D1001094

Military Enterprise and
Technological Change

WORCESTER POLYTECHNIC
INSTITUTE LIBRARY

Military Enterprise and Technological Change

Perspectives on the American Experience

edited by Merritt Roe Smith

The MIT Press
Cambridge, Massachusetts
London, England

First MIT Press paperback edition, 1987

© 1985 by The Massachusetts Institute of Technology

All rights reserved. No part of this book may be reproduced in any form by any electronic or mechanical means (including photocopying, recording, or information storage and retrieval) without permission in writing from the publisher.

This book was set in Linotron 202 Baskerville by Achorn Graphic Services and printed and bound by Halliday Lithograph in the United States of America

Library of Congress Cataloging in Publication Data

Main entry under title:

Military enterprise and technological change.

Bibliography: p.
Includes index.
1. Munitions—United States—History—19th century—Addresses, essays, lectures.   2. Munitions—United States—History—20th century—Addresses, essays, lectures   3. United States—Armed Forces—History—19th century—Addresses, essays, lectures.   4. United States—Armed Forces—History—20th century—Addresses, essays, lectures.   5. Technology—United States—History—19th century—Addresses, essays, lectures.   6. Technology—United States—History—20th century—Addresses, essays, lectures.   I. Smith, Merritt Roe, 1940–
UF533.M55   1985   355.8'2'0973   85-112
ISBN 0-262-19239-X (hard)
      0-262-69118-3 (paper)

Dedication

UF
533
M55
1985
c. 2

For 1st Lt. George B. (Tinker) Pearson, III (1941–1966)
Who believed in the American Dream and died for it in Vietnam

Contents

Preface　*viii*

Introduction　*1*
Merritt Roe Smith

1
Army Ordnance and the "American system" of Manufacturing, 1815–1861　*39*
Merritt Roe Smith

2
The Corps of Engineers and the Rise of Modern Management, 1827–1856　*87*
Charles F. O'Connell, Jr.

3
The Navy Adopts the Radio, 1899–1919　*117*
Susan J. Douglas

4
Ford Eagle Boats and Mass Production during World War I　*175*
David A. Hounshell

(

5

Adjusting to Military Life: The Social Sciences Go to War, 1941–1950   *203*

Peter Buck

6

Military Needs, Commercial Realities, and the Development of the Transistor, 1948–1958   *253*

Thomas J. Misa

7

U.S. Navy Research and Development since World War II   *289*

David K. Allison

8

Command Performance: A Perspective on Military Enterprise and Technological Change   *329*

David F. Noble

9

Technology and War: A Bibliographic Essay   *347*

Alex Roland

Contributors   *381*
Index   *383*

Preface

The idea for this book grew out of a casual conversation. At a 1978 symposium on the American system of manufactures, held at the Smithsonian Institution, a colleague asked me what had been written about the role of the armed forces in American industrialization. I mentioned the literature on the military-industrial complex, but other than that I was both stumped and intrigued. My curiosity led to a bout of library research whose fruits were surprisingly meager. No standard work addressed the topic, and I could find only a few studies treating specific instances of military innovation. Further delving into the subject led to the collaborative effort embodied in the present book.

All but two of these essays were specially commissioned and are appearing in print for the first time. The exceptions are the contributions by David Noble and myself. My essay (originally presented at the Smithsonian symposium mentioned above) is included because it focuses on one of the earliest and most important instances of military enterprise in American history and contains all I had to say on the subject. Noble's essay (somewhat revised from its original appearance in a symposium volume of the American Association for the Advancement of Science) is included because its critical view of the military system seemed a salutary counterweight to the more sympathetic approaches of the other contributors and also because it provides an overview of the author's important research on the development of numerically controlled machine tools. By and large the contributions are the work of younger scholars—most of them historians of technology—whose research on technological innovation has revealed an important military presence.

The topics covered are not intended to be comprehensive,

but they should suggest the inherent richness of our theme and the variety of interpretations that can be brought to bear on it. Other topics that might have been chosen include medical technology and health care, aviation, food processing, and nuclear energy. Our goal, simply stated, is to sample a limited number of important topics, to advance some hypotheses about the meaning of military enterprise, and to suggest paths for future research. We believe that such research can provide new insights into the nature of socioeconomic change and should eventually be integrated into the mainstream of American history and policy studies. We anticipate that a second volume will carry the subject further and will allow a more comprehensive assessment of military enterprise and its influence on the course of American history.

Many people have helped in preparing this book. During the past three years it has gone through several substantial revisions and has profited greatly from the close reading and critiques of three anonymous referees. In addition, Charles Cheape, Colleen Dunlavy, W. David Lewis, Leo Marx, Bronwyn Mellquist, Alex Roland, Barbara Rosenkrantz, Charles Sabel, John Staudenmaier, S. J., and Peter Wallensteen have read portions of the manuscript and offered valuable recommendations. Throughout the project the staff of The MIT Press has provided encouragement and fine editorial support.

An early version of the manuscript formed the agenda for a four-part seminar in the Department of Technology and Social Change at Linköping University (Sweden) during my residence there as a Fulbright scholar in the spring of 1983. More condensed versions were presented to the Program in the History of Science and the Center for Peace and Conflict Research at Uppsala University, the Department of History at the University of Gothenburg, the History of Technology Seminar at the Royal Institute of Technology in Stockholm, the Fondazione Giangiacomo Feltrinelli in Milan, and the U.S. Army Military History Institute at Carlisle Barracks, Pennsylvania. At each of these institutions the audiences were receptive, and their probing questions helped to sharpen our analyses.

Finally, thanks are due to the Smithsonian Institution Press for granting permission to include an updated version of my essay, which originally appeared in *Yankee Enterprise*, a volume edited by Otto Mayr and Robert C. Post.

Merritt Roe Smith
Newton, Massachusetts

Introduction
Merritt Roe Smith

Historical Perspectives

Students of economic development and social change frequently describe the new kind of society that emerged in the late eighteenth and early nineteenth centuries as "industrial capitalism." While this phrase adequately reveals the directive influence exerted by private individuals and firms operating in the marketplace, it tends to deflect attention from the important nexus that exists between government institutions and industrialization. Particularly slighted in this respect is the role of the military as an agent of technological innovation and industrial consolidation, the subject of this anthology.

"Military enterprise," as construed in this book, refers to a broad range of activities through which armed forces have promoted, coordinated, and directed technological change and have thereby, sometimes directly and sometimes indirectly, affected the course of modern industry. The connections we seek to trace are not primarily the applications of technology to weaponry or the special developments that occur under conditions of war. While acknowledging the obvious relationship between technology and warfare, this volume portrays military enterprise as an ongoing process, one that obtains in times of peace as well as times of conflict.[1] Owing to its enduring character, its scale, and its demand for materiel of the highest quality, military enterprise has exerted a powerful influence in deter-

---

1. For similar statements about the continuity of military enterprise, see Clive Trebilcock, " 'Spin-Off' in British Economic History: Armaments and Industry, 1760–1914," *Economic History Review* 22 (1969): 474–90; and Alan S. Milward, *War, Economy and Society, 1939–1945* (Berkeley: University of California Press, 1977), chap. 6.

mining the institutional and technical dimensions of the modern industrial era. Such influence dates back at least to the eighteenth century and ranges from direct involvement in innovative developments to the indirect sponsorship of technological and even behavioral research. Military enterprise thus embraces a wide gamut of activities, and even when restricted to the American experience, as it is here, the subject remains vast and largely unexplored.

Historians recently have studied technological change from four different interpretive perspectives. The oldest and most familiar of these interpretations is that which views technology as a form of expanding knowledge. Here scholarly discourse ranges from narrowly focused "nuts and bolts" studies to more complex treatments of industrial systems. What distinguishes this approach is its primary emphasis on the tangible aspects of technology and the evolving technical and epistemological features of the subject.[2] A second popular mode of interpretation portrays technology as a social force. Studies of this type also tend to place technology at the center of things while emphasizing its "impact" on society.[3] A third perspective views technology as a social product largely shaped by the cultural setting in which it originates and accordingly reflective of the values and ideologies of its creators. Studies of this genre treat innovation in the workplace as well as more broadly defined analyses of technological change in industrial communities, geographic regions, and national cultures.[4] A fourth interpretive perspec-

2. Representative studies are Edwin A. Battison's closely detailed artifactual analysis of "Eli Whitney and the Milling Machine," *Smithsonian Journal of History* 1 (1966): 9–34; Thomas P. Hughes's comparative study of complex *Networks of Power* (Baltimore: Johns Hopkins University Press, 1983); David J. Jeremy's comprehensive treatment of the *Transatlantic Industrial Revolution* (Cambridge, MA: The MIT Press, 1981); and Edwin T. Layton's pathbreaking research on the historical relations of science and technology in "Mirror-Image Twins: The Communities of Science and Technology in Nineteenth Century America," *Technology and Cultur* 12 (1971): 562–80. For important historiographic expositions of technology as knowledge, see Brooke Hindle, *Technology in Early America: Needs and Opportunities for Study* (Chapel Hill: University of North Carolina Press, 1966), pp. 3–28; Hindle, "How Much Is a Piece of the True Cross Worth?" in *Material Culture and the Study of American Life*, ed. Ian M. G. Quimby (New York: W. W. Norton, 1978), pp. 5–20; Edwin T. Layton, "Technology as Knowledge," *Technology and Culture* 15 (1974): 31–41; and Otto Mayr, "The Science-Technology Relationship as a Historiographic Problem," *Technology and Culture* 17 (1976): 663–73.

3. Perhaps the best-known work in this area is Lynn White's *Medieval Technology and Social Change* (New York: Oxford University Press, 1962).

4. See, for example, David F. Noble, *America by Design: Science, Technology and the Rise of Corporate Capitalism* (New York: Alfred A. Knopf, 1977); and Anthony F. C. Wallace, *The Social Context of Innovation* (Princeton: Princeton University Press, 1982). An early and influential historiographic statement of this position is provided by George H.

tive, which ideally should underlie the other three, sees technological change as a social process. This approach recognizes that social and political interactions occur at the earliest stages of invention as well as at the latest stages of development and that they become even more complex when new technologies are taken beyond the research laboratory or experimental shop and introduced into everyday use. In short, interpretations of technology as a social process recognize the fragility of human relationships and the tensions inherent in change. Such a perspective reveals much about the human interactions that occur when new technologies are developed, introduced, and assimilated,[5] but it does not necessarily reveal how human choices influence the form new technologies take.

All four approaches have been employed by historians of military enterprise and all four are represented in this volume.[6] The essays that follow are primarily concerned with the institutional contexts of technological change, how innovations that had been developed under military auspices were transferred to civilian use, and, how, when new technologies were introduced, people in and out of the armed forces responded. An exception, which cautions against too facile an acceptance of military enterprise as a determining agent of industrialization, is David Hounshell's essay about Henry Ford's experience building Eagle boats during World War I. It illustrates the

---

Daniels, "The Big Questions in the History of Technology," *Technology and Culture* 11 (1970): 1–21.

5. Noteworthy examples are Thomas P. Hughes, "Technological Momentum in History: Hydrogenation in Germany 1898–1933," *Past and Present* 44 (1969): 106–32; Edwin T. Layton, *The Revolt of the Engineers* (Cleveland: Press of Case Western Reserve University, 1971); and Bruce Sinclair, *Philadelphia's Philosopher Mechanics* (Baltimore: Johns Hopkins University Press, 1974).

6. For interpretations of military technology as knowledge, see, for example, Edward C. Ezell, *Handguns of the World* (Harrisburg, PA: Stackpole Books, 1981); Bernard Brodie and Fawn M. Brodie, *From Crossbow to H-Bomb* (1962; Bloomington: Indiana University Press, 1973); and I. B. Holley, Jr.'s seminal study *Ideas and Weapons* (1953; Hamden, CT: Archon Books, 1971). For technology as social force, see Carlo M. Cipolla, *Guns, Sails and Empires* (New York: Minerva Press, 1965); Daniel R. Headrick, *The Tools of Empire* (New York: Oxford University Press, 1981); and Dennis E. Showalter, *Railroads and Rifles: Soldiers, Technology, and the Unification of Germany* (Hamden, CT: Shoestring Press, 1975). For technology as social product, see Hugh G. J. Aitken, *Taylorism at Watertown Arsenal* (Cambridge, MA: Harvard University Press, 1960); John F. Guilmartin, Jr., *Gunpowder and Galleys* (New York: Cambridge University Press, 1974); and Merritt Roe Smith, *Harpers Ferry Armory and the New Technology* (Ithaca, NY: Cornell University Press, 1977). For technology as social process, see Robert V. Bruce, *Lincoln and the Tools of War* (Indianapolis: Bobbs-Merrill, 1956); Elting E. Morison, *Men, Machines and Modern Times* (Cambridge, MA: The MIT Press, 1966); and Alex Roland, *Underwater Warfare in the Age of Sail* (Bloomington: Indiana University Press, 1978).

problems that can arise when civilian production technologies are rapidly deployed in wartime. Together, however, the essays clearly document the importance of military enterprise as a force in American history. Although military historians have been looking for themes that will weave their subject more tightly into the fabric of American history, they have largely overlooked this rich aspect of their field.[7] Neither have historians of science and technology paid sufficient attention to the subject.[8] The result is that military enterprise remains one of the least understood and appreciated phenomena in American culture.

Main Themes in the History of Military Enterprise

A major thesis of this book is that military enterprise has played a central role in America's rise as an industrial power and that since the early days of the republic, industrial might has been intimately connected with military might. Whether one looks to the origins of mechanized production or the latest version of the automatic factory, one finds the imprint of military influence. Computers, sonar, radar, jet engines, swept-wing aircraft, insecticides, transistors, fire- and weather-resistant clothing, antibacterial drugs, numerically controlled machine tools, high-speed integrated circuits, nuclear power—these are but some of the best-known industrial products of military enterprise since World War II. And the list can be greatly extended.

Less well known but equally significant is the military's presence at the earliest stages of the industrial revolution. In Europe, as Werner Sombart and others have shown, military demand for arms and munitions exerted an enormous in-

7. In 1975, B. Franklin Cooling chided military historians for their "benighted ignorance" of technology and its relationship to military affairs in "Technology and the Frontiers of Military History" (Paper delivered at the International Commission on Military History Symposium, Washington, D.C., August 18, 1975), p. 9. A condensed and less critical version of Cooling's paper appeared in *Military Affairs* 34 (1975): 206–7. Relevant discussions are also found in Cooling, ed., *War, Business, and American Society* (Port Washington, NY: Kennikat Press, 1977); Cooling, "Military History: A Blending of Old and New," *OAH Newsletter* 12, no. 1 (Feb. 1984): 14–15; Robin Higham, ed., *Guide to the Sources of United States Military History* (Hamden, CT: Archon Books, 1975); Peter Paret, "The History of War," *Daedalus* 100 (1971): 376–96; and Theodore Ropp, "Military Historical Scholarship Since 1937," *Military Affairs* 36 (1977): 68–74.

8. See Jean Missud, "American Science and the Military: An Historiographic Note," *History of Science in America: News and Views* 2 (Dec. 1982): 1–3.

fluence on the development of mining, metallurgy, and machine production.[9] Yet these developments, as significant as they are to understanding the rise of industrial civilization, seem rather sporadic and disconnected when compared with what happened in the United States during the early nineteenth century. Within the relatively brief span of four decades, Americans borrowed, assimilated, and expanded what had been done in Europe over three centuries. Indeed, in the United States three of four major sectors of technological change—interchangeable manufacturing, machine tools, and railroads—were closely tied to military enterprise, while the fourth, textiles, relied heavily on the federal government for tariff protection and other indirect subsidies, as well as direct purchases of cloth. The first two selections address these developments in considerable detail.[10]

The balance of this introduction provides an overview of some of the major aspects of military enterprise in America. I will identify five themes and relate them to the contributions in this volume as well as to a larger body of historical literature dealing with the military's role in the development of industrial capitalism. My aim is to suggest the pervasiveness of military enterprise in the American economy and to assess its social consequences.

Design and Dissemination of New Technologies
Of the many aspects of technological innovation, two of the most important are the establishment of specifications and the execution of designs. They are especially important to military

9. Werner Sombart, *Studien zur Entwicklungsgeschicte des Modernen Kapitalismus*, II. *Krieg und Kapitalismus* (Munich, 1913). Also see Waldemar Kaempffert, "War and Technology," *American Journal of Sociology* 46 (1941). 431–44; Lewis Mumford, *Technics and Civilization* (1934; paperback ed., New York: Harbinger Books, 1963); and William H. McNeill, *The Pursuit of Power: Technology, Armed Force, and Society Since A.D. 1000* (Chicago: University of Chicago Press, 1982). A recent addition to this literature is Anthony F. C. Wallace's fascinating chapter on the British Royal Office of Ordnance and the late seventeenth-century origins of the steam engine in *The Social Context of Innovation* (see note 4 above), pp. 28, 37–61. Also relevant is V. Foley and S. Canganelli, "The Origin of the Slide Rest," *Tools and Technology* 6, no. 1 (1983): 1–7.

10. In addition to the essays by Smith and O'Connell in this volume, see Edwin A. Battison, *Muskets to Mass Production* (Windsor, VT: American Precision Museum, 1976); Editors of the American Machinist, *Metalworking: Yesterday and Tomorrow* (New York: McGraw-Hill, 1977); and Carl E. Prince and Seth Taylor, "Daniel Webster, the Boston Associates, and the U. S. Government's Role in the Industrializing Process, 1815–1830," in *Essays from the Lowell Conference on Industrial History 1980 and 1981,* eds. Robert Weible, Oliver Ford, and Paul Marion (Lowell, MA: Lowell Conference on Industrial History, 1981), pp. 114–27.

enterprise, for at each stage decisions are made which reflect the military's needs and goals and are ultimately embodied in the technology. Although many variables characterize the design process, it is important to emphasize in this context that technologies necessarily reflect the values and aspirations of their makers. Such norms, whether consciously espoused or not, pervade the entire spectrum of development and are particularly important in setting the subsequent course of new technologies. To use Langdon Winner's apt phrase, "artifacts have politics." As products of particular segments of society, technologies are loaded with ideological implications.[11]

By establishing standards and specifications for various goods and contracting with private manufacturers for their production, the military influences the design of many artifacts that eventually enter civilian use. In some cases, like the design of a cooking pouch or weather-resistant shoes, it is difficult to see how military criteria either challenge accepted civilian values or impose unwanted standards on society. In fact, such "consumer goods" as Air Force parkas or dehydrated foods often meet with instant success when transferred to commercial markets. The situation becomes more complex, however, when we shift from industrial products to industrial processes. As David Noble points out with reference to machine tools, the transfer of process-oriented manufacturing technologies from military to civilian use often raises fundamental issues about command and control, especially in connection with work relations on the shop floor.[12]

Probably the most famous modern instance of the transfer of military technology to civilian use is nuclear energy. In their superb study *The Nuclear Navy*, Richard Hewlett and Francis Duncan detail how, between 1953 and 1957, Admiral Hyman Rickover and his naval reactors group oversaw the design and construction of the nation's first civilian nuclear power plant at Shippingport, Pennsylvania. Protracted deliberations between the Atomic Energy Commission, its affiliates, and the power utilities preceded Rickover's selection to head the project. Ultimately the AEC's decision rested with the Navy because it pos-

---

11. Langdon Winner, "Do Artifacts Have Politics?" *Daedalus* 109 (Winter 1980): 121–36.

12. In addition to Noble's chapter in this volume, see his *Forces of Production* (New York: Alfred A. Knopf, 1984) for an extended treatment of the history of numerically controlled machine tools. Smith's chapter below also touches on this question.

sessed the simplest, most thoroughly tested, and most logical reactor technology available at that time. The project succeeded remarkably well. But, as so often happens in new technological applications, the Shippingport undertaking led to inflated expectations among government bureaucrats, utilities representatives, and manufacturers of reactor equipment—so much so that they committed the American electric power industry to "virtually exclusive development of light-water reactor systems."[13] By the late 1950s enthusiasm for a nuclear future was so great that some experts were predicting that energy costs would eventually become negligible in America. That did not happen, of course, nor did the rosy future predicted for nuclear energy come to pass. The reasons for this failure are complex, and they would take us beyond the scope of this introduction. Suffice it to say that although light-water reactors continued to work well on naval vessels, their scaling up for commerical power generation created enormous technical and managerial problems. What seemed rational and controllable for limited military use became problematic and politicized when extended to widespread civilian use.[14]

The design of military products emphasizes performance and uniformity but pays only secondary attention to costs—a fact of military strategy since the early nineteenth century. Referring to aeronautics, Alex Roland points out that "the primary difference over the years seems to be that speed and maneuverability have been most highly valued in military aircraft, while safety, economy, and comfort come first in the civilian field. . . . But as a rule," he concludes, "fundamental advances in one realm have been applicable in the other. . . . There is simply no clear dividing line between civilian and military aviation."[15]

13. Richard Hewlett and Francis Duncan, *The Nuclear Navy* (Chicago: University of Chicago Press, 1974), p. 234.

14. Cf. Irvin C. Bupp and Jean-Claude Derian, *Light Water* (New York: Basic Books, 1978); and Michael D. Stiefel, "Government Commercialization of Large Scale Technology: The United States Breeder Reactor Program, 1964–1976" (Ph.D. dissertation, Massachusetts Institute of Technology, 1981).

15. Alex Roland, "The Impact of War Upon Aeronautical Progress: The Experience of the NACA," in *Air Power and Warfare*, eds. Alfred E. Hurley and Robert C. Ehrhart (U.S. Air Force Academy Military History Symposium 1978; Washington, D.C., 1979), pp. 365–66. For corroborating evidence, see Harvey M. Sapolsky, *The Polaris System Development* (Cambridge, MA: Harvard University Press, 1972), p. 212; Monte D. Wright and Lawrence J. Paszek, eds., *Science, Technology, and Warfare* (U.S. Air Force Military History Symposium 1969; Washington, D.C., 1969), p. 159; and Lloyd J. Dumas, "Innovation Under Siege," *Technology Review* 83 (Nov./Dec. 1980): 9, 76.

With process-oriented technologies, the dividing line between performance and economy is even less distinct. New production technologies such as machine tools almost always have bugs and bottlenecks that prevent them from operating as efficiently as older methods of production. Indeed, the time needed to perfect and assimilate new technologies is often lengthy and costly. The Springfield armory, for example, required over a decade to digest novel techniques for interchangeable manufacture and to produce firearms at costs significantly lower than those made by more traditional craft methods.[16] Yet, despite the considerable costs involved, the military has repeatedly succeeded in getting private manufacturers to adopt innovative but commercially unproven technologies. The reasons for this success lie in the enormous influence the military exercises through its control of an extensive contracting and subsidizing network.

In the nineteenth century armaments contractors were expected to adopt the latest techniques if they wished to retain their contracts. And most of them did. Although contractors had to assume the direct costs of tooling their factories, the military frequently provided them with monetary advances as well as free access to the latest machine patterns, technical drawings, and other information accumulated at the national armories and navy yards. During the 1870s and 1880s, for instance, the Naval Ordnance Bureau, with its intense interest in adopting all-steel breechloading guns and armored vessels, played an instrumental role in getting such firms as the Midvale Steel Company and the Bethlehem Steel Company to acquire the latest open-hearth methods and scale up their plants to produce heavy, high-quality steel forgings. Since European firms like Krupp and Armstrong led the world in innovative practice, Ordnance Bureau chiefs Captain William N. Jeffers (1873–81) and Captain Montgomery Sicard (1881–90) followed an energetic policy of gathering intelligence abroad and rapidly disseminating it at home. These activities, coupled with high contract prices paid to induce firms like Midvale to modernize their plants, resulted in the deployment of a powerful "blue water" navy by 1900. In the process the Navy also helped

---

16. See Felicia J. Deyrup, *Arms Makers of the Connecticut Valley*, Smith College Studies in History, vol. 33 (Northampton, MA: Smith College, 1948); Smith, *Harpers Ferry Armory and the New Technology*; and Paul J. Uselding, "Technical Progress at the Springfield Armory, 1820–1850," *Explorations in Economic History* 9 (1972): 291–316.

to foster the emergence of one of the world's most advanced steelmaking establishments.[17]

In the twentieth century this mode of operation continued, although in many cases the subsidies were more direct. David Noble argues, for example, that during the 1950s the U.S. Air Force "created a market" for numerically controlled machine tools by purchasing, installing, and maintaining "over one-hundred N/C machines in factories of prime contractors." At the same time the Air Force actually paid contractors, aircraft manufacturers, and various part suppliers to learn how to use the new technology. "Not surprisingly," Noble observes, "machine-tool builders got into action, and research and development expenditure in the industry multiplied eightfold between 1951 and 1957." Thus the military inaugurated one of the most important manufacturing technologies of the twentieth century. With the massive financial support afforded by military contracts, cost effectiveness and other economic considerations lost their primacy. The new technology gained a foothold because the Air Force initially shielded it from the competitive rigors of the marketplace.[18]

The importance of nurturing and protecting new technologies bears emphasizing in this context. Although numerous institutions have performed this function, none have done it more often or with greater resources and resolve than the military. This is so, especially since World War II, because a pervasive concern for national security has provided a political shield that, in the words of Mel Horwitch, "protects a program from possible protests emanating from various sectors of society." Horwitch's studies of large-scale technological programs, while only peripherally concerned with military enterprise, provide convincing evidence on this point. The Manhattan Project, the Apollo program, and various weapons systems programs have all enjoyed such protection. "On the other hand," Horwitch notes, "as large-scale demonstration projects

---

17. Richard D. Glasow provides a detailed examination of this subject in "Prelude to a Naval Renaissance: Ordnance Innovation in the United States Navy During the 1870s" (Ph.D. dissertation, University of Delaware, 1978), esp. pp. 121–22, 243–44 ff. Also see B. Franklin Cooling, *Gray Steel and Blue Water Navy* (Hamden, CT: Archon Books, 1979); Walter R. Herrick, Jr., *The American Naval Revolution* (Baton Rouge: Louisiana State University Press, 1966); and Harold and Margaret Sprout, *The Rise of American Naval Power, 1776–1918* (Princeton: Princeton University Press, 1939).

18. David Noble, "Social Choice in Machine Design," in *Studies in the Labor Process*, ed. Andrew Zimbalist (New York: Monthly Review Press, 1979), pp. 25–26.

moved from defense or aerospace toward commercialization goals. . . protective shields that had previously functioned so effectively began to disappear, and with this dissolution the likelihood of managerial success began to diminish considerably." The demise of the American SST project provides a classic example.[19]

Management

The military's seeding and shielding of new technologies points to an important issue: the centrality of management to military enterprise. Most noteworthy is the extent to which historical studies of military-industrial technologies focus on administrative rather than "hard" technical problems. Indeed, scholars purportedly doing technological history often find themselves immersed in entrepreneurial history. And for good reason. Whether one examines the introduction of interchangeable manufacturing in the early nineteenth century or the building of a nuclear submarine in the mid-twentieth, one quickly discovers that technological innovation entails managerial innovation. Technology and management are inextricably connected, although it is clear that under military enterprise the former often provides the impetus for the latter. Changes in technology prompt adjustments in management. Those enterprises that do not adjust usually fail, as Robert Art's study of the TFX (Tactical Fighter Experimental, the original designation of the F-111) project and Edward Ezell's account of the demise of the Springfield armory well illustrate.[20]

What is less clear from the historical record is the degree to

19. Mel Horwitch, "The Role of the Concorde Threat in the U.S. SST Program," (MIT Sloan School Working Paper, May 1982), pp. 1–2. For further elaboration of the shielding concept, see Horwitch, *Clipped Wings: The American SST Conflict* (Cambridge, MA: The MIT Press, 1982); Horwitch, "The Convergence Factor for Large-Scale Programs: The American Synfuels Experience as a Case in Point" (MIT Energy Laboratory Working Paper, December 8, 1982); and Horwitch and C. K. Prahaled, "Managing Multi-Organization Enterprises: The Emerging Strategic Frontier," *Sloan Management Review* 22 (Winter 1981): 3–16. For an interesting counter example about shielding, see Susan Douglas's essay in this volume, pp. 146–47.

20. Robert Art, *The TFX Decision: McNamara and the Military* (Boston: Little, Brown & Co., 1968); Edward C. Ezell, "Patterns in Small-Arms Procurement Since 1945," in Cooling, ed. *War, Business and American Society*, pp. 146–57; and Ezell, "The Search for a Lightweight Rifle: The M14 and M16 Rifles" (Ph.D. dissertation, Case Western Reserve University, 1969). Also see Sapolsky, *Polaris System Development;* Hewlett and Duncan, *Nuclear Navy;* James Webb, *Space Age Management* (New York: McGraw-Hill, 1969); and Alfred D. Chandler, Jr., *The Visible Hand: The Managerial Revolution in American Business* (Cambridge, MA: Harvard University Press, 1977), as well as the essays by Smith, O'Connell, Douglas, and Allison in this volume.

which military enterprise has actually influenced the rise of modern industrial management. Despite the fact that the military is one of the world's oldest bureaucracies, business and economic historians generally are reluctant to assign it credit for managerial innovation. Instead they attribute the rise of modern management to the play of market forces.[21] The evidence presented in this volume, which tends to controvert that view, depicts the military in dynamic interaction with the business community. The contributors thus substantiate Lewis Mumford's claim that the "army is in fact the ideal form toward which a purely mechanical system of industry must tend."[22] Viewed from this perspective, the current relationship between business and the military acquires enhanced significance. When one understands, for example, that Fordism traces its ancestry to the military arms industry of the nineteenth century, one begins to appreciate how deeply military-industrial rationality and centralization are implanted in American culture.[23]

Charles O'Connell's essay on the Army Corps of Engineers and the American railroad system offers particularly strong evidence of the connection between military enterprise and managerial innovation. Likewise, my chapter on the "American system" reveals that the managerial structures and accounting practices developed by the Army Ordnance Department for the national arsenal system spread not only to other privately owned arms factories such as Colt and Remington, but also to other technically related industries. The history of virtually every important metalworking industry in nineteenth-century America—machine tools, sewing machines, watches, typewriters, agricultural implements, bicycles, locomotives—reveals the pervasive influence of military management techniques.[24]

---

21. Notable exceptions are Aitken, *Taylorism at Watertown Arsenal;* and Thomas C. Cochran, *Frontiers of Change: Early Industrialization in America* (New York: Oxford University Press, 1981).

22. Mumford, *Technics and Civilization*, p. 89.

23. See David A. Hounshell, *From the American System to Mass Production, 1800–1932* (Baltimore: Johns Hopkins University Press, 1984).

24. In addition to the essays of Smith, O'Connell, and Noble in this volume, see Battison, *Muskets to Mass Production;* Hounshell, *American System to Mass Production;* and Nathan Rosenberg, "Technological Change in the Machine Tool Industry, 1840–1910," *Journal of Economic History* 23 (1963): 414–43. I. B. Holley's discussion of the introduction of Cost Plus Fixed Fee contracting and its influence on accounting practices in the defense industry during and after World War II provides another important example of the military's impact on modern management. See Holley, *Buying Aircraft: Air Materiel for the Army Air Force* (Washington, D.C.: GPO, 1964), chaps. 15 and 16.

Given these connections, it is not particularly surprising that the earliest experiments with scientific management were conducted by people well acquainted with armory practice and military enterprise.

When Lammot du Pont conducted systemic studies of the movement of materials at his family's gunpowder mills along the Brandywine in 1872, he based his calculations on work norms published in American and French ordnance manuals. Like other firms that relied on large government contracts, the DuPont Company kept itself informed on technical projects sponsored by the military by acquiring numerous tracts and reports of chemical and metallurgical experiments conducted by the Army Ordnance Department. Moreover, ordnance officers often visited the DuPont works to conduct inspections as well as experiments.[25] Similar influences are revealed in the writings of Charles H. Fitch, Henry Metcalfe, and Henry R. Towne, all early advocates of systematic management and all connected with "armory practice" industries. Metcalfe, in fact, was a West Point graduate who served as an ordnance officer at the Springfield armory (1870–75), the Frankford arsenal in Philadelphia (1875–81), and the Watervliet arsenal near Troy, New York (1884–86).[26]

The best known among those fostering more rigorous management techniques is Frederick Winslow Taylor. Taylor's background reveals a connection with armory practice and the machine trade, particularly through his close association with William Sellers, the renowned Philadelphia machine tool builder and foundry owner who kept abreast of developments in the New England firearms and machine tool industries. Tay-

25. Norman B. Wilkinson, "In Anticipation of Frederick W. Taylor: A Study of Work by Lammot du Pont, 1872," *Technology and Culture* 6 (1965): 217. The DuPont Company collections at the Eleutherian Mills Library (Greenville, DE) contain numerous reports of chemical and metallurgical experiments conducted by the Army Ordnance Department during the antebellum and postbellum periods.

26. Charles H. Fitch, "Report on the Manufacture of Interchangeable Mechanism," *Tenth Census of the United States (1880): Manufactures*, vol. 2 (Washington, D.C.: GPO, 1883); Fitch, "Report on the Manufacture of Hardware, Cutlery and Edge Tools," ibid.; Henry Metcalfe, "The Shop Order System of Accounts," *Transactions*, American Society of Mechanical Engineers 8 (1886): 459–68; Metcalfe, *The Cost of Manufactures and the Administration of Workshops, Public and Private* (New York, 1885); Metcalfe, *A Course of Instruction in Ordnance and Gunnery*, 3rd ed. (New York: John Wiley & Sons, 1894); Henry R. Towne, "The Engineer as an Economist," *Transactions*, American Society of Mechanical Engineers, 7 (1886): 429–30; George W. Cullum, *Biographical Register of the Officers and Graduates of the U. S. Military Academy*, 3 vols. (New York: Houghton Mifflin, 1891), 3: 109–10.

lor's interest in scientific management stemmed from his disgust with "soldiering" (the practice of deliberately restricting output) among machinists under his supervision at the Midvale Steel Company and his determination to eliminate it from the shop. Although he is best known for the introduction of job analysis and time-study methods, technological innovations played an important part in his managerial system. Most prominent was his discovery of the properties of chromium-tungsten tool steel and its application to cutting metals. By greatly increasing the speed at which production machines could be operated, the new cutting tools made necessary the design of heavier, more stable machinery. Because managers recognized the importance of high-speed steel and immediately began to adopt it, the new technology enhanced Taylor's reputation and gave him an added opportunity to reorganize work along lines more advantageous to employers. Like so many other instances of military enterprise, Taylor's managerial innovations were closely linked with technological innovations.[27]

Scientific management was an attempt to redefine and control the pace of work by acquiring and controlling knowledge of the production process. Doing time studies, standardizing machine speeds, introducing scheduling cards, and establishing refined accounting procedures and inventory controls accordingly acquired paramount importance. Such methods aimed at consolidating managerial power by seizing control of the shop floor from workers and placing production in the hands of industrial engineers. Through various adroit practices, and particularly through their possession of essential skills, industrial workers had successfully forestalled managerial control of the shop floor throughout the nineteenth century.[28] From the outset of Taylor's reforms, then, a turf-oriented rivalry existed between shop-trained mechanics and professional engineers. The fact that Taylorism faced its sternest tests at the government-owned Rock Island and Watertown arsenals, where bitter strikes occurred in 1908 and 1911 respectively, suggests that economic factors alone cannot explain its popularity among

27. Aitken, *Taylorism at Watertown Arsenal*, chap. 1 (esp. pp. 19–33). Also see Daniel Nelson, *Managers and Workers* (Madison: The University of Wisconsin Press, 1975), pp. 55–61, as well as his biography *Frederick W. Taylor and the Rise of Scientific Management* (Madison: University of Wisconsin Press, 1980).
28. See, for example, David Montgomery, *Workers' Control in America* (New York: Cambridge University Press, 1979); and Daniel Nelson, *Managers and Workers*.

top-level managers. At issue as well were questions of power and authority that were deeply rooted in the command ethos of military-industrial enterprise. These episodes and the investigations that followed raised such a furor that Congress in 1915 passed legislation prohibiting time studies and incentive payments, two of the most hated aspects of Taylorism at the national arsenals. Nonetheless other elements of the Taylor system made deep inroads at government armories and throughout the American industrial system. The degree of the military's commitment to scientific management is perhaps best indicated by the fact, observed by Hugh Aitken, that in 1918 "one third of the members of the Taylor Society were working in the Ordnance Department."[29]

The military's interest in management studies and applications did not abate with Taylorism. Indeed, both world wars witnessed an extension of social research for managerial purposes. The Army's use of intelligence tests during World War I, for example, helped to stimulate general interest in the use of aptitude tests by managers as instruments for matching employee skills to the requirements of different jobs. Similarly, the attitude surveys conducted by the Research Branch of the Special Services Division of the Army in World War II provided postwar sociologists with a striking example of how their science could be used, in the words of one critic, "to sort out and to control men for purposes not of their own willing." The technical achievements in both cases were substantial. The Army's work with intelligence tests in World War I led to significant extensions in the use of factor analysis. The Research Branch's attitude surveys during World War II gave rise to major developments in the theory and practice of scaling, whose potential applications to administrative problems in civilian life were so clear as to cause concern that the whole of sociology would come to be mobilized for solving "managerial problems for industry and the military."[30]

In his chapter on the military and the social sciences, Peter Buck places these activities in a complex institutional frame-

29. Aitken, *Taylorism at Watertown Arsenal*, pp. 231–33, 238; Pat Harahan, "Worker Resistance to Managerial Control of Production at Rock Island Arsenal, 1898–1919" (Paper delivered at the 23rd Annual Missouri Valley History Conference, Omaha, Nebraska, March 7, 1980).

30. Robert Lynd, quoted in John Madge, *The Origins of Scientific Sociology* (New York: The Free Press, 1962), p. 320. I am indebted to Peter Buck for this reference.

work and assesses their paradoxical consequences. Particularly noteworthy is his discussion of why, during the first thoroughly mechanized war in history, Samuel Stouffer and his associates at the Research Branch decided to abandon the study of technology as a critical variable in social analysis. The answer, Buck suggests, lies not so much in their ignorance of technology's significance in human affairs but rather in their fear that such an emphasis would "invite those self-proclaimed 'experts in technology,' the natural scientists, to intervene in the field, and. . . to win out."[31] In short, professional rivalry had a lot to do with the theoretical direction social research took after World War II.

Analogous problems confronted advocates of other management techniques developed under military auspices during and after the Second World War. In the case of operations research, for example, analysts "were hesitant to embrace the social sciences and the arts" for fear of undermining the scientific character and reputation of their emerging field.[32] When systems analysis and its kindred techniques of program budgeting, project management, and cost-benefit analysis came into vogue during the early 1960s, considerable discussion occurred within military circles about their legitimacy as bona fide sciences and their effectiveness as management tools. A good example is the controversy that erupted between Admiral Hyman Rickover and Secretary of Defense Robert S. McNamara over the application of "the systems approach" in decision making.[33] As sophisticated and technically elegant as systems analysis and its companion methods seemed to be, they frequently proved to be flawed in actual practice. The histories of the TFX decision and the C5A jet transport debacle provide the most famous instances, although David Allison detects similar management problems with naval research since World War II.[34]

31. Buck, p. 244 below. Buck's analysis of *The American Soldier* complements Richard H. Kohn's findings in "The Social History of the American Soldier: A Review and Prospectus for Research," *American Historical Review* 86 (1981): 553–67.

32. I. B. Holley, Jr., "The Evolution of Operations Research and Its Impact on the Military Establishment; the Air Force Experience," in Wright and Paszek, eds., *Science, Technology, and Warfare*, p. 95. Also see Robert L. Perry's interesting "Commentary" in the same volume, pp. 110–21.

33. Hewlett and Duncan, *Nuclear Navy*, pp. xiii, 35.

34. Art, *The TFX Decision;* John Newhouse, *The Sporty Game* (New York: Alfred A. Knopf, 1982); A. Ernest Fitzgerald, *The High Priests of Waste* (New York: W. W. Norton, 1972), as well as Allison's essay in this volume.

In a study of operations research during World War II and its subsequent connection with the 1960s "managerial revolution" in the Defense Department, I. B. Holley attributed the problems in defense management not to any inherent shortcoming in analytical technique but rather to the failure of military strategists "to develop an adequate doctrine to optimize its use."[35] The doctrinal determinants of weapons procurement and technological change are important, to be sure. Nonetheless, military management methods received rather severe criticism in the late 1960s for being too inflexible and often ineffective.

In 1969, for example, one critic went so far as to name systems analysis "Hitchcraft" after Charles J. Hitch, a former Assistant Secretary of Defense who had played an influential role in its development.[36] Such a sobriquet was facetious, but it reflected the serious point that certain management practices had indeed been contrived to simulate technological innovation when, in fact, they had little to do with it. A case in point is PERT, a sophisticated computerized program evaluation technique developed during the late 1950s in conjunction with the Polaris submarine project. Described by Harvey Sapolsky as "an alchemous combination of whirling computers, brightly colored charts, and fast-talking public relations officers," PERT nonetheless served an important political purpose in selling the Polaris program to Congress, fending off potential critics from within the Navy Department, and giving assurances to all concerned that the program was in expert hands and would be completed on schedule at projected costs. "It mattered not whether the parts of the [PERT] system functioned or even existed," Sapolsky concludes. "It mattered only that certain people for a certain period of time believed that they did." The ploy worked well, so well in fact that within four years of its announcement 108 of Fortune's top 500 firms in America

---

35. Holley, "Evolution of Operations Research," p. 108. For a full statement of Holley's important thesis about the centrality of doctrine to technological development, see his *Ideas and Weapons* (1953; reprint ed., Hamden, CT: Archon Books, 1971). Elting E. Morison treats an important variant of this thesis in "The War of Ideas: The U. S. Navy, 1870–1890," in Wright and Paszek, eds., *Science, Technology, and Warfare*, pp. 189–96.

36. Ida Hoos, *Systems Analysis in Social Policy* (Westminster: The Institute of Economic Affairs, 1969), p. 21.

had either installed or were planning to install PERT-type systems.[37]

Our historical understanding of the interaction between military enterprise, modern management, and the American business system remains shockingly vague and incomplete. Extensive primary sources exist on such subjects as industrial management, operations research, and systems analysis, but little of substance is known about the contextual history of these methods, the social processes that brought them into being, and the influence they exerted on the commercial sector.[38] Rich rewards await those who investigate the history of military management and its relationship to technological change and industrial development in America.

Testing, Instrumentation, and Quality Control

Two premises underlie modern management: the primary goal is to control the entire job situation; the primary means of achieving control is through standardization. At an administrative level, control is sought by imposing work rules, central planning practices, and uniform accounting procedures. A case in point, alluded to earlier, is scientific management, an administrative strategy that aimed at rationalizing and controlling every aspect of labor on the shop floor. At a technical level, on the other hand, control is sought through the systematic testing of products and processes. Testing is essential to standardization. It is an integral part of military enterprise, and it is here that the military has traditionally exercised its greatest influence in shaping the American industrial system.

Testing and its concomitant, instrumentation, have many variations, all of which aim at regulating the quality of whatever is being produced. At its simplest, testing entails the visual and tactile inspection of, say, a piece of cloth for defects and imperfections. At a higher level it involves the application of precision gauges to determine the accuracy and strength of components being manufactured for assembly into a finished product. Such inspections have a straightforward purpose—to weed out substandard items. At a yet higher level, however,

37. Sapolsky, *Polaris System Development*, pp. 112, 129.

38. A significant exception to this statement is Aitken's *Taylorism at Watertown Arsenal*, a superb work which could well be used as a model for other studies of military management and its relationship to the private sector.

testing becomes more complex. As Edward Constant indicates in his book *The Origins of the Turbojet Revolution,* it "consists of running a complete system—an engine for example—until something breaks, redesigning or strengthening that part, and continuing until something else breaks." Constant adds: "At a more sophisticated level testing involves construction of complex test rigs that are themselves major technological achievements. With such specialized apparatus, data can be collected on the behavior of individual components of systems at various systems performance levels and conditions."[39] Implicit in Constant's description of systemic testing is a feedback concept: items tested reveal information about design and material weaknesses. By closely monitoring the performance of materials under varying conditions, it becomes possible to detect trouble spots and to pinpoint the stage in the manufacturing process where defects occur. Even the most simple of such tests thus yields valuable information about what changes or adjustments are needed to remedy certain production problems.

As one of the largest patrons of the industrial sector, the military, in its role as a tester of products, has often set goals for private manufacturers and thereby influenced the innovative process. The ability to test thus provides one of the main avenues by which the military influences industrial design and production. Many examples of the influence of military enterprise in the sphere of testing can be found in the annals of aviation, electronics, and metalworking. Probably most significant are the services that military-owned and -operated facilities have provided private industry over the years. The tests conducted at these installations often helped to define what became "standard" technological practice. Nineteenth-century developments in metallurgy provide an apt illustration.

In 1841, as the following chapter indicates, the Army Ordnance Department inaugurated the first sustained program for testing ferrous metals in America. From that time until the Civil War a cadre of science-oriented "soldier-technologists" conducted a series of tests and experiments that resulted in the qualitative improvement of cannon as well as other products made of cast iron. After the Civil War the same tradition of testing continued, as the focus of military interest gradually

---

39. Edward Constant, *The Turbojet Revolution* (Baltimore: The Johns Hopkins University Press, 1980), p. 21. Also see Hewlett and Duncan, *Nuclear Navy,* p. 329.

shifted to ordnance applications in steel. An important segment of this work centered at the Watertown arsenal near Boston where "the prototype of a long line of hydraulic testing machines" was installed in 1879. Called the "Emery Testing Machine" after its designer, A. H. Emery of Chicopee, Massachusetts, and built by the Ames Manufacturing Company, a Chicopee firm closely connected with the Springfield armory and the "American system" of manufactures, the unit had a maximum capacity of 800,000 pounds in tension and 1,000,000 pounds in compression and could test materials as heavy as gun forgings or as fragile as horsehair. Between 1881 and 1912 the Emery machine made over 3000 tests, 753 of which were for private firms at nominal fees.[40] One has only to examine the voluminous *Reports of Tests* made at Watertown to appreciate the extent to which the U.S. arsenal influenced the growth of industrial metallurgy, particularly steel, in the United States.[41] When one realizes that Watertown was only one of several Army facilities working in this area and that the Navy Ordnance Bureau made an even larger investment in the testing of materials, the significance of military enterprise in the nineteenth century looms even larger.

During the late nineteenth and early twentieth centuries the military continued and extended its testing work. The Navy, particularly the Naval Experimental Battery near Annapolis, the Naval Torpedo Station at Newport, Rhode Island, and the Washington Navy Yard, conducted hundreds of tests with iron plate, steel forgings, and steel armor. These activities, coupled with the Navy Ordnance Bureau's aggressive role in acquiring the latest technical information from abroad, resulted in numerous improvements in casting, welding, forging, machining and alloying techniques. They also paralleled equally creative work in industrial chemistry, most notably explosives. The list of firms that cooperated with the Navy Ordnance Bureau and profited from its exertions reads like a who's who of American industry in the Victorian era. The Bethlehem Steel Company, Carnegie Steel Company, DuPont Company, Fort Pitt Foundry, Nashua Iron and Steel Works, Naylor and Company, South Boston Iron Company, and West Point Foundry all ap-

40. Judy D. Dobbs, *A History of the Watertown Arsenal* (Watertown, MA: Army Materials and Mechanics Research Center, 1977), pp. 34–35.
41. U. S. Ordnance Department, *Report of the Tests of Metals and Materials for Industrial Purposes . . . at the Watertown Arsenal.* 38 vols. (Washington, D.C.: GPO, 1882–1919).

pear on the list. Most prominent in the bureau's correspondence files was the Midvale Steel Company, Frederick W. Taylor's old firm in Philadelphia. The fact that the great surge of the American steel industry to world leadership coincides with the American Naval Renaissance of the 1880s and 1890s is no accident. This is not to say, however, that military enterprise was the sole reason for the American steel industry's rise to preeminence during the late nineteenth century. But it certainly was a major factor.[42]

Uniformity and Order
Whether one investigates the fabrication of steel plate, the testing of aircraft engines, or the supervision of armory workers, one finds military technology deeply embedded in institutional settings that emphasize uniform behavior as well as uniform production. Uniformity denotes a disposition to order, a desire to have things orderly. Armies have long exhibited uniformity, but the notion takes on added meaning when it intersects with other ideas and institutions associated with the modern era.

Among nineteenth-century American military engineers and ordnance officers, the idea of uniformity became "the great desideratum," a guiding principle in the formulation of military policy toward technology. And it has continued to be a central precept ever since. In this context uniformity means something more than the mechanical ability to produce things with standardized or interchangeable parts. It also refers to a pervasive mental attitude common among soldiers and essential to military enterprise that emphasizes system and order in all things, including labor on and off the shop floor. Not surprisingly, the notion of social and technical order has implications that go far beyond military life. In its ongoing quest for absolute standards and fail-safe solutions to logistical and tactical problems, the military has stood in the forefront of those institutions defining the character of modern life.

As the first essay suggests, French military engineers initially gave the uniformity concept formal expression in America during the early national period. Even more noteworthy, although in need of further study, is the intriguing bond that exists between military enterprise, with its emphasis on uniformity, and other dominant institutions in American, indeed Western,

---

42. See Glasow, "Prelude to Naval Renaissance."

civilization: science, engineering, business, government, religion, and education. All these modern institutions are similar in the sense that they are order-seeking entities with complex bureaucratic hierarchies and organizational networks. They differ, of course, in the ways they exercise power to achieve and maintain order. What matters here, however, is not so much whether one institution is more important than another but rather that all of them share values that underpin industrial civilization as we know it today. That order is a prerequisite for achieving their primary goals of efficiency and control makes these institutions more compatible than incompatible and bestows on each a degree of corporate power and influence far greater than any one could enjoy alone.

Such compatibility has allowed them to pursue and establish social order with considerable effect. By reinforcing one another, these ordering institutions have become the arbiters of national, even international, culture since the eighteenth century. This subtle process has gradually replaced decentralized and more democratic sources of power with highly centralized and more bureaucratic forms of control. The convergence and growth of these institutional forces is a major factor in the rise of the West since the Renaissance. Through them has emerged an intellectual tradition which emphasizes utilitarian values, with their "hard" instrumental approaches to problem solving and their unwavering affirmation of progress. With them are associated "enlightened" ideas of political stability, economic control, religious asceticism, scientific rationality, and educational discipline in the everyday affairs of business, labor, and statecraft. In effect, the bonding of these institutions gives the modern world view coherence. The values they espouse are ultimately reflected in the forms that technology takes. Understanding this search for order thus allows us to comprehend more fully the meaning of technology as a social product. The significance of military enterprise can be appreciated only within this larger social and institutional context.[43]

43. Numerous authors have addressed certain aspects of this question. Some of the most illuminating are Arnold Pacey, *The Maze of Ingenuity: Ideas and Idealism in the Development of Technology* (Cambridge, MA: The MIT Press, 1976); Carolyn Merchant, *The Death of Nature* (San Francisco: Harper & Row, 1980), esp. chaps. 8, 9, and 12; E. P. Thompson, "Time, Work-Discipline, and Industrial Capitalism," *Past and Present* 38 (1967): 56–97; Chandler, *The Visible Hand;* John Higham, "From Boundlessness to Consolidation: The Transformation of American Culture, 1848–1860," *William L. Clements Library* (1969; Bobbs-Merrill Reprint Series in American History, H-414), pp.

Although the theme of uniformity requires further study, it already has provided significant insight into the nature of military-industrial relations and has enlarged our understanding of the role noneconomic factors play in the emergence of novelty in thought and action.[44] Uniformity thus offers a crucial theoretical lens for examining not only a broad range of technical developments over an extended period of time but also the values that condition military enterprise and technological change. It embraces aesthetic as well as political and economic factors and asks how the military ethos influenced the development of engineering practice and industrial institutions. Since many of the same or analogous pressures continue to influence military thought and, through it, the innovative process, the uniformity precept is particularly relevant to the contemporary scene. A fully documented historical study of uniformity as a social product would deepen our understanding of current policy positions that emphasize the need for rationalization in maintaining highly centralized modern military and industrial establishments.

Innovation and Social Processes

As much as expanding bureaucracies with uniform standards and centralizing tendencies have defined the modern era, one quickly discovers that even the most rationally organized and orderly enterprises encounter resistance to technological change. This is as true of the military as it is of civilian institutions. Change roughs up people by disrupting accustomed ways of doing things and redefining power relationships within

---

1–28; Donald G. Mathews, "The Second Great Awakening as an Organizing Process, 1780–1830: An Hypothesis," *American Quarterly* 31 (1969): 23–43; Clifford S. Griffin, "Religious Benevolence as Social Control, 1815–1860," *Mississippi Valley Historical Review* 44 (1957): 423–44; W. David Lewis, "The Reformer as Conservative: Protestant Counter-Subversion in the Early Republic," in *The Development of an American Culture*, eds. Stanley Coben and Lorman Ratner, 2nd ed. (New York: St. Martin's Press, 1983), pp. 80–111; Richard M. Rollins, "Words as Social Control: Noah Webster and the Creation of the American Dictionary," *American Quarterly* 28 (1976): 415–30; Michael B. Katz, "The Origins of Public Education: A Reassessment," *History of Education Quarterly* 16 (1976): 381–407; David Nasaw, *Schooled to Order: A Social History of Public Schooling in the United States* (New York: Oxford University Press, 1979); Robert H. Wiebe, *The Search for Order, 1877–1920* (New York: Hill and Wang, 1967); Samuel P. Hays, *The Response to Industrialism, 1885–1914* (Chicago: University of Chicago Press, 1957); and Noble, *America by Design*.

44. See, for example, Aitken, *Taylorism at Watertown Arsenal*; Noble, *Forces of Production*; and Smith, *Harpers Ferry Armory*, as well as the essays by Smith, O'Connell, and Misa in this volume.

societies. Often it is a painful experience fraught with anxiety and tension. No matter how innocuous new technologies seem to be, they challenge old values. One has only to recall, for instance, that the machinery and institutions of the industrial revolution marked a radical departure from an earlier lifestyle that stressed close personal and communal relationships, harmony between nature and civilization, and the integration of work with the rest of life. Some people readily adjusted to these changes. Others viewed them as destructive of deeply held values. Almost everyone recognized that life under the new industrial regimen would never be the same. Such shifts and their consequences clearly reveal that technological change *is* a social process. Some of the most illuminating instances of this theme in the historiography of technology come from case studies related to military enterprise and technological innovation.[45]

The way people initiate and respond to change depends on who they are, what they do, how long they have been doing it, and how they perceive the potential effects of the novelty in question. Questions of status, tradition, and control thus loom large in any discussion of technological innovation. In some instances, the challenge of change can take on highly personalized overtones, as Elting Morison's discussion of continuous-aim firing at the turn of the twentieth century makes abundantly clear. In the case Morison describes, a brash junior American naval line officer, William Sims, pushed for the adoption of a relatively simple mechanism for increasing the speed and accuracy of gunfire at sea and met formidable resistance from members of the Naval Bureau of Ordnance. Sims eventually succeeded in getting his innovation adopted, but

45. Prime examples are Aitken, *Taylorism at Watertown Arsenal;* David K. Allison, *New Eye for the Navy: The Origin of Radar at the Naval Research Laboratory* (Washington, D.C.: GPO, 1981); Lance C. Buhl, "Mariners and Machines: Resistance to Technological Change in the American Navy, 1865–1869," *Journal of American History* 61 (1974): 703–27; Morison, *Men, Machines and Modern Times,* chaps. 1, 2, and 6; Morison, *From Know-How to Nowhere* (New York: Basic Books, 1974), chap. 8; Alex Roland, *Underwater Warfare in the Age of Sail* (Bloomington: Indiana University Press, 1978); Smith, *Harpers Ferry Armory;* O. Pi-Sunyer and Thomas DeGregori, "Cultural Resistance to Technological Change," *Technology and Culture* 5 (1964): 247–53; O. Pi-Sunyer and Thomas De-Gregori, "Technology, Traditionalism, and Military Establishments," *Technology and Culture* 7 (1966): 402–7; Barton C. Hacker, "Resistance to Innovation: The British Army and the Case Against Mechanization, 1919–1939," *Actes du XIIIE Congrès International d'Histoire des Sciences* 12 (1971; pub. 1974): 61–70; and Richard I. Wolf, "Arms and Innovation: The United States Army and the Repeating Rifle 1865–1900" (Ph.D. dissertation, Boston University, 1981).

only because President Theodore Roosevelt personally inter-
ceded on his behalf and overruled the naval establishment.[46]

Less than a generation later, when Sims himself had become
part of the naval establishment, yet another high-ranking member of the administration intervened to initiate technological
change in the Navy. As David Hounshell suggests, Secretary of
the Navy Josephus Daniels forced upon the Navy's Bureau of
Construction and Repair a scheme dreamed up by Henry Ford
and his engineers to mass-produce fast-cruising, 200-foot submarine chasers to end the German U-boat menace in World
War I. Not a military man, Daniels held a common view that
private inventors such as Thomas Edison and Elmer Sperry
had been responsible for the era's impressive technological
achievements. Hence it was almost axiomatic for him to call
upon Henry Ford rather than his own department to solve the
submarine problem. Ford had gained international fame for
his company's development of assembly line, mass production
techniques used to manufacture the Model T. But as Hounshell
demonstrates, Ford's enormously expensive efforts to manufacture subchasers ultimately faltered mainly because both
Daniels and Ford failed to appreciate that the Navy had developed its own expertise in the design and execution of war ships.
Designed hastily and primarily for easy serial production, the
Eagle boat proved to be both unseaworthy and far more
difficult to make than Ford and his staff envisioned. The fact is
that the Ford Company did not know the first thing about
building steel ships; moreover, shipbuilding proved to be much
different from auto making. Such expertise had been developed through several decades of naval enterprise, just as it had
taken Ford years to conceptualize, introduce, and perfect
methods for the mass production of Model Ts. Ford's mistaken
assumption that he could simply shift from the mass production of automobiles to the mass production of naval vessels is
characteristic of the exaggerated self-confidence and enthusiasm that often comes with great entrepreneurial success.

The examples of continuous-aim firing and the Eagle boat
illustrate how a forceful personality can impose himself on the
course of events and, for better or worse, influence the direction of technological change. In other instances, the process of
innovation is best examined within a larger social-organi-

---

46. Morison, *Men, Machines and Modern Times*, pp. 17–44.

zational context. As Susan Douglas demonstrates, the Navy's adoption of radio communications between 1899 and 1919 was impeded by the service's decentralized structure. The centralizing tendencies inherent in the proposed innovation challenged such hallowed naval traditions as autonomy of command at sea. Only through a tortuous two-decade process of institutional redefinition and realignment could the Navy formulate an effective doctrine for the new technology. In this instance, as in many others, changes in strategy followed upon structural changes associated with the introduction of a new technology.[47]

These and other case studies of technological innovation in the military services reveal a complex tapestry of bureaucratic relationships. Depending on where they are situated in the innovative process, different groups react in different ways to technological change. Interservice rivalry often manifests itself in conflicts over the adoption of new techniques. Witness the tensions that existed within the armed forces over the development and control of missile systems in the late 1950s and early 1960s.[48] At other bureaucratic levels, lines of divergence form between civilian administrators and military officers, as Robert Art describes in *The TFX Decision*.[49] Yet, as bitter as these conflicts can be, the most serious disagreements often occur within different branches of the same service. Here differences between line and staff over what is needed and who should control its development often complicate and sometimes undermine innovative activity. In brief, bureaucratic rivalries litter the ground of technological innovation in the military services. Even within the same technical bureau, staff members may differ over the assignment of priorities to research and development. During the 1950s, for instance, design and production engineers at the Springfield armory differed over the technical development of a new lightweight rifle. In this case, failure to resolve the issue ultimately resulted in the closing of the armory, an action that subsequently lessened the Ar-

47. For important discussions of the relationship of technological innovation to military doctrine, see Holley, *Ideas and Weapons* and David A. Armstrong, *Bullets and Bureaucrats: The Machine Gun and the United States Army, 1861-1916* (Westport, CT: Greenwood Press, 1982).

48. See Sapolsky, *Polaris System Development;* Edmund Beard, *Developing the ICBM* (New York: Columbia University Press, 1976).

49. See note 20 above. Also see Hewlett and Duncan, *Nuclear Navy*, pp. 26, 52, 64, 114.

my's ability to monitor and control American small arms contractors.[50]

As Douglas's treatment of Stanford C. Hooper and the development of the Navy's radio network indicates, age and experience often separated the protagonists in conflicts over the introduction of new technologies. In the case of radio, as in other innovative undertakings, the key person proved to be a relatively young officer who didn't quite fit the mold of his peer group and even displayed a certain impatience with bureaucratic processes that hampered the development of new ideas.[51] Even more pertinent to understanding innovation are the complex, even contradictory, situations that seem to typify the institutional contexts of technological change. Take, for example, the curious position of the Navy's Bureau of Ordnance around 1900. In the case of continuous-aim firing, the bureau is depicted as an institution tenaciously committed to an outmoded technique and unwilling to listen to the suggestions of a precocious junior officer. Yet at the same time the bureau was deeply involved in materials testing as well as the development of innovative steel manufacturing techniques. What accounts for this seemingly paradoxical behavior? Why was the Bureau of Ordnance so obstinate in one case and so open to change in another?

An answer becomes apparent if we consider the contrasting ways in which the projects were initiated. In the case of steelmaking, the impetus to innovation came from inside the bureau. Indeed, the department's reputation in the field went back to the Civil War experiences of one of its heroic figures, Admiral John Dahlgren. With continuous-aim firing the innovative idea came from an obscure line officer outside the bureau. Had Sims been an ordnance officer, an insider, his suggestion doubtless would have been taken more seriously. Engineering-trained ordnance officers of that era tended to be more interested in technical problems of design and production than in less exciting tactical questions of firing accuracy. One's background and position within the naval bureaucracy made a difference. No matter how good an idea might be, if it

---

50. Ezell, "Patterns in Small-Arms Procurement," pp. 148–52. Other examples may be found in Allison, *New Eye for the Navy;* Kent C. Redmond and Thomas M. Smith, *Project Whirlwind* (Bedford, MA: Digital Press, 1980); Barton C. Hacker, "Imaginations in Thrall: The Social Psychology of Military Mechanization, 1919–39," *Parameters* 12 (1982): 50–61; and Hewlett and Duncan, *Nuclear Navy.*

51. Douglas, pp. 154–70 below; and Morison, *Men, Machines and Modern Times,* chap. 2.

came from the wrong person or entered the bureaucracy through the wrong agency, its likelihood of success diminished considerably. Such are the risks and uncertainties that confront technological innovators in large-scale organizations.[52]

Catalysts to Innovation
Closely related to the conditions that restrain technological innovation are those that prompt it. Military enterprise takes on added meaning when viewed from this perspective. A review of the relevant historical literature about military technology suggests, for instance, that able personnel and institutional flexibility, coupled with an infectious enthusiasm for technological change, are important catalysts of technical creativity in the armed forces. Since these factors also exercise a pervasive influence on innovative activity in the private sector, they merit attention.

Elting Morison regards individual personality as of paramount importance in understanding the innovative process. Such innovators characteristically possess a number of special attributes: youthful energy, a "contempt for bureaucratic inertia," a fascination with technically "sweet" ideas, and a gift for discovering new concepts in serendipitous events. Above all, Morison stresses "the interaction of fortune, intellectual climate, and the prepared imaginative mind" in the emergence of novelty.[53]

Other scholars emphasize similar traits but insist on placing them within a larger institutional setting. Rather than dealing with the individual inventor, these writers focus on research or project groups and the translation of their work by certain specially endowed people into the production and installation of equipment. Such a "translator," according to Susan Douglas, was Stanford C. Hooper, the Navy's first Fleet Radio Officer and subsequent director of the service's Office of Communications. Hooper was not an inventor in any technical sense. Rather he excelled at putting things together in different configurations and creating an environment in which the new

---

52. Alex Roland reaches a similar conclusion in *Underwater Warfare*. Morison, *Men, Machines and Modern Times*, chap. 2; Glasow, "Prelude to Naval Renaissance"; and Armstrong, *Bullets and Bureaucrats*, also provide spendid illustrations of the "NIH" ("Not-invented-here") syndrome. A variant of the insider-outsider theme may be found in Hounshell's discussion of Josephus Daniels, Henry Ford, and the Eagle boat.
53. Morison, *Men, Machines and Modern Times*, pp. 25–27.

technology (radio) could take root and eventually flourish. Allowing for certain variations in personal style, the same can be said of Colonel George Bomford and the "American system" of manufacturing, Captain William N. Jeffers and the American naval renaissance, Admiral Harold Bowen and the development of naval research, General Leslie Groves and the Manhattan Project, Admiral Hyman Rickover and the nuclear navy, and Admirals William F. Raborn and Levering Smith and the Polaris system development. Such men, and there have been many like them, pushed the armed services along innovative paths during the nineteenth and twentieth centuries. They emerged as innovators not because of any special technical competence (although many possessed it), but rather because they had broad vision and an entrepreneurial talent for organizing research, development, and production in ways that fostered creativity. This usually meant that they helped to provide a wholesome environment in which people could work effectively. Creating such an environment consisted of something more than the provision of adequate physical facilities and research equipment, however. Often, as in the cases of Bomford and the "American system" and Raborn and the Polaris project, it entailed the establishment of decentralized but closely coordinated structures within a highly centralized military bureaucracy. Such entrepreneurs frequently circumvented standard operating procedures by selecting key staff members from outside the normal chain of command and by using diversionary tactics to attract public attention to one feature of a project while they were emphasizing something else. A good case in point is PERT, the program evaluation technique developed under the Polaris project and discussed above. Another is the Ordnance Department's argument about cost effectiveness with reference to interchangeable manufacturing when, in fact, economy stood relatively low in the department's list of priorities. Like so many other military innovators, Bomford and Raborn were masters of bureaucratic politics. Their impressive records as orchestrators of technological change clearly controvert Lewis Mumford's derisive observation that the military is "the refuge of third-rate minds."[54]

54. Mumford, *Technics and Civilization*, p. 95. Evidence corroborating Douglas's and Morison's observations about able personnel, good fortune, and favorable environments may be found in Glasow, "Prelude to Naval Renaissance"; Hewlett and Duncan, *Nuclear Navy;* Sapolsky, *Polaris System Development;* Aitken, *Taylorism at Watertown Arse-*

In other instances, as David Allison points out in discussing the development of the Sidewinder missile, a favorable environment meant having an organization flexible enough to allow exploratory "hip pocket" research to take place without officially sanctioning or funding it. Institutional flexibility, particularly the ability to alter or circumvent standard bureaucratic procedure when necessary, thus assumes an importance equal to entrepreneurial ability in fostering innovation. Without such flexibility even the most promising ventures can fail to materialize.[55] Integral to this process is an enthusiasm for technological change that often borders on the fanatical. Such passionate commitment has many distinguishing qualities. By far the most common are a compulsive interest in technically elegant problems, an almost childlike fascination for new things, and, of course, a patriotic commitment to national defense. All three traits find adherents outside the military services as well and they therefore serve to encourage openness and interaction between the worlds of science, engineering, business, and defense. Such enthusiasm is a conspicuous feature of the West's ideology of progress, which so often manifests itself in the institutional search for "order" and "system" alluded to earlier.[56]

Historiographic Perspectives

All these themes bear on an old but curiously neglected historiographic debate about the relationship of war to human development. Three scholars in particular—Werner Sombart, Lewis Mumford, and John U. Nef—have shaped the main contours of the discussion. In *Krieg und Kapitalismus,* published on the eve of World War I, Sombart argued that war and the

---

nal; Allison, *New Eye for the Navy;* Redmond and Smith, *Project Whirlwind;* and Horwitch, "Role of the Concorde Threat." Also see the essays of Smith and Allison in this volume.

55. Allison, pp. 315–20 below. Also see Aitken, *Taylorism at Watertown Arsenal;* Allison, *New Eye for the Navy;* Hewlett and Duncan, *Nuclear Navy;* and Sapolsky, *Polaris System Development.*

56. See the essays by Smith and Allison in this volume, as well as Allison, *New Eye for the Navy;* Hacker, "Imaginations in Thrall"; Hewlett and Duncan, *Nuclear Navy;* Horwitch, "Role of the Concorde Threat"; Roland, *Underwater Warfare;* Sapolsky, *Polaris System Development;* and Nancy Stern, *From ENIAC to UNIVAC* (Bedford, MA: Digital Press, 1981). Eugene S. Ferguson provides an insightful examination of technological enthusiasm in "Enthusiasm and Objectivity in Technological Development" (Paper delivered at the meeting of the American Association for the Advancement of Science, Chicago, 1970).

preparations for it spawned capitalism in Europe. He mustered considerable evidence showing that the union of the warlord and the capitalist resulted in the creation of large-scale enterprises designed to meet the increasing pressure of military demands for food, clothing, and the munitions of war. From this impetus issued many of the Western world's great advances in metallurgy, machinery, and standardized production techniques.[57] Such accomplishments, in Sombart's view, tended to outweigh the negative effects of militarism. Indeed, one of his disciples later wrote that "industry learned everything, except invention, from war—organization, discipline, standardization, the coordination of transport and supply, the separation of line and staff, the division of labor."[58]

Although contemporaries recognized the importance of Sombart's book, more than twenty years elapsed before anyone sought either to elaborate or dispute its provocative thesis. One of the first scholars of recognized stature to grapple with Sombart's work was Lewis Mumford, a brilliant iconoclast whose early research focused mainly on the history of architecture, literature, and urban planning. In 1934 Mumford published *Technics and Civilization,* a work of tremendous breadth and insight about the role of technology in human affairs. In it he devoted a chapter to the "Agents of Mechanization," in which he asserted that "militarism forced the pace and cleared a straight path to the development of modern large-scale standardized industry."[59] According to Mumford, armies and factories went hand-in-hand; soldierly habits of thought and practice set the course of modern industrial civilization.

Although Mumford readily acknowledged his intellectual debt to Sombart at many points in *Technics and Civilization,* the general thrust of his argument turned out to be quite different. In contrast to Sombart's positive view of war, capitalism, and technology, Mumford detected a "deterioration of life under the regime of the soldier" owing mainly to the military's tendency to substitute force and rigid uniformity "for patience and intelligence and cooperative effort in the governance of men."[60] He concluded:

57. Sombart, *Krieg und Kapitalismus.*
58. Kaempffert, "War and Technology," p. 443.
59. Mumford, *Technics and Civilization,* p. 87. Also see Mumford's "Authoritarian and Democratic Technics," *Technology and Culture* 5 (1964): 1–9.
60. Mumford, *Technics and Civilization,* p. 94.

The alliance of mechanization and militarization was an unfortunate one: for it tended to restrict the actions of social groups to a military pattern, and it encouraged the rough-and-ready tactics of the militarist in industry. It was unfortunate for society that a power-organization like the army, rather than the more humane and cooperative craft guild, presided over the birth of the modern forms of the machine.[61]

In 1934 Mumford ended his argument on a mildly optimistic note, but subsequently he became more strident in his denunciation of military regimentation and its traumatic effects. Particularly loathesome to him was its association with an elite power complex called "the megamachine" and the vexing problems it posed for mankind. "The murky air of the battlefield and the arsenal blew over the entire field of industrial invention," he wrote in 1970, "and affected civilian life" in ominous ways.[62]

Perhaps his ardent support of Hitler, more than any scholarly shortcomings of his work, prompted severe criticism of Sombart's thesis in the early 1940s by such economic historians as Thomas S. Ashton and John U. Nef. Indeed, at Britain's darkest hour Nef went so far as to associate *Krieg und Kapitalismus* with Nazi aggression. While his argument that wars "interfere with material progress. . . and large-scale enterprise" seemed to carry the day, the intensity of his attack had unfortunate consequences.[63] Historians, perhaps feeling uneasy about studying the relationship between war and technology in an era of nuclear weaponry and "defense Keynesianism," abandoned the vexed issue and turned to more congenial subjects during the postwar period. Except for a handful of economic historians who periodically revived the discussion, the relationship of war and the military to social and economic development failed to attract much scholarly attention.[64] As mentioned earlier, histo-

61. Ibid., p. 96.

62. Mumford, *The Myth of the Machine, Vol. 2, The Pentagon of Power* (New York: Harcourt Brace Jovanovich, 1970), p. 148.

63. John U. Nef, "War and Economic Progress," *Economic History Review* 12 (1942): 37. Also see Nef, *Western Civilization Since the Renaissance* (1950; New York: Harper Torchbooks, 1963); Nef, *Cultural Foundations of Industrial Civilization* (1958; Harper Torchbooks, 1960); Thomas S. Ashton, "The Relation of Economic History to Economic Theory," *Economica* 13 (1946): 84.

64. J. M. Winter provides an excellent historiographic perspective on much of this literature in his introduction to *War and Economic Development* (New York: Cambridge University Press, 1975), pp. 1–10. For entry to a related body of literature on the "military-industrial complex," see B. Franklin Cooling, *War, Business, and American Society;* Cooling, ed., *War, Business and World Military-Industrial Complexes* (Port Washing-

rians of science and technology largely ignored the issue. The same proved true of military historians, who seemed content to confine their interests to the impact of weaponry on warfare. Even William H. McNeill, the most recent entrant into the debate, limits his analysis in *The Pursuit of Power* primarily to weapons as instruments of statecraft and avoids assessing the industrial and social aspects of military innovation. In this respect his focus is narrower than Sombart's, even though his interpretation of military influence is similar. He presents a broad panorama of "the military-technological and political aspects of the rise of European capitalism," their co-evolution, and mutual interaction. Like Sombart, he seeks to show the close connection between military "command economies" and the rise of capitalism, and how in fact they may have been its primary stimulus. He also follows Sombart in criticizing Marxist economic history, though here his argument is neither sustained nor wholly persuasive. In short, McNeill, relying on historical scholarship since 1945, has revived Sombart's thesis and restates it in a way that is more moderate in tone but nonetheless just as sweeping. As Alex Roland indicates in his bibliographic essay in this volume, McNeill's popular work "will likely set the agenda in this field for many years to come."[65]

Assessing Military Enterprise

Quite apart from the serious moral liabilities taken on by militarized societies, the questions remain: Has the influence of the military on civilian industrial development been a good thing? Is military enterprise an effective way of generating new technologies? Has society benefited? The responses to these questions obviously depend on conditions of time, place, and circumstance, to say nothing of individual values. To those who construe military enterprise primarily from the perspective of national security, the answers more often than not are "yes." Indeed some proponents maintain that, given the popularity of

---

ton, N.Y.: Kennikat Press, 1981); Paul A. C. Koistinen, *The Military-Industrial Complex: An Historical Perspective* (New York: Praeger, 1980); Seymour Melman, *Pentagon Capitalism: The Political Economy of War* (New York: McGraw-Hill, 1970); Melman, *The Permanent War Economy* (New York: Simon & Schuster, 1974); Milward, *War, Economy and Society;* and Clive Trebilcock, "British Armaments and European Industrialization, 1890–1914," *Economic History Review* 26 (1973): 254–72.

65. McNeill, *Pursuit of Power,* pp. 116–20, 293, 313, 339; Roland, p. 353 below.

laissez faire ideas in the United States, one of the few ways government can effectively stimulate technological innovation is through the military. The primacy of national defense, they argue, has long served to legitimize research and development efforts that would have otherwise received little if any government support.[66]

With escalating defense budgets in recent years, however, a growing number of economists and political scientists are responding to the above questions with a resounding "no." Echoing the pronouncements of some historians that armies are "pure consumers" that do little or nothing to enhance productivity, they contend that military spending siphons precious human and material resources away from the domestic economy, thus resulting in unmet demands for goods and services and, of course, spiraling inflation.[67] As if this scenario were not enough, the same scholars also detect an equally pernicious effect of military spending on technological innovation. Lloyd J. Dumas argues, for instance, that, owing to the disproportionate allocation of scientific and engineering talent to military research, the rate of commercial innovation has declined appreciably since the late 1960s. "Rising prices and deteriorating relative quality," he observes, "have made U.S.-produced goods increasingly noncompetitive in both world and domestic markets, forcing production cutbacks and plant closures in the U.S. and creating additional inflationary pressure." For Dumas, as well as others, innovation, unemployment, and inflation are not merely linked but laid at the military's doorstep.[68]

66. See, for example, Roland, "The Impact of War Upon Aeronautical Progress." More recent evidence of the military's "agenda-setting power" is presented by Michael Schrage, "Defense Budget Pushes Agenda in High Tech R & D," *Washington Post* (August 12, 1984), pp. F1, F6.

67. See, for example, Joshua Cohen and Joel Rogers, *On Democracy* (New York: Penguin Books, 1983), pp. 35–40, 94–96, 107–10; Lloyd J. Dumas, "Innovation Under Siege," pp. 8–9; Dumas, ed., *The Political Economy of Arms Reduction*, AAAS Selected Symposium 80 (Washington, D.C.: American Association for the Advancement of Science, 1982); Emma Rothschild, "Boom and Bust," *New York Review of Books* 27 (April 3, 1980); Ruth L. Sivard, *World Military and Social Expenditures 1982* (New York: Institute for World Order, 1982); Lester Thurow, "How to Wreck the Economy," *New York Review of Books* 28 (May 14, 1981); John Tirman, ed., *The Militarization of High Technology* (Cambridge, MA: Ballinger Publishing Co., 1984); Lester Thurow, "Which Strategy for Reindustrialization," Technology and Culture Seminar, MIT, March 9, 1981.

68. Dumas, "Innovation Under Siege," p. 8. For an alternative view of the so-called "productivity problem" as a managerial problem, see William J. Abernathy, *The Productivity Dilemma: Roadblock to Innovation in the Automobile Industry* (Baltimore: Johns Hopkins University Press, 1978); Robert H. Hayes and William J. Abernathy, "Managing Our Way to Economic Decline," *Harvard Business Review* (July–Aug. 1980): 67–77; and

Closely connected with the innovation-productivity debate is the argument over whether military-sponsored technologies "spinoff" or "spillover" into the commercial economy with positive social effects. Economists like Dumas are openly skeptical of the concept and cite a 1974 report from the National Academy of Engineering that denies the importance of widespread spillovers from "federally funded programs since World War II."[69] Yet it would be a serious mistake to dismiss the spillover argument as specious. Historians from Sombart to McNeill reveal that important spillovers have occurred. Since they interpret technology either as expanding knowledge or as a social force, they tend to view the spillover process in a positive light. Indeed, much of the evidence presented in this volume corroborates the existence of spillovers in important sectors of the American economy.[70] But the same evidence also suggests that the results of such spillovers are mixed. To appreciate the complexities of the spillover question, it is necessary to dissect the thesis to see where it is supported by empirical evidence and where it is not.

A good starting point is provided by Thomas Misa's essay about the development of the transistor. Misa musters considerable evidence in support of the spillover argument. By sponsoring research, subsidizing engineering development and plant construction, standardizing practice, and disseminating the results of such work, the Army Signal Corps played a central role in defining the structure of the early transistor industry. Indeed, Misa reveals that the Army's programs increased the pace of transistor development to such an extent that the industry actually experienced "a sizable overcapacity" by 1955.

Steve Lohr, "Overhauling America's Business Management," *New York Times Magazine* (January 4, 1981).

69. Dumas, "Innovation Under Siege," p. 9. For a current version of the spinoff argument, see Bruce Steinberg, "The Military Boost to Industry," *Fortune* (April 30, 1984): 42–48.

70. In addition to the essays in this volume and the works of Sombart, Kaempffert, Mumford, McNeill, and Trebilcock cited above, see David Mowry and Nathan Rosenberg, "Government Policy and Innovation in the Commercial Aircraft Industry, 1925–1975" (Unpub. paper, Department of Economics, Stanford University, November 1980); Daniel P. Jones, "From Military to Civilian Technology: The Introduction of Tear Gas for Civil Riot Control," *Technology and Culture* 19 (1978): 151–68; John H. Perkins, "Reshaping Technology in Wartime: The Effect of Military Goals on Entomological Research and Insect-Control Practices," ibid., pp. 169–86; Roland, "Impact of War on Aeronautical Progress"; Constant, *Turbojet Revolution;* Stern, *From ENIAC to UNIVAC;* Hewlett and Duncan, *Nuclear Navy;* Glasow, "Prelude to Naval Renaissance"; and Sapolsky, *Polaris System Development.*

He cautions, however, that a "simplistic 'pump priming' inter-pretation" will not suffice, particularly when one considers the impact the Signal Corps' extensive promotion of diffused tran-sistors had on the industry. The military's emphasis on expen-sive high performance diffusion devices created a tension between its needs and the commercial interests of the Bell Sys-tem, so much so that by 1958 Bell executives began to worry that military requirements "had complicated and perhaps even compromised" their firm's production efficiency. As a result, Misa concludes, the benefit of "the spillover from military to commercial use was incomplete at best."[71]

David Noble's research corroborates Misa's findings. In his study of the development of numerically controlled machine tools, Noble readily acknowledges the Air Force's importance as a promoter of the new technology during the 1950s. But he also points out that the Air Force's design requirements, with their emphasis on sophisticated electronic computer program-ming, proved to be more complex than most manufacturers needed, or could afford. Consequently, when German and Jap-anese machine tool producers began to introduce cheaper and more flexible NC machines and software systems in the 1970s, they quickly dominated the market. In effect Noble argues that a spillover from military to commercial uses occurred in the United States, but in the long run it worked to the detriment of American machine tool builders because NC was designed to satisfy military rather than commercial needs.[72]

His interpretation of technology as a social product and so-cial process leads Noble to some significant observations about the spillover question that more technically oriented scholars often overlook. His criticism of writers who view the military's role in technological development as an "externality" is strongly reinforced by other essays in this book. Equally significant is his treatment of the values inherent in military technologies them-selves and how these values spill over when such techniques are transferred to the civilian sector. He argues, for instance, that as much as new technologies like interchangeable manufactur-ing, numerical control, or containerization suited the military's ideology of command and uniformity, their introduction had

71. Misa, pp. 276, 285, 284, 287, respectively.
72. In addition to his chapter below, see Noble's "Social Choice in Machine Design" and *Forces of Production.*

serious, even "disastrous" consequences for workers. By rationalizing and restructuring work in new ways, these innovations struck at well-established traditions and threatened labor's power to control the shop floor. For engineers and managers, on the other hand, there was a certain aesthetic attractiveness and economic symmetry to rationalization. Above all, the new technologies promised to "shorten the chain of command" between the office and the shop, something which managers and technocrats considered absolutely necessary. Shortening the chain of command through technological innovation thus meant different things to different people. Managers and engineers, like their military counterparts, found the idea advantageous as well as socially constructive. Workers found it noxious. How one reacted to such changes depended on where one stood in the organizational and social hierarchy.

In the final analysis the main question is not whether spillovers have occurred. They most certainly have, and in a number of important areas. The primary question is whether these spillovers have been beneficial in a socioeconomic sense and, if so, to what members of society. There are always winners and losers in the process of technological change. The ability to produce more goods at cheaper costs, for example, is acquired not only through the introduction of more sophisticated manufacturing techniques but also through increased discipline on the shop floor and, unfortunately, structural unemployment. Industrialists and consumers by and large applaud such changes because they seem more economical, hence personally appealing. But these advantages are almost always acquired at the expense of someone else. This is what Noble is driving at when he asks: "Progress for whom and for what?. . . What kind of progress can we, as a society, afford?"[73]

A large body of literature exists about economic growth and industrial development in the United States, but it reveals little about the military's participation. Why? One reason is the persistence of a deeply rooted tradition that sanctions free enterprise and decries government interference in the economy. Yet, as the following essays reveal, the military has been an important agent of technological and managerial innovation. By linking national defense with national welfare, it has sponsored all types of research and development and has served as

73. Noble, p. 346 below.

an important disseminator of new technologies. Just as it helped to inaugurate the industrial revolution in America, it continues to alter the structure of industrial society today. Some aspects of this enterprise are explored in this book. Many more need to be studied. Clearly an assessment of the social value of the military's role in technological change requires a deeper historical understanding of the full range and force of its influence. Although much work remains to be done on the significance of military enterprise in American society, the main outline of the subject is becoming clear. Scholars must recognize the complementarity of military and economic forces as an essential influence on the American industrial system. They also must advance their work on several interpretative fronts. This includes further investigation of the military's contribution to the expansion and impact of technical knowledge. But, above all, it means placing military enterprise in a context that reflects the complexity of technological change as a product of society and its relentless social processes.

1

Army Ordnance and the "American system" of Manufacturing, 1815–1861

Merritt Roe Smith

The mechanization of arms making represents one of the earliest and most important phases of the industrial revolution in America. Private entrepreneurs like Eli Whitney and Samuel Colt are often credited with introducing many of the novel production methods that eventually became known as the "American system of manufactures," but this essay points to a more complex background and assigns an important innovative role to members of the U.S. Army Ordnance Department. Specifically, the essay examines the conceptual and institutional framework the military provided for developing and disseminating the new technology. It also reveals how the manufacturing system reflected the ideology of its military promoters and how factory masters and operatives struggled to define and control the regimen of production during the antebellum period.

The Ordnance Department, with the force under its control, displays . . . energy and skill in the fabrication of arms and other munitions of war. . . . Indeed, the reports . . . exhibit the Army of the United States not in the light in which standing armies in time of peace have usually been regarded, as drones who are consuming the labor of others, but as a body of military and civil engineers, artificers, and laborers, who probably contribute more than any other equal number of citizens not only to the security of the country but to the advancement of its useful arts. (*Annual Report of the Secretary of War*, November 24, 1828)

Government's pivotal role in the antebellum American economy is well known and thoroughly documented.[1] Indeed, all levels of government—federal, state, and local—evolved policies aimed at ensuring domestic stability and promoting economic growth. The magnitude of this involvement ranged widely from direct grants and subsidies to such indirect aids as the provision of banking facilities, tax exemptions, and tariff and patent legislation. Confronted by barriers to expansion, nineteenth-century Americans expected public institutions to enter the economic arena by assuming initial risks, removing bottlenecks, and cultivating resources until private enterprises could become established on a profitable basis. Besides fostering a climate in which private business could organize and flourish, the actual presence of government agents in developing regions had an incalculable psychological impact that spurred public confidence, attracted investment, and stimulated industrial development. Such activities not only served utilitarian purposes but also strengthened popular beliefs in progress, prosperity, and perfectibility. On these ideological foundations rested the viability of republican institutions and the promise of American life.

One of the most visible agents of government enterprise was the United States Army. Due partly to a traditional commit-

---

1. For entry to the literature on this subject, see Robert A. Lively, "The American System: A Review Article," *Business History Review* 29 (1955): 81–96; Sidney Fine, *Laissez Faire and the General Welfare State* (Ann Arbor: University of Michigan Press, 1956), chap. 1; Henry W. Broude, "The Role of the State in American Economic Development, 1820–1890," in *The State and Economic Growth*, ed. Hugh G. J. Aitken (New York: Social Science Research Council, 1959), pp. 4–22; David D. VanTassel and Michael G. Hall, eds., *Science and Society in the United States* (Homewood, IL: The Dorsey Press, 1966), chaps. 8–9; Carter Goodrich, "Internal Improvements Reconsidered," *Journal of Economic History* 30 (1970): 289–311; and Harry N. Scheiber, "Government and the Economy: Studies of the 'Commonwealth' Policy in Nineteenth-Century America," *Journal of Interdisciplinary History* 3 (1972): 135–54.

ment to national security, partly to a romantic attachment to Manifest Destiny, and partly to an abiding concern for conveying a favorable image in a society inherently skeptical of standing armies, members of the military willingly, even eagerly, participated in programs for internal domestic improvement. While the regular Army and Quartermaster Department contributed significantly to the settlement and stabilization of the frontier, the real thrust of military enterprise emanated from three technical bureaus housed in the War Department: the Corps of Engineers, the short-lived Topographical Bureau (1838–63), and the Ordnance Department. These bureaus undertook numerous projects directly affecting the course of diplomacy, science, and industry in America. Precisely because their activities touched so many vital interests of the republic, they occupied a central position in programs for national development, which enabled them to demonstrate their usefulness to society. Yet in the highly partisan ferment of Jacksonian America, not everything they did with internal improvements proved satisfying. Since jobs, contracts, and general regional prosperity weighed in the balance, the bureaus could not escape controversy. Indeed, as public agencies charged with civil as well as military duties, they frequently found themselves in the difficult position of having to serve rival constituencies, with all the associated pressures of congressional scrutiny, sectional jealousy, and pork-barrel politics. That they succeeded in addressing problems of national growth in a tempestuous era testifies not only to their professional zeal but also to their role as catalysts of change and consolidation.

Much has been written about the Topographical Bureau and the Corps of Engineers, whose extensive explorations, geodetic surveys, and construction activities resulted in an impressive fund of scientific data as well as in a wide variety of civil works.[2] A good deal less is known about the exploits of the Ordnance Department, especially its involvement in one of the great technological achievements of the nineteenth century, popularly known as the "American system" of manufactures. The term

2. For the undertakings of the Topographical Corps, see William H. Goetzmann's prize-winning study *Army Exploration in the American West* (New Haven: Yale University Press, 1959). For the Corps of Engineers, see Forest G. Hill, "Government Engineering Aid to Railroads Before the Civil War," *Journal of Economic History* 11 (1951): 235–46; Hill, *Roads, Rails and Waterways: The Army Engineers and Early Transportation* (Norman: University of Oklahoma Press, 1957); and Hill, "Formative Relations of American Enterprise, Government and Science," *Political Science Quarterly* 75 (1960): 400–19.

evidently originated in 1854 when a British military commission investigated the machinery of the United States. In their published report, the commissioners mentioned "the American system" with specific reference to the division of labor and application of machinery in the production of firearms with interchangeable parts. That the expression was printed in lower case and appeared only once in the report did not seem to bother those historians who have seized upon it as a catch-all phrase to describe developments in the American small arms industry prior to the Civil War.[3] Interestingly enough, the ordnance officers and civilian mechanics who developed the system never used the phrase, preferring instead to describe their work as part of a larger program to "insure system and uniformity" in all materiel of war.[4] The subject occupied the minds and efforts of ordnance officials for more than forty years, and this essay seeks to analyze the evolving concept of uniformity and to assess the Ordnance Department's role in facilitating the rise of the new technology.

The year 1815 was one of jubilation in the United States. The War of 1812 had ended and, amid bonfires, cannon salutes, and tolling church bells, citizens everywhere joined in a national celebration the likes of which had not been seen since the Revolution. Republicanism had been vindicated, despotism vanquished, and independence finally achieved. Exhilarated by the public optimism and air of self-congratulation that followed the Treaty of Ghent, the popular press conveyed the feeling that a new era of "Peace and Plenty" was at hand. Slowly but perceptibly, the Revolutionary ideals of liberty and republican virtue were taking on renewed meaning in expectations of progress and prosperity. In such a milieu nothing seemed im-

---

3. The original report appeared in the British *Parliamentary Papers* 50 (1854–55) and is reprinted in Nathan Rosenberg, ed., *The American System of Manufactures* (Edinburgh, Scotland: Edinburgh University Press, 1969), p. 143. James Nasmyth also used the expression in his *Autobiography* (1883) with specific reference to developments at the Springfield armory. Joseph W. Roe later quoted Nasmyth and adopted the term in his pathbreaking history of *English and American Tool Builders* (New Haven: Yale University Press, 1916), pp. 129, 140–41. One of the earliest American writers to refer to the "American system" was Charles H. Fitch in his 1880 census, *Report on the Manufacture of Fire-Arms and Ammunition* (Washington, D.C.: GPO, 1882), pp. 4–5.

4. See Letters Sent–Letters Received and Reports of Inspections of Arsenals and Depots in the Records of the Office of the Chief of Ordnance, Record Group 156, National Archives, Washington, D.C. (hereafter cited as OCO); and Stephen V. Benet, ed., *A Collection of Annual Reports and Other Important Papers, Relating to the Ordnance Department*, 3 vols. (Washington, D.C.: GPO, 1878–90).

possible. Wiser men knew, however, that the country had barely skirted disaster during the war; good fortune rather than military might had ensured the nation's survival. From the beginning faulty arms, insufficient supplies, and tactical errors had plagued the war effort—facts that no one, in 1815, appreciated more than the Secretary of War and his military staff.[5]

Three years earlier, at the urging of the executive branch, Congress had attempted to revamp the cumbrous army supply system by creating three central bureaus—the Quartermaster Department, the Commissary General of Purchases, and the Ordnance Department—to assist the beleaguered War Department in procuring, inspecting, and distributing military equipment. Because the legislation failed either to clarify the relationship between these agencies or to define adequately their respective duties, their activities overlapped at many points during the ensuing war years. These ambiguities resulted in a confusion of authority that perpetuated problems which had plagued the army for decades.[6]

With the end of hostilities, Congress again sought to remedy the situation by passing in February 1815 "An Act for the better regulation of the Ordnance Department." In contrast to the bill that had created this bureau in 1812, the new legislation carefully spelled out the duties and responsibilities of the office and expanded its sphere of authority. Formerly the department's primary mission had consisted of inspecting cannon, proving gunpowder, and supervising the manufacture and storage of gun carriages, munitions, and other equipment at several federal arsenals. Ordnance officers possessed no authority to make contracts, nor did they exercise jurisdiction over the procurement and production of small arms. Under the act of 1815, however, all this changed. In addition to transferring responsibility for the negotiation and supervision of all

---

5. Michael Kammen, "From Liberty to Prosperity: Reflections Upon the Role of Revolutionary Iconography in National Tradition," *Proceedings of the American Antiquarian Society* 86 (1976): 263–72; Russel B. Nye, *Society and Culture in America, 1830–1860* (New York: Harper and Row, 1974), pp. 1–31; Leonard G. White, *The Jeffersonians: A Study in Administrative History, 1801–1829* (New York: The Free Press, 1965), pp. 211–39.

6. U.S. *Statutes at Large*, 2: 732–34; White, *The Jeffersonians*, pp. 215–16, 224–32. For an analysis of small arms procurement before 1812, see Merritt Roe Smith, "Military Arsenals and Industry Before World War I," in *War, Business, and American Society: Historical Perspectives on the Military-Industrial Complex*, ed. B. Franklin Cooling (Port Washington, NY: Kennikat Press, 1977), pp. 25–28.

arms contracts from the Commissary General of Purchases to the Ordnance Department, the bill also placed the national armories at Springfield, Massachusetts, and Harpers Ferry, Virginia, under the latter's immediate command. Equally significant, the reorganization act empowered the chief of ordnance "to draw up a system of regulations . . . for the uniformity of manufactures of all arms ordnance, ordnance stores, implements, and apparatus, and for the repairing and better preservation of the same." For the next forty years this charge became the guiding principle of ordnance policy. Although little noticed at the time, the proviso set the stage for important developments in military technology and the eventual transformation of the American industrial system.[7]

Inquiry into the origins of the uniformity system sheds much light on the processes of technological innovation, but it also raises a fundamental question about the originality of the idea as an American invention. The United States was a developing nation in the early nineteenth century and as such relied on Europe for both manufactured goods and manufacturing techniques.[8] While the immediate need for uniform ordnance sprang from the bitter logistical experience of the War of 1812, the actual formation of the policy emanated from sources deeply rooted in the French military tradition. Since the Revolutionary War, French artillerists and engineers had exercised a pervasive influence on the United States Army. Through them, engineering treatises, testing procedures, arms designs, and educational techniques had made their way to American shores and were assimilated by native officers. Indeed, whenever the United States needed to revise and improve its military program, it looked primarily to the "French system" for appropriate models. This practice continued well into the 1840s.[9]

7. U.S. *Statutes at Large* 3: 203–5. Stanley L. Falk describes these changes as a "Technological Revolution in military arms and equipment" in "Soldier-Technologist: Major Alfred Mordecai and the Beginnings of Science in the United States Army" (Ph.D. dissertation, Georgetown University, 1959), p. 1 and passim.

8. See, for example, Eugene S. Ferguson, "On the Origin and Development of American Mechanical 'Know-How,'" *Midcontinent American Studies Journal* 3 (1962): 3–15; Norman B. Wilkinson, "Brandywine Borrowings from European Technology," *Technology and Culture* 4 (1963): 1–13; Paul J. Uselding, "Henry Burden and the Question of Anglo-American Technological Transfer in the 19th-century," *Journal of Economic History* 30 (1970): 312–37; and David J. Jeremy, "British Textile Technology Transmission to the United States," *Business History Review* 47 (1973): 24–52.

9. For a revealing discussion of the French influence on the U.S. military system, see General Winfield Scott to Secretary of War Lewis Cass, May 24, 1834, Letters Received, Records of the Headquarters of the Army, Record Group 108, National Archives. Also

Among the most influential agents of French technology was Major Louis de Tousard. A graduate of the artillery school at Strasbourg, an aide to Lafayette during the American Revolution, and a skilled practitioner of military engineering, Tousard returned to the United States in 1793 after being convicted of subversive activities by the French National Convention and briefly jailed at the "bloody prison of l'Abbaye." Two years later he joined the newly created Corps of Artillerists and Engineers. He spent the next seven years on special assignments as an inspector of artillery and fortifications. Among other duties, he supervised the construction of several forts along the eastern seaboard, advised founders on the casting and boring of cannon, tested and experimented with iron ordnance, and initiated efforts to convert the garrison at West Point into a full-fledged military academy with a curriculum modeled after the École Polytechnique.[10] At President Washington's request he also initiated research on a three-volume work entitled *American Artillerist's Companion*. This study, published in 1809, synthesized existing knowledge in the field and served as the standard text for American officers for more than a decade. Throughout his treatise Tousard emphasized the need to devise "a system of uniformity and regularity" based on scientific theory and experimentation. "This want of uniformity," he cautioned readers, "impeded for a long time the progress of the French artillery, as it will that of America, unless a similar system is adopted."[11]

---

see U.S. Ordnance Department, *Ordnance Manual for the Use of the Officers of the United States Army* (Washington, D.C.: J. and G. S. Gideon, 1841); Edward C. Ezell, "The Development of Artillery for the United States Land Service Before 1861: With Emphasis on the Rodman Gun" (M.A. thesis, University of Delaware, 1963), pp. 80–82, 122; Peter M. Molloy, "Technical Education and the Young Republic: West Point as America's École Polytechnique, 1802–1833" (Ph.D. dissertation, Brown University, 1975); Benet, *Ordnance Reports*, 1: 18–19, 29, 53–57, 185, 202, 214, 224, 233–34, 510. For references to the "French system of artillery," see William E. Birkhimer, *Historical Sketch of the Organization, Administration, Materiel and Tactics of the Artillery, United States Army* (1884; reprint ed., New York: Greenwood Press, 1968), pp. 225, 235; and Louis de Tousard, *American Artillerist's Companion*, 3 vols. (Philadelphia: C. and A. Conrad and Co., 1809), 1: 200.

10. *Dictionary of American Biography*, s.v. Tousard, Anne Louis de (1749–1817); Norman B. Wilkinson, "The Forgotten 'Founder' of West Point," *Military Affairs* 24 (1960–61): 177–88; Ezell, "Development of Artillery," pp. 79–87.

11. Tousard, *Artillerist's Companion*, 1: v–vi; 2: iii–iv, xiv; Birkhimer, *Historical Sketch of Artillery*, p. 302; William Wade, "Early Systems of Artillery," in *Ordnance Notes, No. 25* (Washington, D.C.: Ordnance Office, 1874), pp. 140–41. Another influential French tract was Colonel Jonathan Williams' translation of De Sheel's *A Treatise on Artillery* (1800).

The "system" to which Tousard referred was the brainchild of General Jean-Baptiste de Gribeauval, who in 1765 had initiated thoroughgoing reforms in the organization and production of French ordnance. Under Gribeauval's direction armorers had developed methods of manufacturing gun carriages and other equipment with standardized parts. Consolidated models, improved techniques, and rigid inspections distinguished the new system. Its strategic significance was soon revealed in the mobility of Rochambeau's light artillery during the American Revolution and later in Napoleon's speedy conquest of Italy.

British observers readily appreciated the tactical advantages of Gribeauval's uniform designs. One officer wrote:

At the formation of their system, they saw the necessity of the most exact correspondence in the most minute particulars, and so rigidly have they adhered to this principle that, though they have several arsenals, where carriages and other military machines are constructed, the different parts of a carriage may be collected from these several arsenals, in the opposite extremities of the country, and will as well unite and form a carriage as if they were all made and fitted in the same workshop. As long as every man who fancies he has made an improvement is permitted to introduce it into our service, this cannot be the case with us.[12]

Such reports prompted the British government to introduce similar changes in carriage construction, an undertaking that

---

12. Captain Ralph Willet Adye, quoted by Tousard, *Artillerist's Companion*, 1: 200; 2: xiii–xvi, xviii. "In order to attain this desired uniformity," Tousard observed, "a certain number of patterns for the length and figure of the guns; for the dimensions and proportions of the carriages and appendages; and for the best manner of training them, was irrevocably fixed by regulations. Exact tables were formed, and accurate scales of the dimensions of every article used in artillery, made and marked with the seal of the government. Models for workmen in wood; patterns, matrices, and dies, for the forges, were made in conformity to those measures to insure the regularity of the principal forms in the carriage work. These were placed in every arsenal, and *officers made accountable for their construction and reception.* Then the expense eventually lessened, evinced the advantage of having adopted a system, and of having introduced economy, precision, and uniformity in this service." It is noteworthy that American ordnance officers used virtually the same language and argument in explaining and defending their plans for uniformity before Congressional committees during the early nineteenth century.

Studies assessing the origins and technological implications of the Gribeauval and British stock-trail systems are needed. Selma Thomas is currently investigating the origins of the French uniformity system. For a preview of her work, see " 'The Greatest Economy and the Most Exact Precision': The Work of Honoré Blanc" (Paper delivered at the Twentieth Annual Meeting of the Society for the History of Technology, Washington, D.C., October 21, 1977). I am indebted to Ms. Thomas for providing a copy of her paper and sharing her research findings on Gribeauval.

culminated around 1794 with the famous "stock-trail" arrangement featuring interchangeable wheels, pintle hooks, and wrought-iron axles (figures 1, 2). These improvements were eventually adopted by France in 1827 and, through French contacts, subsequently influenced American artillery designs during the 1830s. Yet even though the stock-trail system eventually displaced Gribeauval's original artillery plan, his conceptual insights regarding "system and uniformity" made a deep impression on American officers and continued to provide an essential focus for their activities during the early national period.[13]

Gribeauval's influence also manifested itself in another way. As Inspector General of Artillery, he had played an instrumental role in promoting the work of Honoré Blanc, a talented armorer who had developed various labor-intensive methods of manufacturing muskets with uniform parts. Favored by Gribeauval's patronage, Blanc had received approval to undertake experiments with the design and manufacture of small arms at several government installations. By the mid-1780s he had tooled the Vincennes arsenal with novel die-forging, jig-filing, and hollow-milling techniques capable of effecting "the greatest economy and the most exact precision."[14] These innovations held particular interest for American visitors. Thomas Jefferson, while serving as ambassador to Versailles, had visited Blanc in 1785 to witness the way in which the lock components of Blanc's muskets could be randomly selected and interchanged without any fitting or filing. Duly impressed by the demonstration, Jefferson wrote home about the experience and even tried to persuade Blanc to emigrate to the United States. Although the Frenchman declined, Jefferson continued to monitor his activities and even shipped six of Blanc's muskets to Philadelphia in 1789. One person who profited from these diplomatic exchanges was Eli Whitney. Through correspondence and conversation with Jefferson, Secretary of Treasury Oliver Wolcott, and other government officials, Whitney evi-

---

13. Birkhimer, *Historical Sketch of Artillery*, pp. 227–49.

14. Honoré Blanc, *Mémoire Importante sur la fabrication des armes de guerre, à l'Assemblée Nationale* (Paris, 1790), quoted by Thomas, p. 1. For discussions of Blanc's techniques, see William F. Durfee, "The First Systematic Attempt at Interchangeability in Firearms," *Cassier's Magazine* 5 (1893–94): 469–77; Roe, *English and American Tool Builders*, pp. 129–30; Edwin A. Battison, "Eli Whitney and the Milling Machine," *Smithsonian Journal of History* 1 (1966): 11–26; and Merritt Roe Smith, *Harpers Ferry Armory and the New Technology* (Ithaca, NY: Cornell University Press, 1977), pp. 88–9, 91, 232.

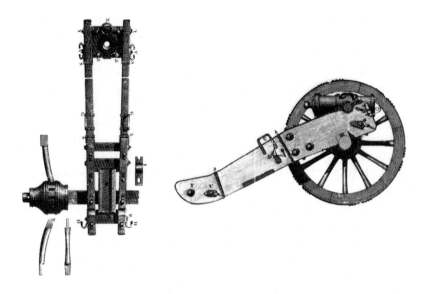

Figure 1
Gribeauval carriage. Note the bracketed form of construction. *Source:* Louis de Tousard, *American Artillerist's Companion*, 1809.

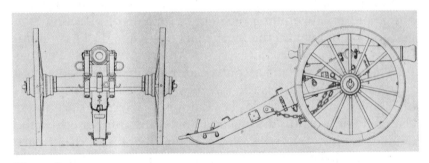

Figure 2
Stock-trail carriage. *Source:* Alfred Mordecai, *Artillery for the United States Land Service*, 1849. Smithsonian Institution Neg. No. 46476E.

dently learned of Blanc's work and tried to emulate it in filling his musket contracts with the War Department. Although his efforts fell short of success, he nonetheless became a zealous advocate of the uniformity principle, popularized the concept, and persuaded many politicians to support policies aimed at standardizing the manufacture of military arms.[15]

All these factors weighed upon the first chief of ordnance, Colonel Decius Wadsworth. As an intimate friend of Whitney, he especially appreciated the importance of uniformity in small arms and had continually championed the New Haven manufacturer's interests at the War Department. As a former member of the Corps of Artillerists and Engineers, he also had worked closely with Tousard, Stephen Rochefontaine, and other Frenchmen in American service. Through them he became well acquainted with French practice, particularly Gribeauval's uniform system of artillery. Hence, upon assuming command of the Ordnance Department in 1812, he openly espoused *"Uniformity Simplicity* and *Solidarity"* and made the phrase his motto. The words poignantly reflected the degree to which he was influenced by French ideas, and thereafter they shaped not only his actions but those of his staff as well.[16]

As early as 1813, Wadsworth had expressed dissatisfaction with the hectic disarray of American ordnance. In language reminiscent of Tousard, he had written a fellow officer about the innumerable variations in artillery and of the urgent need to rectify the situation. In the past, he explained, "every superintendent selected whatever pattern and introduced whatever alteration his fancy suggested." Under present circumstances this could no longer be tolerated:

15. Jeanette Mirsky and Allan Nevins, *The World of Eli Whitney* (New York: The Macmillan Co., 1952), pp. 208, 214, 219–22; Robert S. Woodbury, "The Legend of Eli Whitney and Interchangeable Parts," *Technology and Culture* 1 (1960): 243–44; Edwin A. Battison, *Muskets to Mass Production* (Windsor, VT: The American Precision Museum, 1976), pp. 8–9. For a recent evaluation of Whitney's accomplishments, see Merritt Roe Smith, "Eli Whitney and the American System," in *Technology in America: A History of Individuals and Ideas,* ed. Carroll W. Pursell (Cambridge, MA: The MIT Press, 1981), pp. 45–61.
16. Mirsky and Nevins, *Eli Whitney,* pp. 195–96, 200, 206–7, 261, 267–68, 291; Benet, *Ordnance Reports,* 1: 15–17; C. Wingate Reed, "Decius Wadsworth, First Chief of Ordnance, U.S. Army, 1812–1821," *Army Ordnance* 24 (May–June 1943): 527–30, Part 1; 25 (July–August 1943): 113–16, Part 2; Birkhimer, *Historical Sketch of Artillery,* p. 278; Wadsworth to Secretary of War, August 8, 1812, Letters Sent to the Secretary of War, OCO; Whitney to Roswell Lee, November 10, 1821, Letters Received, Records of the Springfield Armory, Record Group 156, National Archives (hereafter cited as SAR).

The necessity of some regulation to secure *simplicity* and uniformity must be obvious to all; yet men of reflection and experience alone can duly estimate the importance of these two qualities. Every variation in the proportions of pieces of the same calibre exacts a corresponding change in the carriage, and for every distinct calibre will be required not only a suitable carriage but its appropriate equipments and ammunition.

"In a word," he concluded, "unless the number of our calibres and their variations be reasonably reduced, and the whole be settled by some permanent regulation, no possible exertion can give to our artillery that perfection its importance merits and which the public service requires."[17]

Since similar deficiencies existed in the design and manufacture of small arms, Wadsworth called for a comprehensive reform, arguing that all ordnance should be consolidated, systematized, and rigidly regulated. Who would implement these changes? His answer, of course, was "the Ordnance Department." To this end he had actively lobbied to secure passage of the 1815 act reorganizing the department. The language of the bill reflected his influence, and its comprehensiveness provided him with the necessary authority to pursue his cherished goal of "system and uniformity."[18]

As a bachelor with few family obligations, Wadsworth completely immersed himself in professional affairs and made uniformity his special calling. Whether scrutinizing the propositions of private contractors, enforcing departmental discipline, or reorganizing the federal arsenals into those for "construction" and those for "storage and repair," he labored diligently to effect a workable agenda for reform. Nothing seemed to escape his attention. Yet, as much as everyday details absorbed his energy, the key to his policy rested on the introduction of fundamental changes in the government's manufacturing program.[19]

---

17. Wadsworth, May 27, 1813, quoted by Birkhimer, *Historical Sketch of Artillery*, p. 386.
18. Ibid., pp. 277–78; Benet, *Ordnance Reports*, 1: 5–6; U.S. *Statutes at Large*, 3:203–5.
19. See, for example, Benet, *Ordnance Reports*, 1: 19–79. By 1821, the year of Wadsworth's death, the Ordnance Department maintained three arsenals of construction at Watervliet, New York; Pittsburgh, Pennsylvania; and Washington, D.C. Of the twenty-three arsenals reporting to the department in 1855, four (Watervliet, Allegheny, Washington, and St. Louis) were arsenals of construction, six were "arsenals of repair," and thirteen were "arsenals of deposit."

Wadsworth always had a special attachment to artillery and, as chief of ordnance, he most often thought and spoke of uniformity in that context. Even before the reorganization of 1815, he had begun to consider means of improving artillery. By 1817 he had proposed the adoption of the British stock-trail system, the effectiveness of which he had witnessed during the War of 1812. To a bureaucracy officially committed to Gribeauval's designs, however, the time was not propitious for such a drastic change. Officers in his own bureau expressed uncertain feelings about the proposal. This uneasiness, coupled with a bitter jurisdictional dispute with the artillerists, forced Wadsworth to satisfy himself with piecemeal efforts to regularize calibers and improve the manufacture of field and siege-garrison equipment under the Gribeauval plan. It was, to be sure, an unsatisfactory arrangement, one that Wadsworth deeply resented. But, given the posture of his peers, he realized that more thoroughgoing reforms would have to wait until petty jealousies and traditions had dissipated. As a result, the real thrust of his uniformity policy centered on parallel developments in the manufacture of small arms.[20]

The first indication that the Ordnance Department intended to take action occurred in June 1815, when Wadsworth called a special meeting at New Haven, Connecticut, to discuss the problems of standardizing firearms and to formulate an appropriate strategy. Present were Superintendent Roswell Lee of the Springfield armory, Superintendent James Stubblefield of the Harpers Ferry armory, former Springfield superintendent Benjamin Prescott, and Wadsworth's old friend and confidant Eli Whitney, who hosted the gathering. After several days of deliberation, they agreed that uniformity should first be applied to the manufacture of muskets and thereafter extended to all military sidearms. To this end Wadsworth assigned Stubblefield the task of making several pattern pieces at Harpers Ferry based, appropriately, upon the French model musket of 1777. With the completion of this phase, the plan called for further preparations and tests of pattern muskets and rifles and their eventual manufacture as regular products at the national armories. If the experiment proved successful, the group

20. Ibid., 1: 29, 34–36, 53–55, 58–61, 71–72; Reed, "Decius Wadsworth," Part 1, p. 530, Part 2, p. 115; Birkhimer, *Historical Sketch of Artillery*, pp. 230–37, 260–62; Ezell, "Development of Artillery," pp. 95–100; Falk, "Soldier-Technologist," pp. 231–33; Wade, "Early Systems of Artillery," pp. 141–42, 148.

decided, the program would then be extended to arms made by private contractors.[21]

To his credit, Wadsworth understood that the Ordnance Department could not devise a complex engineering strategy and then simply withdraw as a passive observer of the undertaking. Success depended on closely monitoring and orchestrating every stage of the plan. Since he was overworked and beginning to feel the painful effects of a malignant tumor on his arm, he also knew that he had to select a deputy with the necessary skill and determination to oversee the new program and stand firm in the face of opposition. That person was Lieutenant Colonel George Bomford, an extremely able officer who was a graduate of West Point and who had served since 1812 as Wadsworth's principal assistant (figure 3). To Bomford fell the responsibility of implementing the uniformity system in America.

Wadsworth and Bomford made strange bedfellows. Although both men came from well-to-do backgrounds and had powerful political connections, no two persons could have been more different in habit and demeanor. Wadsworth was outspoken, Bomford reserved. Wadsworth's mercurial temperament contrasted sharply with Bomford's calculated restraint, just as Bomford's flamboyant lifestyle clashed with Wadsworth's puritan austerity. They frequently disagreed on issues, yet both men realized that despite their differences they complemented and strengthened each other. What united them was a zealous commitment to the idea of uniformity and a single determination to see the system introduced on a large scale. Although the cantankerous chief would not live to see the culmination of the mechanical ideal, his ideas would be carried forward and elaborated by his protégé. As Wadsworth's designated successor, Bomford would head the Ordnance Department for more than twenty years (1821–42), during which he would witness every important development in the firearms industry. Indeed, from an administrative standpoint, Bomford would do more than any other person to make the uniformity system an American reality.[22]

21. Wadsworth to Colonel George Bomford, May 15, June 13, 1815, Stubblefield and Lee to Wadsworth, March 30, 1816, Correspondence Relating to Inventions, OCO; Lee to Stubblefield, August 6, 1816, Letters Sent, SAR.

22. For further information on Bomford (1780–1848), see Smith, *Harpers Ferry Armory*, p. 153.

Figure 3
Lieutenant Colonel George Bomford (1782–1848), chief of the Ordnance
Department from 1821 to 1842. *Source: Records of the Columbia Historical Society* of Washington, D.C., vol. 13 (1910). Smithsonian Institution Neg. No.
81-193.

As architects of the new system, Wadsworth and Bomford
pushed relentlessly for its speedy introduction at the national
armories. Initially neither officer contemplated serious problems. Their timetable called for completion of the pattern arms
by the fall of 1815, with full-scale production of the new
"Model 1816" musket to begin at both armories before the turn
of the year (figure 4). When Stubblefield failed to deliver the
patterns on time, however, Wadsworth began to realize that
neither acceptance nor success would come easily. "It is a Pity
that so much Tardiness is manifested in deciding on the Model
of the Musket," he wrote Bomford in August 1815. "The Delay
looks to me as if the Model I had proposed, or rather the basis
which I had laid down for constructing the Model was not
altogether approved of." This suspicion was confirmed when
the patterns finally reached the Ordnance Department three

Figure 4
U.S. Flintlock Musket Model 1816 (pattern piece dated 1822) manufactured at the Springfield armory. *Source:* Collections of the Division of Armed Forces History, National Museum of American History. Smithsonian Institution Neg. No. 77-3318.

months later, only to be found deficient in uniformity as well as design.[23]

Sobered by the experience, Wadsworth ordered the preparation of new pattern pieces—this time at both national armories. Moreover, he directed the superintendents to cooperate with one another not only in preparing the patterns but also in all matters related to management and manufacturing on the uniformity principle. This instruction, repeated time and again during the next fifteen years, struck a responsive chord with Springfield's Lee, who seized the opportunity to improve interarmory communications by befriending and establishing an ongoing dialogue with Stubblefield, the Harpers Ferry superintendent. Through correspondence and periodic visits, both men began to transmit information related to annual inventories, accounting methods, wage rates, and general shop procedures. At Lee's initiative they also exchanged men, machinery, and raw materials, a practice which soon spread to other establishments and had important technological ramifications throughout the industry. Not surprisingly, Lee's resourcefulness, outgoing manner, and disciplined attitude made him one of the most valued members of the American arms-making community; no one appreciated this more than the chief of ordnance. Under Lee's energetic direction, Springfield quickly emerged as a leading metalworking center and pivotal clearinghouse for the acquisition and dissemination of technical information (figures 5, 6). Yet, precisely because he was making the armory "a credit to the Government and an ornament to the nation," he frequently found himself em-

23. Wadsworth to Bomford, August 1815, Correspondence Relating to Inventions, OCO.

Figure 5
Roswell Lee, superintendent of the Springfield armory from 1815 until his death in 1833. *Source:* Charles W. Chapin, *Sketches of the Old Inhabitants and Other Citizens of Old Springfield,* 1893.

Figure 6
The upper waterworks at the Springfield armory, 1830. *Source:* Claud E. Fuller, *Springfield Shoulder Arms, 1795–1865,* 1931. Smithsonian Institution Neg. No. 48619.

broiled in affairs that extended far beyond his normal range of activities as a factory master.[24]

Between 1816 and 1819, the Ordnance Department continually reminded Stubblefield and Lee of the need for uniformity in all things and admonished them whenever their efforts fell short of success. Since the craft-oriented armorers of Harpers Ferry resented the intrusion of ordnance officers and often balked at their insistent demands for innovation and change, Stubblefield most frequently received the departmental reprimands. Lee also felt these pressures because Bomford expected him to manage Springfield as well as encourage improvements at Harpers Ferry. At times this dual burden became so intense that he threatened to resign his post and seek employment elsewhere. On other occasions he vented his frustration while urging his superiors to be more patient. A typical instance occurred in November 1817, after Bomford had complained for a third time in two years about "the want of uniformity" at the national armories and demanded an immediate remedy. "It is difficult for a Pattern Musket to be made by any one to *please every body,*" Lee responded. "*Faults* will *really exist* and *many imaginary* ones will be *pointed out.* . . . It must consequently take some time to bring about a uniformity of the component parts of the musket at both Establishments." Lee's continued pleas for "time, patience & perseverence" failed to make much of an impression at the Ordnance office. Captivated by the engineering ideal of uniformity and convinced of its urgency, Bomford wanted results, not excuses, and continued to voice criticism whenever the need arose.[25]

Despite these encounters, Lee never wavered in his allegiance to "the grand object of uniformity." Like Bomford and Wadsworth, he understood that the venture's success depended on systematic procedures and regularized controls and that the Ordnance Department attempted to maintain these standards by closely scrutinizing armory operations. The term "system" meant something very real to these men. Far from being a chimerical symbol, it involved specific regulation of the

24. Stubblefield and Lee to Wadsworth, March 30, 1816, Correspondence Relating to Inventions, OCO; Lee to Stubblefield, August 6, 1816, November 10, 20, December 18, 1817, June 18, July 11, 1818, Letters Sent, SAR; *American State Papers: Military Affairs,* 2: 553.

25. Lee to Stubblefield, November 20, 1817, to Bomford, September 11, 1821, Letters Sent, SAR; Lee to Senior Officer of Ordnance, November 20, 1817, Letters Received, OCO.

total production process from the initial distribution of stock to the final accounting of costs. Above all it embraced the coordination of manufacturing activities between two widely separated arms factories. Like a complex machine—an analogy often used by ordnance officers—the correct functioning of the whole depended on the proper relationship of its parts. Coordination and control thus actuated the systems concept. Much as he disliked the incessant prodding of his military superiors, Lee appreciated the significance of their actions. His recognition that uniformity could be achieved only through constant vigilance and scrutiny by a central bureau accentuated the entrepreneurial role of the Ordnance Department and made him a faithful expediter of its policy.[26]

Ordnance officials relied on two methods of monitoring armory operations. One dealt primarily with fiscal affairs and involved the keeping of accurate accounts; the other addressed problems of quality control and involved the careful inspection of finished firearms. Both methods assumed critical importance in developing and expanding the uniformity system and reflected the influence of uniformity as an organizing concept.

Because every government agency stood accountable to Congress for funds received and expended, the department introduced sophisticated bookkeeping methods at a very early date. Beginning in 1816 all officers in charge of federal arsenals and armories were required by regulations to submit quarterly returns detailing the work performed at their respective posts. Based on standard double-entry bookkeeping, these returns included abstracts and receipts of expenditures for buildings, raw materials, and plant equipment, as well as detailed inventories of work performed and property on hand at the end of each period. The inventories identified and tabulated the types of arms manufactured or repaired and listed components still in progress and the amount of stock—coal, iron, oil, and the like—remaining. In addition, the reporting officer transmitted monthly payroll accounts that recorded the name of each armorer, the type of work he performed, the piece rate for each task, and his total wages. At the end of the fiscal year, the superintendent submitted an annual report that summarized the quarterly returns, tallied production, and pre-

---

26. *American State Papers: Military Affairs*, 2: 543–44, 552–53; Lee to Asa Waters, June 8, 1819, Letters Sent, SAR.

viewed plans for the coming year. After review by the chief of ordnance and his staff, these records were sent to the Treasury Department for final auditing and approval.[27]

Accurate accounts not only served to justify expenditures and appropriations before a cost-conscious Congress, they also permitted the Ordnance Department to evaluate armory and arsenal operations. By examining the accounts of Springfield and Harpers Ferry in 1820, for instance, Wadsworth discovered that piece rates were much higher at the latter establishment. As a result, he ordered a 12.5 percent wage reduction. In like manner, the department used accounts to compare the prices paid for supplies and raw materials at various posts and to alter procurement practices when appropriate. Even more important, accurate bookkeeping provided a means of controlling and coordinating arms inventories throughout the arsenal network, a measure that preceded similar methods in the railroad industry by more than two decades. Precise information on the location, distribution, and condition of equipment among widely scattered arsenals served a strategic purpose. In times of emergency such knowledge enabled officials to direct arms and munitions shipments where they were most needed; in peacetime it allowed them to replenish depleted stocks and dispose of unserviceable weapons. For these reasons the chief of ordnance paid particular attention to the returns of outlying installations and sought constantly to improve their accuracy. One benchmark was achieved around 1829 with the complete standardization of all armory and arsenal accounts. Interestingly enough, however, the armories made little effort until the late 1830s to calculate production costs based on allowances for interest, insurance, and depreciation. Even then ordnance officials used the data only to make rough comparisons of work done at Springfield and Harpers Ferry and not to achieve more effective control over internal production processes. Because they operated in a guaranteed market with output quotas more or less fixed by congressional appropriations, they evidently saw no need to explore further the managerial implications of accurate costing.[28]

27. See payrolls and accounts of U.S. armories and arsenals, 1816–50, Second Auditor's Accounts, Records of the United States General Accounting Office, Record Group 217, National Archives.

28. Smith, *Harpers Ferry Armory*, p. 152; Felicia J. Deyrup, *Arms Makers of the Connecticut Valley: A Regional Study of the Economic Development of the Small Arms Industry, 1798–*

Just as accurate bookkeeping enabled the Ordnance Department to survey the general spectrum of production, so regular
inspections served to control the precision of arms being produced. Compared with accounting methods, however, those
for inspection evolved slowly. This is not surprising given that
few ordnance officers originally contemplated the manufacture
of fully interchangeable weapons. Wadsworth himself seemed
ambivalent about the need for interchangeability when he first
formulated the department's policy in 1815. Instead, uniformity was a flexible, open-ended undertaking that developed
from a rather crude emphasis on the similarity of parts, to
more exacting criteria during the 1820s, to truly interchangeable standards two decades later. Yet even here the picture of a
steady evolutionary path toward mechanical perfection is misleading because it obscures retrogressive tendencies that existed in the arms industry throughout the antebellum period.
Technological innovation doubtlessly moved forward during
these decades, but it was accompanied by temporary setbacks as
armorers attempted to digest new mechanical techniques and
effectively integrate them into the total production process.
Under these circumstances the quality of arms delivered to the
government often varied, even among muskets purportedly of
the same model. To remedy these defects, the Ordnance Department devised regularized checks to measure the work of
the national armories and private contractors. From its endeavors emerged an inspection program that improved quality
controls and encouraged the further introduction of mechanized techniques. For this reason inspection methods often
paralleled other key technological changes in the industry.[29]

In 1816, the inspection of finished firearms varied not only
from one armory to another but also within different branches
of the same armory. For the most part the procedure consisted
of making qualitative comparisons with a pattern arm and its
parts. That is, an inspector discovered work defects mainly by

1870, Smith College Studies in History, vol. 33 (Northampton, Mass.: Smith College,
1948), pp. 49–51, 131; Bomford to Lee, July 25, 1829, Letters Sent, OCO; Lee to
Thomas B. Dunn, January 16, 1830, Letters Sent, SAR. On the managerial innovations
of the Springfield armory and trunkline railroads, see Alfred D. Chandler, Jr., *The
Visible Hand: The Managerial Revolution in American Business* (Cambridge, MA: Harvard
University Press, 1977), pp. 72–75, 79–121; O'Connell essay in this volume.

29. Gene S. Cesari, "American Arms-Making Machine Tool Development, 1789–
1855" (Ph.D. dissertation, University of Pennsylvania, 1970) and Deyrup, *Arms Makers*
document these changes. Also see Fitch, *Report on Fire-Arms*, p. 4.

eye rather than by instrument. The basic criteria for inspecting the lock or firing mechanism of a musket, for example, consisted of taking the piece apart and examining its parts to make sure that they had been made in a "workmanlike" manner. If the inspector found the components properly shaped, filed, and finished, he reassembled the piece and tried its action to determine if it functioned smoothly and gave "good fire." If the lock passed these tests, it received the inspector's stamp of approval. Other than a caliper to check the exterior dimensions of the gun barrel and two plugs to verify the bore, no gauges were used during the inspection. The subjective nature of this process, which made it exceedingly difficult to maintain uniformity among muskets of the same model, indicates why the chief of ordnance so frequently expressed dissatisfaction with the arms made at the national armories between 1815 and 1821.[30]

The first significant advance in inspection procedures occurred at the Springfield armory around 1818. There, under the watchful eye of Roswell Lee, master armorer Adonijah Foot and several other workmen developed a method of gauging musket components both during and after the manufacturing process. Although the procedure needed to be perfected, by 1819 it was far more sophisticated than inspection procedures at other armories. Within two years Lee could report to Bomford with considerable pride that "our Muskets are now substantially uniform." "Yet," he quickly added, "I am sensible that considerable improvements are yet to be made to complete the system of uniformity throughout all the Establishments."[31]

Long before Lee made this pronouncement, Bomford had been thinking about the larger implications of the Springfield undertaking. Throughout the experiment he had kept in close touch with Lee, continually urging him on. Once the endeavor ended, he applauded the result and in the summer of 1821 announced his intention to introduce Springfield's gauging standard not just at Harpers Ferry but among the private contractors as well. This decision signalled the end of craft-oriented inspection procedures and the beginning of a new mechanical tradition. From that time on, hardened steel gauges

---

30. Stubblefield to Bomford, November 9, 1821, Letters Received, OCO.

31. *American State Papers: Military Affairs*, 2: 544; Lee to Bomford, September 11, 1821, Letters Received, OCO.

gradually replaced human skill in the testing and evaluation of ordnance.[32]

By 1823, the Springfield standard stood in place. In addition to designing an improved pattern musket—sometimes designated the Model 1821—Lee and Stubblefield, acting under Bomford's directive, had prepared and distributed six sets of gauges to various musket contractors. Of the "go-no go" variety, each set consisted of ten different pieces which verified the lock mechanism, the bore and exterior of the barrel, the fall of the stock, the size of the bands, the diameter of the ramrod, and the length and width of the bayonet. To guard against defects produced either by faulty workmanship or wear in the working gauges, Bomford introduced quarterly reinspections of sample weapons produced at both public and private establishments. Different individuals were designated for this duty, although the task most frequently devolved upon the master armorers at Springfield and Harpers Ferry. They carried out the inspections with a master set of gauges, making written reports in which they compared and evaluated the work of the different armories. If any deficiencies existed, they immediately notified the chief of ordnance who in turn enjoined those responsible to remedy the situation. At the same time Bomford established a special reference collection of military firearms at Washington for purposes of further comparison and study. He also began to apprise private manufacturers that the issuance of future arms contracts would depend on their current performance, especially the degree to which they updated their operations and cooperated with the department in sharing new inventions or other relevant information. Heeding this injunction, major contractors like Marine T. Wickham, Brooke Evans, and Nathan Starr began almost immediately to adopt new machinery and other labor-saving techniques from the national armories.[33]

Most promising was the work of John H. Hall, a gifted New Englander who in 1811 had patented a breechloading rifle (figure 7), which, with Bomford's support, he had been produc-

32. Lee to Bomford, September 11, 1821, OCO; Lee to Stubblefield, September 29, 1821, Letters Sent, SAR; Stubblefield to Lee, October 12, 1821, Letters Received, SAR.

33. Stubblefield to Bomford, November 9, 1821, Stubblefield and Lee to Bomford, December 4, 1821, Letters Received, OCO; Lee to Stubblefield, December 15, 1821, to Bomford, December 18, 1821, November 23, 1822, to Eli Whitney, November 13,

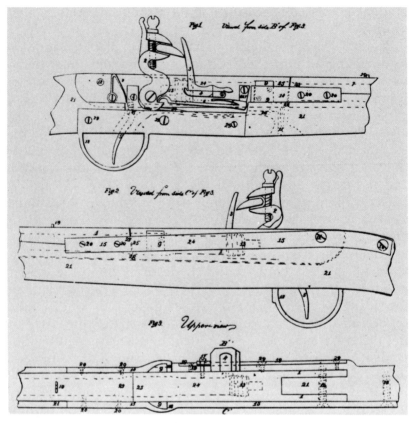

Figure 7
Drawings for John H. Hall's patent of May 21, 1811. Smithsonian Institution Neg. No. 81-195.

ing under special contract at Harpers Ferry since 1819. With the aid of over sixty-three inspection gauges and an impressive stable of machinery, Hall conclusively demonstrated in 1826 that his rifles could be made with interchangeable parts, the first of their kind in America. Eight years later Simeon North (figure 8), an equally talented private contractor from Middletown, Connecticut, added another dimension to the evolv-

1822, Letters Sent, SAR; Stubblefield to Lee, November 17, 1821, August 2, 1822, Letters Received, SAR; Bomford to Stubblefield, August 15, 1827, Letters Sent, OCO; Deyrup, *Arms Makers*, pp. 56–59. Also see Asa Waters to Bomford, April 6, 1824, Letters Received, OCO; Bomford to Nathan Starr, November 19, 1827, Letters Sent, OCO; Starr to John H. Eaton, July 10, 1829, Correspondence Relating to Inventions, OCO; James Baker to Lee, April 6, 1826, Marine T. Wickham to Lee, February 18, 1829, Letters Received, SAR.

Figure 8
Simeon North (1765–1852). Smithsonian Institution Neg. No. 48698.

ing pattern of precision when he adopted Hall's gauges and succeeded in making rifles whose parts exchanged with those made by Hall at Harpers Ferry. Together Hall and North provided tangible evidence of what could be accomplished by adopting uniform practices at two widely separated factories. Having personally witnessed the Hall-North operations, Bomford concluded that the uniformity principle had indeed reached a new level of refinement, and he became even more adamant in his conviction that one day all armories would achieve in a larger, more efficient manner what Hall and North had done on a limited scale. That day would arrive in the mid-1840s, when the national armories and private contractors began to produce the Model 1841 percussion rifle and the Model 1842 percussion musket. These were the first fully interchangeable firearms to be made in large numbers anywhere, one of the great technological achievements of the modern era (figures 9, 10, 11). Yet, as impressive as they were, these accomplishments tended to obscure fundamental ambiguities and tensions associated with the introduction of the uniformity system.[34]

Throughout the 1820s chronic problems plagued and frustrated the Ordnance Department's policy. For one thing, conflicting opinions existed over the need for mechanization as well as the importance of uniformity. At the main armory at Harpers Ferry, for instance, managers as well as artisans held tenaciously to preindustrial traditions, grumbled about the introduction of new regulations, and vilified Bomford and his "visionary" schemes. Similar feelings also existed among some contractors, although they were less willing to oppose the uniformity policy for fear of losing their contracts. These attitudes bothered Bomford and his staff, but what troubled them more were the lapses that continued to appear in the quality of firearms. Due no doubt to the increasing rigor of inspections, the department received numerous reports of flaws, subterfuges, and shoddy workmanship during the 1820s. By 1827 the problem had become so serious at Harpers Ferry that Lee pleaded with Stubblefield to take "measures . . . to remedy the deficiency." He cautioned, "The variation between our Arms seems to be greater now than at any time previous. . . . I hope we shall be able to make some improvement on this point. If not, we shall surely be censured. Let us exert ourselves to the utmost."[35]

---

34. These developments are fully treated in Smith, *Harpers Ferry Armory*, pp. 184–251, 280–92. Increasingly sophisticated gauging standards continued to parallel mechanized manufacturing developments throughout the antebellum period. During the early 1840s, for instance, sixty gauges—many of them multiple gauges—were used in the inspection of contract arms (see figure 11). By 1854 their number had increased to nearly 100, and during the Civil War the Springfield armory would use over 150 gauges in testing the accuracy of component parts. See Deyrup, *Arms Makers*, p. 146; and Rosenberg, *The American System*, pp. 143, 187–90. The accuracy of these methods received an unexpected test in 1852 when, as a result of a flood at Harpers Ferry, 9000 percussion muskets with unmarked parts were stripped, cleaned, and randomly reassembled "with every limb filling and fitting its appropriate place with perfect exactness." For further information, see Benet, *Ordnance Reports*, 2: 536.

35. Lee to Stubblefield, October 31, 1827, Letters Sent, SAR. For a sampling of inspection reports on contract as well as national armory firearms, see Inspection report of Philip Hoffman, February 24, 1823, Adonijah Foot to Bomford, June 10, 1823, Extracts from the confidential report of Colonel Archer, 1824, Elisha Tobey to Lee, October 27, 1827, Letters Received, OCO; Lee to Bomford, June 6, 1826, Joseph Weatherhead to Bomford, May 22, 1827, Letters Sent, SAR.

Figure 9
U.S. Percussion Rifle Model 1841 (pattern piece) designed and manufactured at the Harpers Ferry armory. *Source:* Collections of the Division of Armed Forces History, National Museum of American History. Smithsonian Institution Neg. No. 73-312.

Figure 10
U.S. Percussion Musket Model 1842 (pattern piece) manufactured at the Springfield armory. *Source:* Collections of the Division of Armed Forces History, National Museum of American History. Smithsonian Institution Neg. No. 73-313.

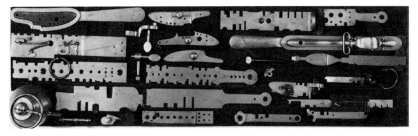

Figure 11
Inspection gauges for the U.S. Rifle Model 1841. An entire set consisted of sixty gauges and six "appendages" (screwdrivers, bench hammer, and punch). *Source:* Collections of the Division of Armed Forces History, National Museum of American History. Smithsonian Institution Neg. No. 62468.

No one could deny that Lee had always exerted himself to the utmost, performing services far beyond the normal call of duty. Between 1823 and 1830 special assignments ranging from service on a commission to investigate the establishment of a western armory to twice filling in as acting superintendent at strife-ridden Harpers Ferry had kept him away from Springfield for more than fifteen months. Indeed, in 1823 and again in 1827 he had spent more time away from the armory than at it. But this was not all. In addition to acting as Bomford's chief troubleshooter, Lee also bore responsibility for overseeing the inspection and delivery of all contract arms made in New England. This was a particularly thankless job that involved him in numerous disputes with contractors and hampered his ability to maintain the sort of friendly relations and cooperative contacts necessary for assimilating new techniques at Springfield.

By 1830 these and other activities had sapped Lee's energy and impaired his health. He therefore asked to be relieved of supervision of the contract service. Recognizing that Lee was ill and that he needed to spend more time attending to affairs at Springfield, Bomford acceded to the request by creating the office of Chief Inspector of Contract Arms and then appointed a promising lieutenant named Daniel Tyler to the post. The new office placed all small arms contractors under Tyler's oversight. Tyler performed his job well and the quality of contract arms began to improve.[36] Encouraged by the success, Bomford sought to introduce even more thoroughgoing administrative reforms aimed at consolidating control over the rest of the innovative but erratic industry. Before he could do so, however, an important organizational change had to be made.

In 1821 the Ordnance Department had lost its status as an independent military bureau when Congress, in the interest of economy, had merged it with the Corps of Artillery. The reorganization produced an uneasy union fraught with distrust and ill feeling. Bomford especially chafed under the new arrangement as a "Brevet Colonel of Artillery on Ordnance Service," a

---

36. *American State Papers: Military Affairs*, 2: 731; Smith, *Harpers Ferry Armory*, pp. 171–74, 178–81; William Riddal to Lee, December 21, 1829, Letters Received, SAR; Lee to James Carrington, August 4, 1830, Letters Sent, SAR; Tyler to Bomford, January 8, March 7, 1831, January 7, 1832, Letters Received, OCO; Deyrup, *Arms Makers*, pp. 59–65.

title which he considered diminutive. Although the change did not impair developments in the manufacture of small arms, it did retard artillery technology as the two factions jealously guarded their prerogatives and quarreled with one another over the determination of long-range policies. Such rivalry in the field of artillery resulted in stagnation. Young artillerists resented ordnance assignments as an unproductive and unnecessary interruption of their professional careers. Ordnance specialists, on the other hand, became particularly restive about being submerged in an amorphous organization that obfuscated their identity. During the 1820s they repeatedly petitioned the Secretary of War to redress the situation. A revision finally came in April 1832, when Congress enacted a bill reestablishing the Ordnance Department as a separate army bureau and returning it to a status comparable to that which had existed in 1815.[37]

The reorganization act of 1832 gave Bomford the authority he needed to effect further managerial changes. As head of an agency charged with administering the procurement, inspection, and distribution of arms over vast geographic distances, he had spent more than a decade addressing problems of a scale and complexity unknown to most comtemporaries in the business world. What is more, he had learned from experience that no individual could personally oversee such widely dispersed yet critically related activities. Since the magnitude of these tasks was so great, the only viable alternative was to develop a decentralized managerial structure through which he could delegate responsibility to subordinates. Daniel Tyler's success as Chief Inspector of Contract Arms had demonstrated the effectiveness of this stategy, so in June 1832 Bomford sought to complement that office and expand channels of communication by appointing Lieutenant Colonel George Talcott as "Inspector of Armories, Arsenals and Depots." That Bomford chose Talcott for this duty indicates the importance of the office. At the time of his appointment Talcott commanded the Watervliet arsenal near Troy, New York, and was the second ranking officer in the department. In 1848, he would succeed Bomford as chief of ordnance. In fact, from 1832 until the Civil War the inspectorship of armories and arsenals would be filled

---

37. U.S. *Statutes at Large* 3: 615, 4: 504; Falk, "Soldier-Technologist," pp. 141–44.

by senior ordnance officers and all but one would become chiefs of ordnance.[38]

As his title indicated, Talcott's main assignment involved the regular inspection of all federal arsenals and armories followed by written reports to his superior, the chief of ordnance. These reports touched on subjects ranging from the condition of buildings and machinery to the general discipline of the establishments. Most particularly they compared the operations of different installations, scrutinized their accounts, and evaluated the performance of the superintendents, paymasters, and other supervisory personnel. As internal documents not intended for public distribution or use, the reports provided remarkably candid appraisals which the chief of ordnance doubtlessly found valuable. In 1832, for instance, Talcott wrote forthrightly of Harpers Ferry that "the machinery at this armory is far behind the state of the manufacture elsewhere and the good quality of their work is effected, at great disadvantage, by manual labor." "Uniformity of work," he emphasized, "is scarcely attainable by manual labor." Three years later he modified his evaluation. "This Armory is much improved since my inspection of 1832—a great advance is perceptible in the introduction of new machinery. Nevertheless," he quickly added, "much remains to be done to bring it up to the point at which the Springfield Armory stood at that time." Since Springfield was so well managed, it invariably became the model against which the performances of Harpers Ferry and other government arsenals were measured.[39]

Besides reporting on conditions at the institutions they visited, the inspectors of armories and arsenals also recommended various organizational and administrative changes. Moreover, they possessed the authority to make unilateral decisions, as in 1836 when Talcott directed the superintendent at Harpers Ferry to store rejected gun barrels for future use rather than to destroy them. Such orders, of course, were subject to review by the chief of ordnance, but he rarely reversed a decision.[40]

38. Benet, *Ordnance Reports*, 1: 235–36.

39. Talcott to Bomford, December 15, 1832, "Inspection of Harpers ferry Armory," July 17–25, 1835, Reports of Inspections of Arsenals and Depots, OCO.

40. Talcott, "Inspection of Harpers ferry Armory," November 3, 1836, Reports of Inspections of Arsenals and Depots, OCO.

Throughout the antebellum period the primary mission of Talcott and his successors centered on the evaluation of production methods and maintenance of uniform standards at all government installations. Indeed, their attempt to impose uniformity in all areas of arms manufacture helped to define the problems and unite the efforts of the inspector of armories and arsenals, the Chief Inspector of Contract Arms, and even the chief clerk in charge of accounts at the Ordnance Department. By evaluating the work of outlying installations and providing a continual flow of information, these officers relieved the chief of ordnance of a heavy burden. In effect, they formed a bureaucratic team of middle managers whose separate but interrelated activities vastly improved the internal rhythm of the ordnance enterprise.[41]

The publication in 1834 of an official set of "Ordnance Regulations" formally codified the institutional arrangements that Bomford and Wadsworth had erected during the previous twenty years. In addition to defining the functions of the department, the regulations carefully delineated lines of authority and communication between the chief of ordnance, his representatives, and working personnel in the field. Everyone from prestigious inspectors to the lowliest arsenal custodians thus gained a clear understanding of his respective role and responsibilities within the larger organizational context. At the national armories, for example, the regulations set standards for uniform accounting and manufacturing practices and established explicit guidelines for such other matters as employee housing, travel allowances, and work discipline. Similar provisions were established for arsenals, depots, and private contractors. Indeed, nothing seemed to escape the regulations' comprehensive embrace. A decided emphasis on order, control, and accountability distinguished the new rules, which reinforced the Ordnance Department's long-standing commitment to uniformity. By issuing such an elaborate set of instructions and revising them when needed, Bomford and his colleagues formalized an administrative network, which greatly facilitated day-to-day operations and clearly identified them with long-range goals. Beyond this, the regulations brought the department public recognition as a central institution of technological

41. See the file of Reports of Inspections of Arsenals and Depots, 1832–60, OCO.

change within the government. In doing so, they paved the way for further elaborations of the uniformity system.[42]

A clear indication of the department's growing stature and consolidation appeared in 1839 with the establishment of an investigatory body known as the Ordnance Board. The idea for such a commission was not new. On numerous occasions between 1818 and 1839, the Secretary of War had assembled special boards of officers from different branches of the service to investigate and report on model arms, artillery designs, new inventions, administrative changes, and other related matters. What distinguished the new board of 1839 was its convergent membership and its permanence as an ordnance institution. Consisting entirely of ordnance officers, it became the ongoing research and development branch of the Ordnance Department and, within this limited realm, shared important decision-making powers with the bureau chief. Its functions were twofold: "to standardize and systematize Army ordnance equipment" and to supervise "the testing, evaluation, and design of this equipment." "More than anyone else," one writer observes, the members of the board "were responsible for the development of American military technology during the twenty-two years before the Civil War."[43]

Once established, the Ordnance Board continued the investigations of its predecessors and initiated a number of new ones. The most significant of these undertakings concerned the introduction of a uniform system of artillery. Since 1832 the War Department had made a concerted effort to advance this branch of the service. By 1839 a specific agenda had emerged. Among other things, the Secretary of War had sanctioned the introduction of uniform designs and nomenclature for a new field artillery, approved the adoption of stock-trail carriages in place of the outdated Gribeauval plan, and agreed to a complete revision of siege-garrison and seacoast artillery. The introduction of stock-trail gun carriages presented the fewest

---

42. *Regulations for the Government of the Ordnance Department* (Washington, D.C.: Francis P. Blair, 1834). Also see U.S. Ordnance Department, *Regulations for the Government of the Ordnance Department* (Washington, D.C.: Jacob Gideon, Jr., 1839); U.S. Ordnance Department, *Ordnance Manual for the Use of Officers of the United States Army* (Washington, D.C.: J. and G. S. Gideon, 1841); William A. DeCaindry, *A Compilation of the Laws of the United States Relating to and Affecting the Ordnance Department* (Washington, D.C.: GPO, 1872).

43. Wade, "Early Systems of Artillery," pp. 142–43; Falk, "Soldier-Technologist," pp. 286–87.

problems, since the basic technology for their uniform production already existed in the form of woodworking machinery developed during the 1820s at the national armories. The greatest difficulty centered on finding more uniform methods of making cannon. Interestingly, the solution to this problem involved the Ordnance Board in a protracted series of investigations aimed at determining the "uniform" properties of iron. In this respect the qualitative concept of uniformity not only informed production practices but also fostered new approaches to problem solving within the department. This, in turn, had another important ramification. Formerly ordnance officers had concerned themselves primarily with promoting and monitoring the uniformity system. By the 1840s they were beginning to become directly involved in the everyday problems of actually building and improving ordnance equipment. Besides denoting a new degree of military involvement in the production process, the endeavor marked the final stage in the rise of the uniformity system.[44]

Derivation of suitable design principles for the new artillery hinged on the production of better gun metal. Since 1801 civilian founders had supplied the Army with cast-iron cannon which, for the most part, had served their purposes well. However, so little was known about the physical properties of iron that, whenever cannon burst, no one could adequately explain why. This gap in technical knowledge, coupled with the fact that most European nations refrained from using iron as a construction material, had prompted several military commissions to recommend during the 1830s that the Army revert to bronze for light field artillery. Although iron founders who held ordnance contracts firmly opposed the idea and lobbied against it, the Secretary of War decided in favor of bronze in 1841, after the Ordnance Board, recently returned from a nine-month tour of inspection in Europe, persisted in its recommendation. While this action settled the fate of light artillery, a bothersome problem remained. Since bronze was costly and could be used economically only in small artillery pieces,

44. These developments are fully documented by Falk, "Soldier-Technologist," pp. 228–85; and Birkhimer, *Historical Sketch of Artillery*, pp. 243–47, 262–65, 280–83. For information on the adaptation of woodworking machinery at the Springfield armory to the manufacture of carriages at the Allegheny arsenal, see Thomas Blanchard, "An Estimate of the Cost of Machinery for making Gun Carriages for the U. States," February 2, 1830, Letters Received, OCO.

the Ordnance Board had no choice but to continue using iron for heavier siege-garrison and seacoast guns. In reaching this decision, everyone realized that the successful design and construction of large guns depended on producing improved castings which would give safe and durable performance. To attack this problem the board recommended and the War Department approved a special program of industrial research.[45]

The quest for more uniform foundry practices spanned nearly two decades and addressed several separate but related problems associated with the strength of materials. Investigations began in the spring of 1841 when William Wade, a former ordnance officer who had become a proprietor of the Fort Pitt Iron Works in Pittsburgh, returned to government service as a civilian "attending agent" at the private foundries. His position was analogous to the Chief Inspector of Contract Arms; his duties, as defined by the Ordnance Board, included the introduction of more uniform production procedures among the contractors as well as the inspection of finished cannon. While executing these tasks, Wade reported to his old friend Bomford who in 1841 had turned over direction of the Ordnance Department to Colonel Talcott in order to devote full time to the design and improvement of large seacoast artillery known as "Columbiads." During the next ten years and intermittently until his retirement in 1854, Wade spent countless hours conducting comparative trials of cannon, building various gauges and testing machines, and examining fractured samples of iron in an effort to establish correlations between their tensile, torsional, and transverse strength, their specific gravity, and the durability of artillery when subjected to continuous fire.[46]

In carrying out these carefully documented studies, Wade received valuable assistance from several junior officers engaged in similar projects. Most notable was Captain Alfred Mordecai, a West Point graduate and "soldier-technologist" of

45. Ibid.; Ezell, "Development of Artillery," pp. 107–31.

46. Captain Benjamin Huger to Bomford, January 28, 1841, Letters Received, OCO; U.S. *Statutes at Large* 5: 512; Benet, *Ordnance Reports*, 2: 35–36; Ezell, "Development of Artillery," pp. 125–33. The results of Wade's work are documented in U.S. Ordnance Department, *Reports of Experiments on the Strength and Other Properties of Metals for Cannon* (Philadelphia: Henry C. Baird, 1856), pp. 11–313. For his unpublished papers, see Reports of Experiments, Tabular Reports of the Inspection of Cannon, Reports Relating to Tests Made on Cast Iron Cannon, and Reports on the Manufacture of Heavy Ordnance, OCO. Edwin Layton places Wade's investigations in larger historical context in his perceptive "Mirror-Image Twins: The Communities of Science and Technology in 19th-Century America," *Technology and Culture* 12 (1971): 571–72.

considerable scientific merit who worked with Wade between 1842 and 1844. Mordecai subsequently achieved prominence for adapting French methods of testing and measuring the explosive force of gunpowder by substituting ballistic and cannon pendulums for older eprouvette mortars. His meticulous reports, published in 1845 and 1849, provided incontrovertible evidence linking the efficiency of gunpowder to its granulation and density. This knowledge, the first of its kind in America, paved the way for further advances in the manufacture of explosives. It also found valuable applications in the planning of fortifications and in the design of cannon, projectiles, and small arms.[47]

The most imaginative series of experiments occurred at the Pikesville arsenal near Baltimore under another protégé of Bomford and Wade, Lieutenant Louis A. B. Walbach. While painstakingly testing and classifying samples of iron from 2808 cannon during the mid-1840s, Walbach detected "a striking relation between the strength and density of the metal in the same gun" and "the peculiarity of the *color of structure*, and the form and size of the crystals" in the fractured sample. Although he believed these properties could "serve to point out the quality of the iron," certain discrepancies in the observed data prevented him from making a definitive statement about their proper relationship. This bottleneck set the stage for an added dimension in ordnance research.[48]

In his final report of 1847, Walbach suggested that chemical analysis might provide an answer to the anomalies he had encountered in testing the physical properties of iron. He therefore sought and eventually received permission to establish a "Chemical Laboratory" at Pikesville. In 1849 he engaged Professor Campbell Morfit of the University of Maryland to study the effects of carbon, sulphur, phosphorous, and other elements on the quality of iron. Working in collaboration with his former teacher, James C. Booth of Philadelphia, Morfit be-

47. Falk, "Soldier-Technologist," pp. 302–4, 338–51, 354–60, 403–4, 526–27. Also see Mordecai's *Report of Experiments on Gunpowder, Made at Washington Arsenal, in 1843 and 1844* (Washington, D.C.: J. and G. S. Gideon, 1845); and his *Second Report of Experiments on Gunpowder, Made at Washington Arsenal, in 1845, '47, and '48* (Washington, D.C.: J. and G. S. Gideon, 1849).

48. Benet, *Ordnance Reports*, 2: 60–75, 138–47, 185, 195–204; Ordnance Department, *Reports of Experiments on Metals for Cannon*, pp. 325–46. Walbach's original correspondence and reports are found in Reports Relating to Tests Made on Cast Iron Cannon and Reports on the Manufacture of Heavy Ordnance, OCO.

tween 1851 and 1855 submitted three reports which adduced metallurgical evidence clarifying and correcting the earlier findings of Wade, Mordecai, and Walbach.[49] Together with the concurrent work of Captain Benjamin Huger, Lieutenant Thomas J. Rodman, and other members of the Ordnance Board, these investigations yielded criteria for evaluating gun metal both before and after the casting process. Although much remained to be learned about the physics and chemistry of iron, the knowledge gained from the experiments enabled founders to exercise far more control over the quality of their castings. Much to the delight of the Ordnance Board, their products significantly improved. Equally important for the growth of modern engineering, the investigations clearly demonstrated the advantages of systematic research over cut-and-try empiricism. By treating the technology of metalworking in this way, members and associates of the Ordnance Department made science a permanent part of the Army's research and development program and provided a model for manufacturers in the private sector. In doing so, they stood in the vanguard of those who were beginning to shed craft traditions for a more rigorous approach to the solution of technical problems. This shift of focus and its translation into everyday practice became an essential part of the uniformity system. By the mid-1850s, the impact of the new technology could be seen not only in specialized production techniques like Rodman's sophisticated hollow-casting process but also in the complete standardization of American small arms and artillery.[50]

"My ability," Major Alfred Mordecai confided to his brother in June 1861, "consists in a knowledge and love of order and system, and in the habit of patient labor in perfecting and arranging details; and my usefulness in the Army arises from the long continued application of these qualities to the specialties of my habitual business." Although Mordecai had recently resigned from the service to avoid choosing sides in the fratricidal struggle that enveloped the nation, he could reflect with pride

49. Benet, *Ordnance Reports*, 2: 527–29; Ezell, "Development of Artillery," pp. 133–37. For Morfit's reports, see Ordnance Department, *Reports of Experiments on Metals for Cannon*, pp. 371–428.

50. Falk, "Soldier-Technologist," pp. 385–87. Rodman's investigations (1843–60) are detailed by Ezell, "Development of Artillery," pp. 137–57. Also see Rodman, *Reports of Experiments on the Properties of Metals for Cannon, and the Qualities of Cannon Powder* (Boston: Charles H. Crosby, 1861).

on his many exploits as a "soldier-technologist." For twenty-nine years he had labored diligently as a member of the Ordnance Department. During that time he had served more than two decades on the Ordnance Board, played a leading role in the development of uniform artillery, witnessed the rise of interchangeable manufacturing in the firearms industry, conducted hundreds of experiments with gunpowder, breechloading rifles, time fuses, and other novel weapons, superintended two major arsenals of construction, and administered the affairs of the entire department as a special assistant to the bureau chief. Somewhat ironically, his distinguished service record did not insure rapid promotion, but it brought other professional rewards. In 1853 he served as judge of exhibits at New York's Crystal Palace exposition. The same year he was elected to membership in the prestigious American Philosophical Society and invited to join the equally prestigious American Association for the Advancement of Science. These honors certainly acknowledged Mordecai's many individual achievements, but in a larger sense they also recognized the institutional accomplishments of the Ordnance Department. At both levels they symbolized the growing link between military innovation and the civilian world of science and industry.[51]

Few colleagues could match Mordecai's breadth of experience and impressive record of achievement. Yet his "love of order and system" and "habit of patient labor" reflected a pervasive spirit that had inspired and guided the ordnance enterprise since its inception. Order and system, patience and perseverance, pragmatism and perfectibility, simplicity and solidarity: these value-laden precepts formed a common intellectual bond between members of the department and gave them a keen sense of professional identity. From this ethos had emerged their intense commitment to pursue a policy of uniformity and to see it materialize in the design of special gauges and machinery for the manufacture of all types of arms and equipment.

Scholarly treatments of entrepreneurship are usually confined to individuals or groups operating within the private business arena. This focus is understandable in view of the importance that is assigned to risk taking. Yet even though

---

51. Mordecai to Samuel Mordecai, June 2, 1861, quoted by Falk, "Soldier-Technologist," pp. 593–94. Also see Falk, pp. 409–10.

antebellum ordnance officials cannot be labeled entrepreneurs in the strictest sense of the word, they nonetheless exercised important entrepreneurial functions in managing the armory-arsenal complex, in making critical decisions, and—not to be taken lightly—in lobbying and accounting for congressional appropriations. The introduction of the uniformity system provided the context for these activities, involving officers like Mordecai in a complex innovative process that moved through at least six interactive stages of problem identification, idea response, invention, research and development, introduction into use, and diffusion to other users. Each stage formed a particular pattern that impressed itself on the structure of the whole, but the one that best reflected the department's peculiar style of innovation was the process of diffusion.[52]

As early as 1815 the chief of ordnance and his staff had recognized the importance of disseminating new techniques. Without the rapid assimilation of machine and gauging processes, the coordination and control necessary for uniform production would have proved well-nigh impossible. Continual communication and cooperation therefore became essential elements in the department's strategy. Resultant information transfers soon manifested themselves in published reports of tests and experiments and in the day-to-day operations of the armaments network.

Nowhere was the emphasis on communication and cooperation stronger than in the small arms industry. There, as indicated earlier, private contractors as well as federal administrators regularly corresponded, visited, and assisted one another in working out the basic configurations of the uniformity system. To insure that the diffusion process would not be hindered, the Ordnance Department also insisted that the national armories open their shops to visitors, who could make drawings, borrow patterns, and obtain other information pertinent to their special interests. At the same time, the department had an implicit understanding with all arms contractors that they had to share their inventions with the national armories on a royalty-free basis if they wished to continue in government service. In this way a number of novel metal- and woodworking

---

52. Thomas P. Hughes presents a thoughtful analysis of these stages in "Inventors: The Problems They Choose, the Ideas They Have, and the Inventions They Make," in *Technological Innovation: A Critical Review of Current Knowledge*, eds. Patrick Kelly and Melvin Kranzberg (San Francisco: San Francisco Press, 1978), pp. 166–82.

techniques that had originated in the private armories of persons like Simeon North became part of the public domain. Such an "open door" policy, while accentuating the public-service orientation of the Ordnance Department, explains why so few crucial machines and machine processes were actually patented during the antebellum period. The same policy also accounts for the highly integrated character of the American firearms industry as well as for the relative speed with which manufacturers adopted the new technology.

Although ordnance officials well understood the importance of diffusing mechanized techniques within the arms industry, they did not anticipate the far-reaching consequences the process would have for other sectors of the economy. By the mid-1840s the manufacturing methods aggregated at the national armories had begun to spread beyond private firearms factories like the Robbins & Lawrence works in Windsor, Vermont, to all sorts of factories and machine shops producing metal products. To a certain extent this change manifested itself through former ordnance officers who had assumed managerial posts with various firearms, foundry, and railroad businesses. Far more frequently, however, armory practice was transmitted by workmen who had received their early training at one of the public or private arms factories and subsequently moved to new positions as master machinists and production supervisors at other manufacturing establishments. Typical in this respect was Jacob Corey MacFarland, a skilled machinist who left the Springfield armory around 1845 to become the foreman of the Ames Manufacturing Company's machine shop at Chicopee, Massachusetts.

The experience of the Ames Company well illustrates how firms availed themselves of the new technology. Established in 1834 by the brothers Nathan P. and James T. Ames, the company became one of the earliest in the United States to manufacture a standard line of machine tools. As a mixed enterprise operating in a limited market, the firm undertook special jobbing contracts and made a wide variety of millwork, mining equipment, cutlery, small arms, cannon, statuary, and other metal products. Yet, despite their obvious technical skill and versatility, the Ames brothers were basically copyists rather than innovators. And for good reason. Chicopee was only a short distance from Springfield, where the brothers had ready access to patterns and drawings owned by the national armory.

Aided by a number of former Springfield workmen, their stock of machinery clearly reflected this influence.[53]

Other transitional establishments—Colt, Robbins & Lawrence, the Sharps Rifle Company, E. Remington & Sons, George S. Lincoln & Company, and the Providence Tool Company, to name but a few—followed similar practices, thereby acquiring the latest armory know-how and relieving themselves of lengthy time lags and costly expenditures associated with learning-by-doing. Frequently borrowed tool and machine designs had to be adapted to different uses. This, in turn, led to further improvements and inventions. Coupled with the aggregate growth of factory production, such activities explain why the late antebellum period witnessed so many significant advances in machine design and production.

From this complex tapestry of cooperation, diffusion, and innovation, one can discern a mechanical genealogy that directly links the Springfield armory and the New England arms industry with the rise of the modern machine tool industry. Together these institutions and their cadres of remarkably mobile mechanics fostered a quasi-educational phenomenon one writer aptly describes as "technological convergence." Through them armory practice spread to technically related industries and by the late 1850s could be found in factories making sewing machines, pocket watches, railroad equipment, wagons, and hand tools. From these beginnings it was only a matter of time before the new technology found applications in the production of typewriters, agricultural implements, bicycles, gramophones, cameras, automobiles, and a host of products associated with the mass production industries of the twentieth century. Interestingly, machining processes rather than precision instrumentation constituted the most important transfer from the arms industry. Since precision production was expensive, most businessmen contented themselves with manufacturing highly uniform but not necessarily interchangeable parts. Only the government could afford the luxury of complete interchangeability.[54]

53. Smith, *Harpers Ferry Armory*, pp. 288–90; Cesari, "American Arms-Making," pp. 92–131. Other examples are provided by David A. Hounshell, "The System: Theory and Practice," in *Yankee Enterprise: The Rise of the American System of Manufactures*, eds. Otto Mayr and Robert C. Post (Washington, D.C.: Smithsonian Institution Press, 1981), pp. 127–52.

54. Nathan Rosenberg discusses the concept of technological convergence in his seminal article on "Technological Change in the Machine Tool Industry, 1840–1910,"

The naive enthusiasm with which many mid-nineteenth-century writers celebrated the advent of the machine tends to obscure the inherent difficulties that accompanied the early industrialization of America. Mechanization, despite its many practical benefits, exacted a costly social and psychological toll. The very process of factory innovation uprooted a large segment of the nation's population, catapulting it into a vastly different, bureaucratically organized environment. None felt the impact of these changes more forcefully than the people who lived and worked in the mill communities. Yet as much as churchlike factories and whirling machines defined and symbolized the new order of things, what really distinguished the age were the unsettling changes effected by the division of labor. In this respect, the experience of the national armories was no different from other large industrial establishments in the United States.

When labor was mechanized and divided in nineteenth-century arms factories, individual work assignments became more simplified while the overall production process became more complex. Coordinating and controlling the flow of work from one manufacturing stage to another therefore became vital and, in the eyes of factory masters, demanded closely regulated on-the-job behavior. Under these conditions the engineering of people assumed an importance equal to the engineering of materials. As conformity supplanted individuality in the workplace, craft skills and other preindustrial traditions became a detriment to production. To those who planned and orchestrated the uniformity system, such changes, though initially unanticipated, seemed fully compatible with their ideas of rational design and with their paternalistic concepts of Christian stewardship. To those who worked under the system, the new regimen represented a frontal assault on valued rights and privileges. The resulting confrontations between labor and management over the control of work highlight fundamental tensions that seethed beneath the glowing

Journal of Economic History 23 (1963): 423–30. For further documentation, see David A. Hounshell, *From the American System to Mass Production, 1800–1932* (Baltimore: Johns Hopkins University Press, 1984); Deyrup, *Arms Makers*, pp. 117–28, 146–59; Cesari, "American Arms-Making," pp. 92–264; Battison, *Muskets to Mass Production*, pp. 16–19, 31; Roe, *English and American Tool Builders*, pp. 128–44, 164–215; Robert A. Howard, "Interchangeable Parts Reexamined: The Private Sector on the Eve of the Civil War," *Technology and Culture* 19 (1978): 633–49.

veneer of industrial achievement. Therein lay the subtle nuances of technology as a social process.[55]

As the general regulations of the Ordnance Department sought to coordinate and control the federal network of armories and arsenals, so new formal work rules aimed at governing the daily activities of their employees. The process began in 1816 when Roswell Lee, at the behest of the chief of ordnance, issued a set of twenty-three regulations for the Springfield armory. The very nature of these rules suggests the sorts of problems Lee and his predecessors had experienced in getting workers to abandon customary habits and reorder their lives around values of diligence, thrift, and responsibility. Among other things, the regulations forbade scuffling, playing, fighting, gambling, drawing lotteries, drinking "ardent spirits," and making "any indecent or unnecessary noise" during working hours. Besides enjoining the armorers "not to begin, excite or join in any Mutinous, riotous or Seditious conduct," the rules also stipulated that "due attention is paid to the Sabbath and no Labor, Business, amusement, play, recreation . . . or any proceeding incompatible with the Sacred Duties of the day will be allowed." Those who transgressed these norms were chastized and, in some cases, fined. Repeat offenders were fired and blacklisted, their chances of finding employment elsewhere in the region virtually eliminated. Not surprisingly, these and other sanctions defined a pattern of paternalism at Springfield that closely paralleled similar practices in the New England textile industry. And, as in the textile mills, enforcement of the ordnance regulations proved difficult.[56]

Lee's success in making Springfield a showplace of American industry owed much to his ability to impose regulations without alienating his labor force. With his death in 1833, however, the situation changed. Discipline relaxed under his successor, John Robb. By 1841 a special commission appointed to examine "the

55. Cf. Paul Faler, "Cultural Aspects of the Industrial Revolution: Lynn, Massachusetts, Shoemakers and Industrial Morality," *Labor History* 15 (1974): 367–94; Alan Dawley, *Class and Community: The Industrial Revolution in Lynn* (Cambridge, MA: Harvard University Press, 1976); Smith, *Harpers Ferry Armory;* Thomas Dublin, *Women at Work: The Transformation of Work and Community in Lowell, Massachusetts, 1826–1860* (New York: Columbia University Press, 1979); Anthony F. C. Wallace, *Rockdale: The Growth of an American Village in the Early Industrial Revolution* (New York: Alfred A. Knopf, 1978); Jonathan Prude, *The Coming of Industrial Order* (New York: Cambridge University Press, 1983).

56. Undated armory regulations, ca. 1816, Letters Received, SAR; Lee to Levi Dart, October 10, 1816, to Bomford, November 2, 1816, Letters Sent, SAR.

condition and management" of the armory noted many irregularities in the daily operations of the establishment. In their published report the members expressed "surprise . . . that no regular hours were established for labor." They noted that "every mechanic, working by the piece is permitted to go to his work any hour he chooses, and to leave off at his pleasure." Furthermore,

in some instances the machinery at the water-shops has been kept running for the accommodation of a single mechanic; and in most of the visits of the board, though made in hours usually devoted to labor, these shops were found nearly deserted. The reading of newspapers during the ordinary hours of labor appears to be so common a practice as not to be deemed improper.

Indeed, the report bristled, "the reading was continued even during the inspection of the board." That many armorers could labor only a few hours daily and still earn relatively high wages of forty to sixty dollars a month convinced the commissioners that "there has been great looseness in the management of the armory; and although the machinery has been brought to a high degree of perfection, and the work done is generally creditable to the mechanics, yet these results have been attained by a lavish expenditure of the public money." Such conditions, they concluded, "could not exist in a private [business], and ought not to be permitted in a public establishment."[57]

The irregularities detected at Springfield seemed to fade in significance when compared with affairs at Harpers Ferry. For years the Potomac armory had suffered the reputation of being locally controlled, shamefully abused, and flagrantly mismanaged. It also held the dubious distinction of employing one of the most independent and troublesome labor forces in the country. Attempts to introduce work rules during the 1820s generally went unheeded. "Workmen came and went at any hour they pleased," one officer recalled, "the machinery being in operation whether there were 50 or 10 at work." Along with these practices, armorers claimed the privileges of keeping frequent holidays, transferring jobs at will, drinking whiskey on the premises, and selling their tools "as sort of a fee simple inheritance." They also boasted that anyone who interfered with these rights could expect the same fate that befell Thomas

57. Benet, *Ordnance Reports*, 1: 401–3.

Dunn, the one superintendent who had adopted and rigorously enforced Lee's regulations. In 1830, after only six months in office, Dunn was murdered by a disgruntled armorer. Everyone knew that the workers determined their own standards of conduct and that they did so with the connivance and support of community leaders and local politicians. "Every way considered," a newly appointed master armorer wrote in amazement to a friend, "there are customs and habits so interwoven with the very fibers of things as in some respects to be almost hopelessly remitless."[58]

Ironically the very innovations that made the uniformity system possible contributed to the perceived "labor problem" at Springfield and Harpers Ferry. The introduction of self-acting machinery enabled armorers paid by the piece to complete more work in less time. Rather than trying to maximize their incomes by making optimal use of the new technology, however, most of them chose to limit their output to customary levels and carry home approximately the same monthly wages. Such practices, unhampered by managerial constraints, gave them the leisure to pursue other occupations such as farming and allowed them to maintain the relatively high level of piece rates previously paid for hand labor. Thus, instead of lending itself to the rationalization of production, unregulated piece work became the means by which armorers controlled their wages and working hours while retaining other preindustrial traditions in the midst of rapid technological change. From the beginning, then, the new technology held out the possibility of an alternative form of organization.

Committed as they were to "stability of things and stability of mind" in all undertakings, members of the Ordnance Department deplored the instability that characterized labor practices at the national armories. As ranking officers in the department, Bomford and Talcott felt especially embarassed and frustrated about the situation. On numerous occasions during the 1830s, they had cautioned superintendents about the lack of internal discipline and had urged them to institute reforms. Beyond

58. Major John Symington to Captain William Maynadier, July 12, 1849, Edward Lucas, Jr. to Bomford, August 29, 1839, Letters Received, OCO; Talcott, "Inspection of the Harpers ferry Armory," July 17–25, 1835, Reports of Inspections of Arsenals and Depots, OCO; Benjamin Moor to Major Rufus L. Baker, May 5, 1831, Letters Received, Allegheny Arsenal Records, Record Group 156, National Archives. For further documentation, see Smith, *Harpers Ferry Armory*, pp. 252–304.

this, however, they could do little to enforce the regulations because they resided so far away from the armories and because the highest officials at Harpers Ferry and Springfield—superintendents, paymasters, and master armorers—were civilians who held their appointments through the machinations of the patronage process. Accordingly, political interests and considerations informed their actions as managers. Since their local power and support depended on maintaining good will in their respective communities, they rarely flaunted their authority or did anything that threatened to jeopardize their relations with the armorers. Like good politicians, they catered to the interests of their constituents and, with regard to work regulations, left well enough alone.[59]

Thwarted in many attempts to control the civilian superintendents, ordnance officials had long been looking for an opportunity to initiate sweeping administrative changes at the national armories. That time arrived in 1841 with the inauguration of President William Henry Harrison. When the newly appointed Secretary of War, John H. Bell, announced his intention to introduce a thorough agenda of reform and called upon the chief of ordnance for advice, Bomford recommended replacing civilian superintendents with ordnance officers at the armories. Bell agreed to the plan and Bomford immediately ordered two of his most experienced subordinates to take command at Springfield and Harpers Ferry. Although patronage-conscious politicians felt threatened by the arrangement and roundly denounced it as "full of mischief in *all respects,*" the secretary nonetheless remained firm. The department's "new reign" began on April 16, 1841.[60]

During the next thirteen years military superintendents exercised exclusive control over the internal operations of both national armories. At their insistence, workers gradually abandoned the task-oriented world of the craft ethos and reluctantly entered the time-oriented world of industrial capitalism. That the large-scale manufacture of interchangeable firearms paralleled this change was no mere coincidence. Early on ordnance officers had recognized the importance of work rules, clocked

59. *American State Papers: Military Affairs,* 2: 553.
60. Bomford to Bell, April 6, 1841, Bell to Bomford, April 1, 1841, Major Henry K. Craig to Bomford, April 1, 1841, William B. Calhoun to Bell, April 1, 1841, Letters Received, OCO; Bomford to Craig, April 2, 1841, Letters Sent, OCO; Benet, *Ordnance Reports* 1: 385–86, 389, 394–97.

days, and regularized procedures in stabilizing the complex human and physical variables present in the workplace. Experience had taught them that there was no other alternative—a factory discipline characterized by rigid bureaucratic constraints had to be inculcated and absorbed by all employees. Only in this way could the delicate parameters of the uniformity system be maintained.

Such considerations, of course, gave little solace to the many armorers whose work ways were being altered. Time and again pieceworkers and inspectors complained about having to keep regular hours. Time and again they grumbled about the relentless pressures imposed by the new administration as well as the rigor with which it enforced the rules. While older artisans bemoaned the disappearance of traditional skills, other armorers protested against the installation of time clocks and the lowering of piece rates. All these feelings were reinforced by kinsmen and neighbors who customarily distrusted strangers and resented outside meddling in their local affairs. Because the culprits were military men, politicians from both regions continually fanned discontent by publicly attacking the "despotism and oppression" of military rule.[61]

Not surprisingly, members of the Ordnance Department denied these charges and repeatedly asserted that the change from civilian to military superintendents "produced a great, if not entire reformation of the abuses formerly existing" at the armories. As Bomford's successor, Talcott became deeply involved in the dispute and often found himself clarifying and defending the department's position. "The regulations governing the workmen are *not changed*," he wrote the Secretary of War on one occasion, "they are *merely enforced*." Indeed, he declared, "The *real ground of opposition* to the present mode of supervision is well known to be this":

The men have been paid high prices & were in the habit of working from 4 to 6 hours per day—& being absent whole days or a week. At the end of a month their pay was generally the same in amount as if

61. Statement of John H. Strider to Secretary of War, July 26, 1848, Major John Symington to Talcott, June 29, 1946, August 4, 12, 1848, May 12, 1849, May 21, 1851, Letters Received, OCO; Benjamin Moor, "Objections to the Military Superintendencies of the National Armories," James H. Burton Papers, Yale University Archives; Charles Stearns, *The National Armories* (Springfield: G. W. Wilson, 1852); Benet, *Ordnance Reports*, 1: 431. Labor relations at the national armories are discussed by Deyrup, *Arms Makers*, pp. 164–66; and Smith, *Harpers Ferry Armory*, pp. 268–74, 298–300.

no absence had occurred. They are now required to work full time and during fixed hours (according to old regulations) and the master of the Shop keeps a time account showing the time *actually spent in labor*. Here is the *great oppression* complained of. At the end of a month the *quantity of labor performed*, or *product*, and the *time* during which it is effected, are seen by a simple inspection of the Shop books. The degree of diligence used by each man is also known and hence results a knowledge of what is the *fair price* to be paid for piece work!!! The Armorers may attempt to disguise or hide the truth under a thousand clamors—but this is the *real cause* of their objections to a Military Superintendent. He enforces the Regulations which lay bare their secret practices (frauds—for I can use no better term).[62]

Although Talcott's spirited rejoinders delineated the sources of antagonism and ill feeling at the armories, there was no denying the close association that existed between military discipline and work discipline. What pained him most was the charge of "despotism and oppression," terms which harkened back to the Revolutionary era and still conveyed deep meaning in Jacksonian America. Such epithets portrayed the Ordnance Department as a threat to the enduring ideals of republicanism. Yet as Talcott and his colleagues constantly asserted, nothing could have been farther from the truth. Like other industrialists, they thoroughly agreed with an 1846 pronouncement by New York's Matteawan Company that systematic control of the work process was necessary "to convince the enemies of domestic manufactures that such establishments are not 'sinks of vice and immorality,' but, on the contrary, nurseries of morality, industry, and intelligence." Beyond this, they firmly believed that an orderly and well-regulated work environment would not only promote efficiency but also instill values conducive to the moral growth and well-being of the country. Viewed in this context, the enforcement of armory discipline had a much larger cultural purpose.[63]

The identification of godliness with productivity melded perfectly with the guiding principles of the Ordnance Department. By amalgamating the sacred and the profane, members of the department sought to clothe utilitarianism in the garb of re-

62. Talcott to John C. Spencer, May 17, 1842, Letters Received, OCO. Also see Colonel Henry K. Craig to Secretary of War, October 28, 1851, November 2, 1852, November 11, 1853, Letters Sent to the Secretary of War, OCO; Benet, *Ordnance Reports*, 1: 395–97, 431, 501; 2: 368–70, 436, 445, 532–37, 543–45.

63. "Manufacturing Industry of the State of New York," *Hunt's Merchants' Magazine* 15 (1846): 371–72; Benet, *Ordnance Reports*, 1: 50, 221, 431.

publican simplicity and thus legitimize their efforts. In their minds the uniformity system meant something more than a means of production; it represented a barometer of national development, an achievement in which Americans could take pride and from which republican institutions could draw strength and vitality. In this sense, uniformity not only stood for material progress through technology but also for the maintenance of virtue in an era of developing industrialism.

It is impossible to specify exactly the economic costs of introducing the uniformity system. Since the new technology evolved in increments closely associated with daily manufacturing activities, existing records simply do not distinguish between developmental and production costs. An admittedly rough approximation of expenditures for special tools, equipment, raw materials, labor, administration, and shop space as well as interest, insurance, and other hidden capital costs suggests that the development of uniform standards required a total investment exceeding two million dollars. Although private contractors expended time and money in developing new techniques, only the federal government could have financed such a massive undertaking. That it did so over a forty-year period underscores the importance not only of capital but also of longevity as a key ingredient in the evolution of complex technological systems. What the government provided, in addition to large infusions of money, was an ongoing bureaucratic organization within which the new technology—itself a bureaucratic phenomenon—could evolve. That the innovation transcended both individual limitations and the confines of isolated geographic environments was due largely to agency of the Army Ordnance Department.

2

The Corps of Engineers and the Rise of Modern Management, 1827–1856
Charles F. O'Connell, Jr.

Due mainly to the research of Alfred D. Chandler, Jr., we know that the railroads were largely responsible for creating modern industrial management. Chandler contends that the railroads, because of the unique size and complexity of their operations, developed totally new administrative practices quite independent of other institutions in American society. In this essay Charles O'Connell challenges Chandler's thesis by pointing out that members of the U.S. Army Corps of Engineers played an instrumental role in building a number of key railroads and consequently introduced military methods of managing them. As the railroads grew and adopted new administrative procedures, the military influence receded. O'Connell nonetheless maintains that military management exerted a formative influence on the development of railroad management. In addition to documenting the connection between technological and managerial innovation, O'Connell's piece sheds considerable light on both the diffusion of knowledge and the social interactions that occur as innovations take shape.

The United States Army contributed significantly to national development during the first half of the nineteenth century.[1] In addition to controlling new territory and policing the frontier, the Army encouraged some of its best educated and most experienced officers to pursue careers that included little in the way of traditional military service. One group of officers in particular became involved in large business ventures that afforded them the opportunity to make a number of important contributions to the emerging field of industrial management. These were the engineers who served some of the nation's early railroads between 1827 and the mid-1850s. The organizational and administrative experience these officers brought to the railroads had a great influence on the early development of the art of management in the United States.

As the railroads evolved and expanded, they began to exhibit structural and procedural characteristics that bore a remarkable resemblance to those of the Army. Both organizations erected complicated management hierarchies to coordinate and control a variety of functionally diverse, geographically separated corporate activities. Both created specialized staff bureaus to provide a range of technical and logistical support services. Both divided corporate authority and responsibility between line and staff agencies and officers and then adopted elaborate written regulations that codified the relationship between them. Both established formal guidelines to govern routine activities and instituted standardized reporting and accounting procedures and forms to provide corporate headquarters with detailed financial and operational information, which flowed along carefully defined lines of communication.

---

1. The historiography of the military's involvement in the development of the United States is far too voluminous to detail here. Excellent descriptive bibliographies are available in Robin Higham, ed., *A Guide to the Sources of the United States Military History* (Hamden, CT: Archon Books, 1975) and its supplement, published in 1981. Some standard works that provide a good overview of the topic are William H. Goetzmann, *Army Exploration in the American West, 1803–1863* (New Haven: Yale University Press, 1959); William H. Goetzmann, *Exploration and Empire* (New York: Alfred A. Knopf, 1966); Francis P. Prucha, *Broadaxe and Bayonet* (Lincoln, NB: University of Nebraska Press, 1953); Francis P. Prucha, *Sword of the Republic* (New York: Macmillan Publishing Co., 1969); Forest G. Hill, *Roads, Rails, and Waterways* (Norman, OK: University of Oklahoma Press, 1957). For discussions of the Army Ordnance Department's managerial and technological accomplishments at the Springfield and Harpers Ferry armories, see Alfred D. Chandler, Jr., *The Visible Hand* (Cambridge, MA: The Belknap Press of the Harvard University Press, 1977), pp. 72–75; Merritt Roe Smith, *Harpers Ferry Armory and the New Technology* (Ithaca, NY: Cornell University Press, 1977); and Smith's chapter in this volume.

As the railroads assumed these attributes, they became America's first "big business."[2]

Were these similarities coincidental, or did the railroads use the United States Army as a management model? While some historians have suggested that the railroads may have made limited use of the Army as a source of organizational and administrative expertise, the alternate hypothesis, which denies the Army a role in the early evolution of railroad management, has gained widespread acceptance.[3]

This essay suggests that the United States Army provided a management model for three of the nation's important early railroads. As the Army grew, institutionalized many of its support services, and assumed responsibility for a range of diverse organizational activities, its leaders formulated management procedures designed to ensure the most efficient and economical execution of its mandated functions. These developments proceeded at an uneven pace, dependent for the most part on the nation's changing military situation and the different political and economic philosophies of the faction in power. Despite occasional setbacks, by the mid-1820s the Army had adopted complex bureaucratic management procedures more advanced than the administrative practices used by contemporary busi-

2. Alfred D. Chandler, Jr., is largely responsible for developing the concepts that enable historians to trace the role of the railroads as management pioneers, and this theme appears in much of his work. In addition to *The Visible Hand,* see, for example, "The Beginnings of 'Big Business' in American Industry," *Business History Review* (hereafter *BHR*) XXXIII, 1 (Spring 1959): 1–31; *Strategy and Structure* (Cambridge, MA: The MIT Press, 1965); *The Railroads: The Nation's First Big Business* (New York: Harcourt, Brace & World, 1965); "The Railroads: Pioneers in Modern Corporate Management," *BHR* XXIX, 1 (Spring 1965)· 16–40; "The Railroads: Innovators in Modern Business Administration" (with Stephen Salsbury), in Bruce Mazlish, ed., *The Railroads and the Space Program* (Cambridge, MA: The MIT Press, 1965), pp. 127–62; "The United States: Seedbed of Managerial Capitalism," in Alfred D. Chandler, Jr., and Herman Daems, eds., *Managerial Hierarchies* (Cambridge, MA: Harvard University Press, 1980).

3. For other views on the "army-as-model" idea see, for example, Thomas C. Cochran, *200 Years of American Business History* (New York: Delta Publishing Company, 1977), who suggests that some industries tried the Army model but rejected it because they found it too cumbersome and complicated. Harold C. Livesay, *Andrew Carnegie and the Rise of Big Business* (Boston: Little, Brown & Co., 1975), notes that the Army had experience moving men and material over long distances and that the railroads adopted elements of military organization and terminology. Even Lewis Mumford, who otherwise takes a rather dim view of the military (it is usually "the refuge of third-rate minds") acknowledges that the Army is an ideal structural model for industry to copy; see *Technics and Civilization* (New York: Harbinger Press, 1963). Professor Chandler is the leading exponent of the alternate view. See Chandler and Salsbury, "The Railroads: Innovators," p. 160 and Chandler, *Visible Hand,* p. 95 for two explicit rejections of the Army's possible role as a railroad management model.

nesses. That the Army's procedures were potentially useful to the nation's business community was not recognized until early railroad managers faced operational and administrative problems similar to those the Army had already encountered. Forced by organizational crises to adopt new business methods between 1827 and the mid-1840s, the railroads often turned directly to the Army for management procedures. Even as late as the mid-1850s, former Army officers in railroad service, working in concert with their civilian colleagues, applied some of the fundamental principles and procedures of military administration to the railroads they served. In the process they helped to shape the course of managerial developments in an industry that became the acknowledged leader in the field of business administration in the United States.

Until about 1818 social, political, and economic forces placed America's national military establishment in a precarious position. Opponents of the military argued that the institution strained the nation's meager resources, and they bolstered their arguments by conjuring up the specter of a tyrannical standing army subverting the rights of free Americans. Largely because of those criticisms, Federalist attempts to place the national military establishment, and especially the Army, on a more rational, stable footing were unsuccessful. Despite periodic reform efforts, military administration remained fragmentary, decentralized, and inefficient. The field army and the staff bureaus were poorly organized and inadequately supported. Army morale and performance suffered accordingly.[4]

The War of 1812 demonstrated the cost of neglect and mismanagement. The Army's performance during the war, especially at the command and staff level, did little credit to either the organization or the nation. The administrative incompetence that plagued the war effort did, however, offer an opportunity to the reform-minded nationalists who had helped push the nation into the war. Out of crisis and near disaster came the ideas and the innovative spirit to begin redressing the organizational and managerial problems that had afflicted the Army for

4. Information on the United States Army during the colonial, Confederation, and early national periods is available in Douglas E. Leach, *Arms for Empire* (New York: Macmillan, 1967); Don Higginbotham, *The War of American Independence* (New York: Macmillan, 1971); Richard H. Kohn, *Eagle and Sword* (New York: Free Press, 1973). Russell F. Weigley's *History of the United States Army* (New York: Macmillan, 1967) is the standard institutional history of the Army.

decades. Nonexistent or ineffective civilian leadership stifled military reformers within the War Department from March 1815 through October 1817, when the situation changed dramatically.[5]

During John C. Calhoun's tenure as Secretary of War (October 1817 to March 1825), the United States Army worked to create a more comprehensive organizational hierarchy. It also created the procedural systems needed to make its new organization function more effectively. Though most of the reforms applied at this time were borrowed from concepts suggested or tried during the previous forty years, this reform effort enjoyed a degree of long-term success because of the nation's changing political environment and because of the energy and talents of Calhoun and his military subordinates. Calhoun served in the House of Representatives during the War of 1812 and witnessed firsthand the chaos and mismanagement that characterized the nation's war effort. A proponent of military preparedness even before he accepted the cabinet portfolio from James Monroe, Calhoun brought new direction to military administration in the United States. His efforts produced the essential elements of the military system the Army used until 1903.[6]

In 1818 the War Department supervised a variety of projects across the continent. It controlled the field army and the staff bureaus that supported the combat arms. It also administered the construction of a string of coastal fortifications, conducted a coast survey, explored and mapped the trans-Mississippi west,

5. A brief yet comprehensive history of the "Second War for Independence" is Harry L. Coles, *The War of 1812* (Chicago: University of Chicago Press, 1965). Histories of the war that focus on local areas frequently detail military management within narrow limits. An especially useful study of this type is Louis L. Babcock, *The War of 1812 on the Niagara Frontier* (Buffalo, NY: Buffalo Historical Society, 1927).

6. Calhoun is the subject of several biographies, most of which tend to concentrate on his legislative career. Perhaps the best examination of his career as Secretary of War is Charles M. Wiltse, *John C. Calhoun, Nationalist, 1787–1828* (Indianapolis: Bobbs-Merrill, 1944). Also useful is Leonard D. White, *The Jeffersonians: A Study in Administrative History, 1801–1829* (New York: Macmillan, 1951). Other useful works include Gerald M. Capers, *John C. Calhoun—Opportunist: A Reappraisal* (Gainesville, FL: University of Florida Press, 1960); Arthur Styron, *The Cast-Iron Man: John C. Calhoun and American Democracy* (New York: Longmans, Green & Co., 1944). Calhoun's general and specific reform proposals emerge most clearly in his official reports and his public and private correspondence. For the former, see *American State Papers, Class V: Military Affairs*, 7 vols. (Washington, D.C.: Gales and Seaton, 1832–1861), especially volumes 1 and 2 (hereafter cited as *ASP-MA*). For Calhoun's correspondence, see Robert L. Meriwether and W. Edwin Hemphill, eds., *The Papers of John C. Calhoun*, 13 vols. to date (Columbia, SC: University of South Carolina Press for the South Caroliniana Society, 1959 . . .) (hereafter cited as *PJCC*).

handled Indian affairs, fabricated arms and ammunition, and distributed bounty lands and pensions to veterans. By the time Calhoun moved on to the vice-presidency in 1825, Congress had directed the War Department to assume responsibility for the continuation of the National Road and to provide engineering assistance to internal improvements projects as well.[7] At the same time, House and Senate military committees scrutinized Army expenditures with extreme diligence because the consensus for increased military spending and preparedness was fragile and the nation's economic condition remained precarious. Both proponents and foes of military spending believed that Congress had to ensure that Army appropriations were expended with care. Congress therefore bombarded the secretary's office with constant requests for a variety of detailed estimates and periodic reports.[8]

Faced with the problems of exercising effective control over diverse, scattered operations while providing Congress with the detailed cost estimates and reports it demanded, Calhoun and his subordinates scrutinized the Army's organizational requirements and formulated specific structural and procedural plans for meeting these needs. Calhoun's managerial philosophy stressed centralization of authority, strict personal and financial accountability, and the value of an elaborate, functionally specialized administrative hierarchy following comprehensive written regulations.[9] He convinced Congress to approve the changes needed to effect his goals because the new system promised to satisfy both Congressional demands and his own managerial expectations. The system of accountability Calhoun installed, and the bureaucratic means by which this accountability was enforced, would be among the most important parts of the military management system later borrowed by the business community.

7. Hill, *Roads, Rails and Waterways*; Weigley, *History of the United States Army*; White, *The Jeffersonians*; L. D. Ingersoll, *History of the United States War Department* (Washington, D.C.: Francis B. Mohun, 1879); William B. Skelton, "The United States Army, 1821–1837: An Institutional History" (Ph.D. dissertation, Northwestern University, 1968); and Carlton B. Smith, "The U. S. War Department, 1818–1842" (Ph.D. dissertation, University of Virginia, 1967) provide information on the Army's activities during this period.
8. White, *The Jeffersonians*, pp. 94–107 discusses Congressional efforts to supervise governmental activities in general. In addition, the numerous reports reprinted in *ASP-MA* demonstrate the volume and range of detailed information Congress demanded specifically from the War Department.
9. White, *The Jeffersonians*, p. 246; Calhoun to Charles Ingersoll, December 14, 1817, *PJCC*, II, pp. 16–17.

Under Calhoun's leadership the War Department expanded and strengthened the military staff system. Seven staff departments—Adjutant and Inspector General, Quartermaster, Subsistence, Pay, Medical, Ordnance, and Judge Advocate General—and a Corps of Engineers provided a variety of technical, administrative, and logistical support services, theoretically leaving the Secretary of War free to concentrate on matters of national and corporate strategy and policy.[10] With the assistance of his staff officers, Calhoun prepared detailed departmental regulations that defined the specific functions, duties, and responsibilities of each of the staff departments and established a complicated set of standardized reporting and accounting procedures. Calhoun was pleased with the reception accorded the staff regulations, but he recognized that the field army was still largely unaffected by his reform efforts. He therefore readily allowed Brigadier General Winfield Scott to prepare a new set of general regulations for the Army as a whole, a project Scott first proposed in September 1818.[11]

Scott's *General Regulations for the Army,* completed in July 1821 and revised in early 1825, was the capstone of Calhoun's reform efforts. The first comprehensive management manual published in the United States, the 1821 edition contained 79 articles, comprising approximately 400 pages of text, in which Scott described the organizational theory that shaped the Army and laid out in great detail the rules and procedures that governed all phases of military life, from the most convoluted questions of relative rank and the division of responsibility between line and staff officers to the most mundane details of camp life and personal hygiene. The regulations required all

10. Calhoun laid out his suggestions for staff reforms and explained the rationale behind these suggestions in a report to Senator John Williams on February 5, 1818, in National Archives Record Group 107, Records of the Office of the Secretary of War, "Reports to Congress by the Secretary of War," Microcopy 220, Roll 1, Vol. I, p. 436. He also discussed his ideas with his military subordinates: Calhoun to Major General Jacob Brown (commanding general of the Army), December 17, 1817, *PJCC,* II, pp. 22–23. The Calhoun correspondence shows that he frequently relied on ideas suggested by his military subordinates when he formulated specific reform plans and regulations; see, for example, Brigadier General Winfield Scott to Calhoun, December 16, 1817, *PJCC,* II, p. 21.

11. Events surrounding Calhoun's efforts to provide comprehensive staff and general regulations can be reconstructed from material in his papers and in National Archives Record Group 107, Records of the Office of the Secretary of War, Microcopy 6, "Letters Sent, Military Affairs, 1800–1861," Rolls 10, 11, and 12; Microcopy 22, "Registers of Letters Received by the Secretary of War, 1800–1860," Rolls 9–19; Microcopy 221, "Letters Received by the Secretary of War, Registered Series, 1801–1860," passim.

units, not just the staff bureaus, to compile regular monthly, quarterly, and annual reports, summaries, and inventories. They also laid out the internal lines of authority and communication that regulated the flow of information within the Army. Scott's work included as separate articles the staff regulations Calhoun had promulgated earlier.[12]

The theme of personal and financial accountability pervaded the regulations. The reasons for this institutional obsession with accountability were clear. The War Department was one of the most complex organizations in the United States. To keep the military bureaucracy functioning efficiently, to meet internal and external demands for information, and to safeguard the reputation of the Army and its officers, the War Department worked constantly to ensure that it could trace the professional activities of its personnel. Line and staff officers reported where they were, what they were doing, and how they were spending the funds allotted to them. This information allowed the central office to compare, for example, the managerial talents of individual officers. It also provided the department with the means to exercise a high degree of centralized control over the Army's functionally specialized, decentralized operations. No contemporary American business faced such a complex set of managerial problems. As the Army developed and refined its administative techniques, it created a bureaucratic management structure with a degree of complexity that was without precedent in the American business community.

The War Department based its accounting system on a series of periodic reports submitted by its officers. A member of the Corps of Engineers maintained a daily log and register and completed four monthly, eight quarterly, and three annual reports and estimates, which the Washington office used to trace expenditures, prepare estimates, and evaluate the performance of its officers in the field.[13] The chief engineer spent much of

12. The 1821 edition of the regulations is U. S. War Department, *General Regulations for the Army; or, Military Institutes*, compiled by Major General Winfield Scott (Philadelphia: M. Carey & Sons, 1821). It is also available in *ASP-MA*, II, pp. 199–267, although this draft does not include the final versions of the staff regulations, articles 66–73. The 1825 edition is U. S. War Department, *General Regulations of the Army; or, Military Institutes*, revised by Major General Winfield Scott (Washington, D.C.: Davis & Force, 1825). Unless otherwise noted, all references to the regulations cited below will be drawn from the 1825 edition.

13. See *General Regulations*, 1825, pars. 890–3, 897, 915, pp. 169–70, 177. The blank forms are included on pp. 179–99. An engineer submitted the following reports:

his time examining these accounts and frequently returned reports that seemed inadequate. Colonel Alexander Macomb, chief engineer from 1821 to 1828, returned individual reports that omitted required information. He occasionally sent back complete sets of returns that did not meet the standards he and the War Department demanded.[14] One officer did not furnish his reports in duplicate, so Macomb referred him "to the 897 paragraph of the general army regulations, to which you will be pleased to attend."[15] Macomb summarily rejected requests for extended deadlines, and in some instances he refused to approve vouchers for what he deemed unnecessary purchases, which meant that the officer bore the cost of the item himself.[16]

The Engineer Department also applied its regulations to the civilian "survey brigades" the Army employed to supplement its limited manpower resources. In June 1825 James Shriver led his brigade of civilians into west central Ohio to survey the route for an extension of the National Road. Shriver reported that his team found it almost impossible to use the forms the Engineer Department supplied. Colonel Macomb replied that the official forms were "conceived to be more full and distinct, and consequently better adapted to the fulfillment of the pur-

---

Form
1   Annual Summary and Estimate
2   Monthly Inventory of Construction Material on Hand
3   Quarterly Progress Report and Inventory
4   Quarterly Inventory of "Equipage and Appurtenances"
5   Quarterly Provisions Expenditures
6   Quarterly Forage Expenditures
7   Detailed Annual Estimate
8   Annual Work Schedule
9   Quarterly Funds Estimate
10 Standard Receipt Form
11 Monthly Payroll
12 Quarterly Report of Expenditures
13 Quarterly Abstract of "Accounts Current"
14 Monthly Per Diem Claim Form

14. Macomb to Major John J. Abert, December 14, 1824, in National Archives Record Group 77, Records of the Office of the Chief of Engineers, Records of the Topographical Bureau, "Letters Sent Relating to Internal Improvements Under the Act of April 30, 1824," p. 44 (hereafter cited as "Letters Sent, Internal Improvements," RG 77); Macomb to Lieutenant Colonel Stephen H. Long, April 6, 1825, "Letters Sent, Internal Improvements, 1824–1830," RG 77, p. 125.

15. Macomb to Captain William G. McNeill, April 14, 1826, "Letters Sent, Internal Improvements, 1824–1830," RG 77, p. 159.

16. See, for example, Macomb to Major John J. Abert, August 22, 1826, "Letters Sent, Internal Improvements, 1824–1830," RG 77, p. 304; Macomb to Major Stephen W. Kearney, August 24, 1826, "Letters Sent, Internal Improvements, 1824–1830," RG 77, pp. 306–7.

poses for which they are intended" than the sample Shriver forwarded, but he told Shriver to use those forms his team could master. Macomb warned him, however, that the Engineer Department "desired that those prescribed may be adopted as soon as they shall be understood."[17] Two of Shriver's colleagues, Jonathan Knight and Caspar Wever, also had difficulty adapting to the War Department's methods. Macomb advised Wever:

None of your accounts correspond with the forms prescribed by the regulations, altho' they are made out with as much distinctness probably as is customary with accounts relating to ordinary transactions of a private nature. It would be desirable, however, that they should be made hereafter agreeably to the forms prescribed.[18]

Despite Congressional indifference or hostility, personal jealousy, and organizational conflicts, Calhoun successfully implemented a coherent system of military management in the United States between 1818 and 1825.[19] The institution Calhoun took over in late 1817 reflected the fears, prejudices, and weaknesses of American society. The Army he left in 1825 was well on its way to becoming a modern, reasonably effective corporate bureaucracy, more advanced and better organized than any contemporary social, political, or business organization in the United States.

One measure of the impact of the Calhoun reforms is the relative success with which the Army conducted its affairs during the 1820s. As the officers assigned to the staff bureaus became more proficient in the execution of their duties, the War Department found itself better able to administer the field army and to control its other responsibilities. Perhaps the key indicator was the performance of the supply system: the troops received more and better food, clothing, arms, and ammunition more regularly than ever before. The new methods did not, of course, completely solve the Army's administrative problems. They were, however, a substantive improvement over previous efforts. Even so, few executives outside the War

17. Macomb to Shriver, June 21, 1825, "Letters Sent, Internal Improvements, 1824–1830," RG 77, pp. 92–93.

18. Macomb to Wever, December 16, 1825, "Letters Sent, Internal Improvements, 1824–1830," RG 77, in National Archives Microcopy M-65, Roll 1, p. 231.

19. The political context of these reforms is discussed in detail in the Calhoun biographies cited above, especially Wiltse, *Calhoun—Nationalist*, in the introductions to the appropriate volumes of the Calhoun papers, and in White, *The Jeffersonians*.

Department or the government were aware of the success of the Calhoun reforms. The military staff bureaus confined their work to military functions, while the civilian business community exhibited little desire to learn more about Army administrative procedures.[20]

The work of the Corps of Engineers in the 1820s gave the Army the opportunity to demonstrate its administrative competence to a wider civilian audience. During the decade the Corps applied military operating procedures to a variety of civilian projects and therefore implicitly demonstrated to some of the nation's economic leaders the potential utility of the Army's administrative system. As a result, when some of these entrepreneurs found that their traditional business methods could not cope with the organizational and administrative demands of their firms in the late 1820s, they turned to the Army for advice and to its bureaucratic system as a model.

The Army's direct involvement with private internal improvements projects began with the passage of the General Survey Act of April 30, 1824. Congress, facing criticism that the federal government was not doing enough to promote internal improvements, directed the War Department to make its engineers available to private or locally sponsored transportation projects. From 1824 to 1827 most of the officers assigned under the act conducted canal surveys. Attention then turned to a new transportation technology, the railroad, and in 1827 the Engineer Department assigned three civilian and eleven military engineers to the service of the newly organized Baltimore & Ohio Railroad Company.[21]

20. For a more detailed examination of the impact of the Calhoun-era organizational and administrative reforms on one staff bureau, see Erna Risch, *Quartermaster Support of the Army: A History of the Corps, 1775–1939* (Washington, D.C.: Quartermaster Historian's Office, Office of the Quartermaster General, 1962); and James A. Huston, *The Sinews of War: Army Logistics, 1775–1953* (Washington, D.C.: Office of the Chief of Military History, U. S. Army, 1966).

21. Hill, *Roads, Rails and Waterways* is the standard treatment of the Corps of Engineers' involvement with internal improvements. The Corps also gained valuable experience from an administrative standpoint during its involvement with the nation's fortification program. The administrative history of this effort can be reconstructed from correspondence and reports in National Archives Record Group 77, Records of the Office of the Chief of Engineers, especially "Letters Received, 1819–1866." While the engineers themselves gained useful managerial experience during their work on this program, this expertise was not transferred to civilian contractors. The civilians the engineers dealt with as part of this work were usually small local concerns that supplied materials to the project. These small firms had no need to experiment with the military's administrative methods because traditional business management techniques worked well enough for them. The exchange of information had to wait until the business community was ready to use the resources available in the Army.

The government engineers began their work in July 1827. On April 5, 1828, they presented their survey report to the B & O Board of Directors. From the outset the members of the survey brigades used Engineer Department accounting and reporting procedures for their submissions to the directors. The government engineers used a reporting system that enabled them to account for the public and private funds they expended and to provide technical information to the railroad in a standardized format. Since the B & O Board of Directors did not suggest any more appropriate system, the brigades used the military procedures they were most familiar with. The directors neither questioned nor challenged the adoption of Army practices. As a result, B & O survey operations were conducted according to technical and administrative standards Army engineers used on similar military projects. The teams used regulation military surveying and mapping methods as they examined various routes. They forwarded reports to the Board of Directors according to the schedule prescribed by the Engineer Department. They used Engineer Department forms to record and transmit their findings and reports. The first management procedures used by the Baltimore & Ohio Railroad were thus standard United States Army Engineer Department practices. The information available to the railroad's management was essentially the same information, submitted in the same format, as the data used by the Corps of Engineers to conduct its operations.[22]

The B & O Board of Directors selected a route from among the alternatives surveyed by its engineers and turned its atten-

---

22. This discussion of the Baltimore & Ohio is based on material in National Archives Record Group 77, Records of the Office of the Chief of Engineers, specifically material in the records of the Topographical Bureau and the Board on Internal Improvements, including Microcopy 65, "Letters Sent, Office of the Chief of Engineers, Relating to Internal Improvements, 1824–1830;" Microcopy 505, "Registers of Letters Received by the Topographical Bureau of the War Department, 1824–1866;" and Microcopy 506, "Letters Received by the Topographical Bureau of the War Department, 1824–1866." Two of the officers most closely involved with the project left a memoir of their activities: Stephen H. Long and William G. McNeill, *Narrative of the Proceedings of the Board of Engineers of the Baltimore and Ohio Rail Road Company, From Its Organization to Its Dissolution, Together With an Exposition of Facts, Illustrative of the Conduct of Sundry Individuals* (Baltimore: Bailey & Francis, 1830), although it must be used with caution, since it is a one-sided summary of the controversy that surrounded the board. It is most useful for the correspondence reprinted therein. Information on the procedures the B & O engineers used on the surveys is available in Stephen H. Long, *Rail Road Manual, or a Brief Exposition of the Principles and Deductions in Tracing the Route of a Rail Road* (Baltimore: William Woody, 1829). These two volumes and, unless otherwise noted, all other contemporary published railroad material, are available in the Economics and Finance Department Library of the Association of American Railroads in Washington.

tion to the problems of building the line. To supervise construction activities, the board created a "Board of Engineers of the Baltimore & Ohio Railroad Company." The members of this board included company president Philip E. Thomas, Lieutenant Colonel Stephen H. Long, and Jonathan Knight. Long was an experienced Army engineer and head of one of the survey brigades; Knight came to the B & O from the National Road, where he worked under the supervision of the Engineer Department. One of the Board of Engineers' first acts was to appoint Caspar Wever the superintendent of construction. Wever too was a veteran of the National Road and a former employee of the Engineer Department. The first management hierarchy of the B & O was composed almost exclusively of men familiar with the principles and procedures that made up the Army's management system.[23]

The Board of Engineers soon met to consider the problems of creating an administrative system to control the railroad's survey and construction operations. As early as April 1828, as the railroad company faced the challenges of moving from the planning to the construction stage, Captain William G. McNeill, the head of the second military survey brigade assigned to the B & O, discussed with Thomas the outline of such a system. Thomas in turn suggested that McNeill draft a set of formal written regulations for the railroad. The resulting regulations were "similar to those which govern generally in the U.S. Engineer Department." The similarity was more than superficial, since McNeill later noted that "when I thought applicable, I have transcribed literally from the printed regulations of the U.S. Engineer Department."[24]

Thomas, a Baltimore banker, was intrigued by the managerial possibilities inherent in McNeill's regulations. Not long after receiving McNeill's report, Thomas asked Army Chief Engineer Alexander Macomb to forward any information he could relating to "what course the Rail Road company should pursue to [record?] the amount advanced on account of ser-

23. The formation of the Board of Engineers is discussed in Long and McNeill, *Narrative*, pp. 18–21, and in the "First Annual Report of the Board of Engineers of the Baltimore and Ohio Rail Road, September 30, 1828," in *Second Annual Report...of the Baltimore and Ohio Rail Road*, 1828. Also see Herbert H. Harwood, Jr., *Impossible Challenge: The Baltimore & Ohio Railroad in Maryland* (Baltimore: Barnard, Roberts & Co., 1979). Harwood is primarily a technical study, but it does contain some useful insights into the organization of the railroad.
24. McNeill to Thomas, May 1, 1828, in Long and McNeill, *Narrative*, pp. 31, 34.

vices, &c." Macomb responded by forwarding a copy of the Engineer Department regulations. He suggested that if the railroad formally adopted the general system of reports, estimates, and accounts the Army used, the directors would "at all times have the means of judging accurately of the state of their affairs and the progress of their work."[25]

The Army's influence began to be felt. In early June 1828, Thomas directed the Board of Engineers to establish an official written code of regulations to govern the operations of the B & O Engineer Department. The board assigned the task to Captain McNeill. On June 10, 1828, the Board of Directors adopted McNeill's "Regulations for the Engineer Department." McNeill admitted that the new railroad regulations corresponded "as nearly as the nature of the two services would admit, to those adopted in the United States Engineer Service."[26]

The B & O Engineer Department Regulations of 1828 closely followed the form and drew heavily on the content of Article 67 (the Engineer Department regulations) of the 1825 edition of Scott's *General Regulations of the Army*. They laid out the organization of the department and defined the duties and responsibilities of its officers. The system of personal and financial accountability was in all major respects identical to the system used by the Corps of Engineers. Acknowledging that the railroad lacked suitable alternative management structures or procedures in its early days, the engineer regulations even directed that reporting officers use Army report forms "until other forms shall be prescribed." In several instances McNeill simply took complete sections from Article 67 and used them verbatim in the B & O regulations.[27]

The specific textual connections between McNeill's regulations and Scott's publication are noteworthy, but perhaps more

25. Thomas to Macomb, May 17, 1828, "Register of Letters Received, Topographical Bureau, 1824–1866," "T" Letter 518, RG 77, M-505, Roll 1; Macomb to Thomas, May 20, 1828, "Letters Sent, Internal Improvements, 1824–1830," RG 77, M-65, Roll 2, p. 85.

26. Long and McNeill, *Narrative*, p. 29; the regulations themselves are in the *Narrative*, Appendix I, pp. 11–16. The subtitle of the regulations gives some clue as to the managerial problems McNeill felt were most pressing: "Organization of the Engineer Department of the Baltimore and Ohio Rail-Road Company, providing a system of government for the same; and for securing a strict accountability in the financial operations of the Baltimore and Ohio Rail-Road Company, and economy in the disbursement of its funds."

27. The comparison described here is between McNeill's regulations, *Narrative*, Appendix I, pp. 11–16, and *General Regulations for the Army*, 1825, Article 67, pp. 169–76. B & O Regulations para. 6 is an almost exact quotation from *General Regulations* par.

important was the "philosophical" precedent the regulations established. The B & O now had a set of regulations that, like Article 67, explicitly established accountability and hierarchy as fundamental organizational principles that would shape future managerial reforms. These concepts were not completely alien to the American business community, but there were few, if any, business organizations that could boast of such a detailed management hierarchy or such advanced bureaucratic management procedures.[28]

McNeill's 1828 regulations for the B & O Engineer Department represented the railroad's first attempt to formulate and codify an operational structure and a set of specific administrative procedures applicable to all railroad employees. While these regulations applied only to the technical operations controlled by the B & O Board of Engineers, they established organizational and administrative precedents that would shape later managerial developments in the B & O and other railroads. The B & O recognized that the fundamental managerial techniques laid down in 1828 worked well; and as the railroad expanded and modified its management procedures over the years, it continued to rely on basic systems established during its early days. The B & O was also the first of the major trunk lines to begin operations. Its activities were widely publicized in the railroad periodicals that appeared, including the *Railroad Advocate* and the *American Railroad Journal*. Later railroads, including the other trunk lines, normally examined the B & O's management procedures before making administrative changes of their own.[29]

915; B & O par 8 comes directly from *General Regulations* par. 892. Also compare:

B & O Regulations	with	*General Regulations*
par. 2		pars. 890, 904, 906, 907
par. 4		par. 889
pars. 5, 7		pars. 891, 897, 909

28. This assertion relies heavily on Chandler's work. See note 2, above.

29. The movement of individuals between the railroad lines, the work of specialized publications such as Henry Varnum Poor's *American Railroad Journal* and the *Railroad Advocate,* and the efforts of the lines themselves to gather information about their contemporaries and competitors fostered the flow of management information. Since the B & O was one of the first major railroads, it was a subject of much interest. See Alfred D. Chandler, Jr., *Henry Varnum Poor: Business Editor, Analyst, and Reformer* (Cambridge, MA: Harvard University Press, 1956). The "railroad tour" became an almost mandatory part of the process of making structural and procedural changes on new and established lines. Both the Western and the Pennsylvania, discussed below, were avid students of the state of the management art on other railroads.

McNeill's effort was not without its flaws, although these were not immediately obvious. He prepared the regulations in haste, reflecting the desire of all parties involved to get the project started with dispatch. The regulations were not based on any detailed, rational examination of the factors that influenced the railroad's activities, but this criticism must be mitigated by the knowledge that the novelty of the railroad required a high degree of administrative adaptation at this stage of its development. From a narrow technical perspective, the regulations allowed little freedom of entrepreneurial action, especially if they were enforced with the rigor Calhoun applied. The regulations ignored the social, political, aesthetic, and psychological conditions that influenced the railroad's route and the design and construction of its facilities. McNeill assumed that the same drive for simplicity and economy that shaped his work as a military engineer also influenced railroad decision making. Unfortunately, utility, economy, and accountability, the cornerstones of McNeill's managerial principles, while recognized as generally desirable goals, occasionally assumed positions of secondary importance when the railroad made decisions. McNeill's apparent inability or unwillingness to recognize the many factors that influenced civilian management behavior ultimately contributed to the clash of personalities and philosophies that rent the B & O management hierarchy in 1829.[30]

Throughout the last half of 1828, Long and Wever argued repeatedly and with growing bitterness over how McNeill's regulations would be interpreted and applied, and by whom. Long favored a strict interpretation of the regulations and demanded that Wever, who, according to the regulations, was an agent of the Board of Engineers, obey them without demur. Thomas, Knight, and the members of the Board of Directors often sided with Wever in these disputes. The civilians were more willing to give the superintendent of construction a longer rein. Like the officers, they too wanted results, but they also believed that the complicated procedures the officers demanded only delayed the work. After a struggle between Long and Wever over Wever's accounts, the Board of Directors noted that

---

30. McNeill's position is stated in the *Narrative*, pp. 114–45, 155, and 197–99. After a tour of the road in the spring of 1828, both McNeill and Knight complained about excessive embellishments added to the work, which they saw as proof of "bad taste and want of judgement." *Narrative*, p. 145.

Col. Long's *education* and *pursuits* had no doubt induced him to be more particular in matters of this sort, than the welfare and interests of the Railroad service required, and that in the present case the rigid formality he wished to observe might be dispensed with.[31]

These disputes were rooted squarely in the Army engineers' "education and pursuits," as the B & O Board of Directors recognized. Wever and the other civilian engineers on the project represented the craft tradition in engineering. They based their work on intuition and experience and were willing to accept higher costs if it allowed them to work with materials (stone for bridges, for example) they knew were safe, solid, and able to withstand the pressure the rolling stock would create. Long and McNeill, on the other hand, represented a new school of engineering that relied on scientific methods and mathematical analysis to determine more precisely the needs and capabilities of planned structures. They calculated what was required and built accordingly, often using techniques and materials that produced cost savings. They touted these savings in reports that were calculated to appeal to a cost-conscious board, but the presentations also revealed an unwavering concern for proper, analytically based engineering methods, system, regularity, uniformity, and order. These issues, with cost control, played the most crucial role in determining how the officers responded to the technical and managerial challenges they faced. It seems likely, however, that the cost savings the officers promised were intended to make an unusually complicated management system more palatable to a Board of Directors used to less formal procedures. From Long and McNeill's perspective, what mattered most was the system.[32]

---

31. Long and McNeill, *Narrative*, p. 139

32. For a more detailed treatment of the role of the Army officers' belief in the value of a systematic approach to the technical and managerial problems they faced, see Daniel H. Calhoun, *The American Civil Engineer: Origins and Conflict* (Cambridge, MA: The Technology Press, MIT, 1960). Also see Long, *Rail Road Manual,* which lays out some of the detailed formulas and procedures an academically trained engineer could be expected to use in the course of his railroad work. Harwood, *Impossible Challenge*, p. 15, notes that the railroad was part commercial artery, part monument in the eyes of some of those involved with the B & O. A finished stone viaduct was a more impressive, if more costly, monument, a fact apparently lost on Long, McNeill, and some of their military colleagues. This issue formed the background of the numerous clashes described in Long and McNeill's *Narrative*. Neither side showed any real willingness to compromise or to try to understand the other side's feelings or objectives. The officers came across as haughty, self-righteous, and supercilious; the civilians appear deceitful, petty, spiteful, and occasionally dishonest, but one must consider the perspective from which the book was written.

Despite these controversies, the railroad's civilian managers did recognize that some elements of the system the officers used could save the railroad from public embarrassment. In February 1829 the Maryland legislature asked the railroad for a report of its activities and expenditures. Survey records were available, but construction records, the responsibility of Wever, were incomplete. Thomas "set forth in glowing colors the evils that were to be apprehended from the failure to meet the demand of the legislature, and give them all the information required." Long told Thomas that the system the officers were trying to introduce "would have kept the board in constant preparation for any emergency of this kind." According to Long, Thomas promised "that he would do anything in his power to introduce such a system, if Col. Long would help him out of his present difficulties." Long compiled the required data from copies of reports the assistant superintendents of construction (three lieutenants from the Corps of Engineers) kept after they submitted their periodic summaries and estimates to Wever, as required by the railroad's regulations.[33]

After more than a year of conflict, the B & O Board of Directors decided that it could no longer tolerate the Army officers in positions of managerial responsibility. It dissolved the Board of Engineers in January 1830. The Army reassigned its officers, and the B & O staffed a new management hierarchy. Jonathan Knight became chief engineer, while Caspar Wever retained his position as superintendent of construction. The Army engineers were gone, but their managerial legacy remained.[34]

In 1830 the B & O Board of Directors adopted new by-laws and regulations for the railroad. Primarily the work of Jonathan Knight, this new set of regulations owed much to its Army and B & O ancestors, and it contained provisions that would have pleased Long and McNeill. The regulations formalized a system of accountability and authority that resolved many of the interpretative and procedural conflicts that plagued the operations of the Board of Engineers. The new regulations also gave the B & O a more rational executive chain of command. While the railroad simplified its organizational structure, the administrative system it adopted was more

---

33. Long and McNeill, *Narrative*, pp. 127–29.
34. Ibid., p. 341.

complex, as befit the growing complexity of the railroad's operations.[35]

Knight's *Rules and Regulations* did not acknowledge its debt to McNeill's regulations or to Article 67, but the lineage is unmistakable.[36] That Knight saw the need to rewrite the regulations is telling in itself, since it suggests the degree to which the leadership of the B & O had learned that formal written procedures enabled it to exercise a degree of control over the railroad's operations that might not have been possible with the less formal methods other contemporary businesses used. As it gained a more accurate picture of the specific operational and organizational problems it faced, the B & O came to appreciate more fully the potential of the military management system. Questions of cost control and accountability focused managerial attention on the reporting system that was one of the key elements of McNeill's efforts. The railroad's management recognized that this complicated system of detailed reports was a source of information that enabled the railroad to coordinate and control its activities and to maintain a close watch on costs as well as cash flow.

The language the railroad used to describe the benefits of its management procedures to its stockholders recalled the terms Calhoun used to sell his system to Congress a decade before. In 1830, as Knight prepared his regulations, the Board of Directors noted that "a well digested plan for the future government and regulation of all the operations of the company may now, and should without delay, be adopted, in which shall be clearly defined, the duties and responsibilities of the officers and agents employed in its service."[37] The next year the directors reported that "the system organized for the regulation of the business of the Company, and for the government of its agents, has, in its operation, been found to be efficient and practical, clearly defining the duties of the several officers and maintain-

35. Baltimore & Ohio Railroad Company, *By-Laws, and Rules and Regulations of the Baltimore and Ohio Railroad Company,* 1830.

36. A comparison of Article 67 of the *General Regulations,* 1825; McNeill's "Regulations for the Engineer Department," 1827; and Knight's *By-Laws, Rules and Regulations,* 1830, demonstrates the links among the three. We see the same attention to detail, an almost identical textual organization, and many of the same procedures appearing in each. Compare, for example, Article 10 of Knight's *Rules and Regulations* with McNeill's paragraph 5, both dealing with a system of regular activity reports, and Knight's Article 12 with McNeill's paragraphs 5 and 8 on accountability. In the latter case, McNeill copied directly from Article 67, and Knight copied directly from McNeill.

37. *Fourth Annual Report,* B & O Railroad, 1830, p. 11.

ing a strict responsibility in every department." They also noted with pleasure that company operations were conducted with "harmony and fidelity," even though these operations required "numerous agents, whose duties rendered it necessary they should be dispersed over an extensive district of country."[38] Knight added his own view of the evolution of the railroad's management procedures. "It is only in practice," he wrote, "that the defects of a system become manifest . . . it becomes our duty to note the defects, investigate the causes, and as far as practicable, to provide against the possible recurrence of these evils."[39]

A management system overtly tied to Army methods served the B & O well until 1847, when operational problems became more pressing than technical ones. On February 10, 1847, the B & O Board of Directors approved a new organizational manual that was the first significant step in the process of administrative reform on the trunk-line railroads during that critical period in the late 1840s and early 1850s when they emerged as the nation's first big business. In preparing this manual, Knight's successor as chief engineer of the B & O, Benjamin H. Latrobe, examined management systems then in use on other lines, a luxury not available to McNeill or Knight in the late 1820s. Many of the lines Latrobe looked at were in New England, where the military influence on railroad managerial developments was strong. Furthermore, New England railroads of the 1840s also reflected managerial developments borrowed from the B & O itself. Ironically, the railroad's search for what it thought were more sophisticated managerial models brought it into contact with systems rooted in the B & O's own Army-based approach modified to accommodate expanding operations.[40]

McNeill played the key role in carrying the thread of military influence to the New England railroads. After leaving the B & O in 1830, he worked for a number of railroads, including

---

38. *Fifth Annual Report*, B & O Railroad, 1831, p. 7.

39. Ibid., p. 49.

40. See Baltimore & Ohio Railroad Company, *Organization of the System of Service of the Baltimore & Ohio Railroad, Under the Proposed New System of Management*, 1847. For a discussion of this "new" system see Chandler, *Visible Hand*, pp. 99–101; Chandler, "The Railroads: Pioneers," pp. 22–26; and Chandler and Salsbury, "The Railroads: Innovators," pp. 135–37.

several in eastern Massachusetts; and in 1836 he went to work for the Western Railroad as a consulting engineer. The Western also hired one of McNeill's former associates on the B & O, Captain William H. Swift, to fill the position of resident engineer. Both men signed service contracts with the railroad on March 18, 1836.[41]

Survey operations began in early April 1836. Construction work progressed smoothly, and on December 21, 1841, the Western inaugurated through service from Boston to Albany in cooperation with the Boston & Worcester Railroad. From the outset the Western's management demonstrated a continuing concern for problems of fiscal accountability and control. Money was tight, and the management was constantly on the lookout for waste and extravagance. The railroad's charter also required the line to file periodic reports with the state, so detailed operating information was required on a regular basis. The Board of Directors therefore worked to install "a system of accountability, which alone can secure the efficient and economical administration of the business of the Corporation."[42]

The exact contributions Swift and McNeill made to this management system are less clear than McNeill's work on the B & O. Like the B & O, however, the Western adopted from the outset procedures similar to those used by the United States Army. Company president Thomas B. Wales was aware of the officers' business methods, since the subject had been discussed before the railroad hired them, and he allowed Swift a free hand in establishing basic accounting and reporting procedures. Swift sought the approval of the Board of Directors as he developed the outlines of the management system, but neither Wales nor the Board of Directors found any reason to be dissatisfied with his work.[43] When a special committee of the

---

41. The history of the Western Railroad is described in Stephen Salsbury, *The State, The Investor, and the Railroad: The Boston & Albany, 1825–1867* (Cambridge, MA: Harvard University Press, 1967). Edward H. Kirkland, *Men, Cities and Transportation*, 2 vols. (Cambridge, MA: Harvard University Press, 1948) is an excellent survey of railroad development in New England. The contracts and correspondence between the interested parties are available in the Records of the Western Railroad Corporation, 1836–1864, Boston & Albany Manuscript Collection, Baker Library, Harvard University, Cambridge, Massachusetts, especially "1836–1838 Letterbook" and "Clerk's Files," Files 24 and 26.

42. Western Railroad Corporation, *First Annual Report...of the Western Railroad Corporation*, 1836, p. 11.

43. See, for example, Swift to Wales, May 30, 1836, and Wales to Josiah Quincy, June 21, 1836, in Western Railroad Corporation, "Letterbook, 1836–1838," pp. 34–36.

board examined the company's operations in detail in 1837 and again in 1838, it expressed its general satisfaction with what company officials had created.[44]

Before 1839 Congress and the War Department were willing to allow their officers to accept employment with the railroad firms to which they were assigned and still maintain their military rank. This policy enabled the government to retain the services of officers like Swift and McNeill, who might otherwise have resigned in favor of far more lucrative positions in private industry. On July 5, 1838, however, Congress repealed the General Survey Act of 1824 and ordered the officers back to their units or their assigned duty stations.[45] McNeill was unaffected by this policy change because he had previously resigned his commission, but the Western faced the prospect of losing Swift. The railroad petitioned Congress to exempt Swift from the recall, but to no avail. Swift decided to remain in the Army and resigned from the Western on July 5, 1839. McNeill assumed Swift's duties temporarily, but he was too involved in other projects to devote his full attention to the Western. Since the railroad insisted that its operations be supervised by a full-time engineer, the Board of Directors hired George W. Whistler, another former Army engineer, as its chief engineer in October 1839.[46]

One of Swift's last acts in the service of the Western Railroad was to serve as a member of the committee formed in April 1839 to plan the organization of the transportation department. George Bliss, the Western's agent and a member of the committee, reported that the plan the group submitted to the Board of Directors was largely Swift's work.[47] Formalized as

44. Western Railroad Corporation, Board of Directors "Minutebook, 1836–1841," pp. 79–81; *Fourth Annual Report*, WRR, 1839, pp. 24–32.

45. See *ASP-MA*, vol. VI, pp. 104–6 for the Congressional report that advocated recalling these officers. As Congress added duties to the Engineer Department, the Corps of Engineers itself became an advocate of recall. See Gratiot to Poinsett, October 30, 1837, in National Archives Record Group 77, Records of the Office of the Chief of Engineers, "Communications to the Secretary of War and to Congress, 1836–1840," vol. I, pp. 216–17.

46. Western Railroad Corporation, Board of Directors "Minutebook, 1836–1841," pp. 137, 158, 169; "Clerk's Files," 1839, File 24. McNeill, Swift, and Whistler were all related, through Whistler's marriages to Mary Swift and Anna McNeill. The Whistler-McNeill union produced several offspring, including William Gibbs McNeill Whistler and the noted American artist James Abbot McNeill Whistler, who immortalized his mother in a famous painting. See Albert Parry, *Whistler's Father* (Indianapolis: Bobbs-Merrill, 1939).

47. George Bliss, *Historical Memoir of the Western Railroad* (Springfield, MA: Samuel Bowles & Co., 1863), p. 90.

Transportation Department regulations in 1840, the plan created a department under the control of the chief engineer, who was ex officio general superintendent as well. He was assisted by two masters of transportation, who in turn supervised maintenance and operating crews and station workers.[48]

Swift's experiences as an Army engineer and as an employee of the B & O (he had been a member of McNeill's survey brigade) shaped his response to the organizational demands he faced on the Western Railroad. What Swift did, in essence, was to apply the basic principles that had formed the regulations controlling engineering operations to the problems of transportation activities. The new regulations took many of the reporting procedures, for example, that the engineers used and, after modifying the forms to reflect different managerial requirements, applied them to a department that supervised a completely different kind of activity. The same hierarchical, bureaucratic structure that characterized the engineer department was now also applied to the transportation branch.[49] The transportation department regulations thus represented a further step in the evolution of railroad management practice and demonstrated the continuing influence of military procedures.

Events soon demonstrated that the system was still far from perfect. A series of accidents, culminating in a head-on collision between two passenger trains on October 5, 1841, focused public and corporate attention on the railroad's operating problems. On October 16 the Board of Directors appointed a committee to examine the operational and administrative practices of other railroads and to draw up such regulations "as shall . . . best promote the safety of the traveller and the security of the trains."[50]

Bliss again credited a military officer (or, more precisely, a former officer) with proposing most of the reforms the committee suggested in its report. According to Bliss, the commit-

48. Western Railroad Corporation, Board of Directors "Minutebook, 1836–1841," pp. 154–56; "Report of a Committee on Running the Road," June 12, 1839, "Clerk's Files," 1839, File 21; *Transportation Department Regulations,* 1840, in AAR Library, Washington, D.C.

49. See WRR *Transportation Department Regulations,* 1840. Compare these to the B & O regulations discussed above, especially the 1827 and 1830 editions cited in notes 26 and 35, above. The similarity between these regulations is even more striking when one considers the relatively few changes that had been made within the intervening decade.

50. Salsbury, *Boston & Albany,* pp. 182–86; Western Railroad Corporation, Board of Directors "Minutebook, 1836–1841," p. 306.

tee relied on George W. Whistler—the new chief engineer and yet another alumnus of McNeill's B & O survey brigade— "for the rules and regulations for the trains, and for systematizing the operations of employees of every grade."[51] While neither the military's organizational theory nor its administrative procedures offered the railroad precise guidelines for responding to the series of events that propelled reform, Whistler drew upon his military experience and his work with other systems based on the military's management model for a pattern of organizational and administrative procedures of proven utility to a railroad. The investigating committee's "Report on Avoiding Collisions and Governing the Employees" of November 30, 1841, is usually seen as one of the seminal works in the evolution of railroad management, although the committee noted repeatedly that Swift's operational system was fundamentally sound and that the changes the committee suggested were little more than amendments to the rules and regulations then in force. Indeed, the committee cautioned the board that adopting the "Twelve Articles" as if "they were new subjects of arrangement," might "excite the wonder of the public that measures for their safety had been so long delayed."[52]

One measure of the success of the 1841 reforms is that managerial procedures stabilized on the Western, despite Whistler's resignation on May 17, 1842, to become chief engineer of the Moscow to St. Petersburg railroad project in Russia. Swift returned to the Western in 1851 to serve a two-year term as president. In 1868 the Western and the Boston & Worcester combined to form the Boston & Albany Railroad, which became one of the most successful of the New England regional lines.[53]

The contributions military engineers made to management developments on the Western were perhaps fundamentally more important than those they made on the B & O. The focus of organizational development on the B & O was technical matters; the basic question facing Long and McNeill was how to build a functional, durable road at the lowest possible cost. McNeill faced the same questions at the Western, but as his administrative goals and methods were known before he ac-

51. Bliss, *Historical Memoir*, p. 90.
52. "Report on Avoiding Collisions and Governing the Employees," November 30, 1841, in Western Railroad Corporation, "Clerk's Files," 1841, File 104, pp. 1–3.
53. See Salsbury, *Boston & Albany* for details on the post-1842 period.

cepted the job, he was able to provide the procedural "answer" from the outset. The key to the military's contribution to the Western was the ability of former officers to translate the techniques used to administer engineering projects into operational procedures for managing the complex daily business of a railroad.

Direct military involvement with the railroads ended in 1839, but indirect contact continued. The railroads offered prestigious and financially rewarding careers to engineers dissatisfied with the Army. Such an officer was Herman Haupt, who, because of a clerical error, was able to resign his commission within weeks of his graduation from West Point in 1835. After working as a railroad engineer and a teacher, Haupt came to the attention of J. Edgar Thomson, chief engineer of the Pennsylvania Railroad. In late 1847 Thomson hired Haupt as his "Principal Assistant."[54]

Haupt so impressed Thomson with a route survey he completed soon after joining the line that Thomson decided to assign him to a more ambitious project. As work on the Pennsylvania Railroad's first division continued, Thomson recognized that it was time to consider the railroad's future management needs. In February 1849 Thomson sent Haupt on a tour of New York and New England railroads with instructions to "examine their Stations, Depots, Shops, Roads, in fact everything connected with their business operations." On his return Haupt prepared a plan of organization and management for the Pennsylvania, complete with "forms and blanks for every branch of the service." On March 7, 1849, Thomson submitted Haupt's "Report on General Systems and Policy for the Pennsylvania Railroad" to the Board of Directors.[55]

Haupt's organizational plan divided responsibility for the operation of the road among four departments—Transportation, Maintenance of Way, Motive Power, and Maintenance of

54. James A. Ward, *That Man Haupt: A Biography of Herman Haupt* (Baton Rouge: Louisiana State University Press, 1973); Francis A. Lord, *Lincoln's Railroad Man: Herman Haupt* (Rutherford, NJ: Farleigh-Dickinson University Press, 1968); Frank A. Flowers, "Reminiscences," in Herman Haupt, *Reminiscences of Herman Haupt* (Milwaukee, WI: Wright & Joys Co., 1901).

55. Ward, *That Man Haupt*, p. 26; Flowers, "Reminiscences," pp. xvii–xviii; *Fourth Annual Report of the Pennsylvania Railroad Corporation*, 1850, p. 43; Pennsylvania Railroad Company, Board of Directors Minutebook No. 1, March 30, 1847 to September 4, 1851, currently held by the Legal Department of the Penn Central Corporation, Philadelphia, p. 155.

Cars—under a superintendent of transportation. The plan laid out a detailed reporting and accounting system that demanded daily, weekly, and monthly reports from employees at all levels. Haupt's plan also contained a significant innovation: it created a General Transportation Office whose business consisted solely of "regulating all the other offices on the line, and in securing accuracy and uniformity in their accounts."[56]

Haupt and Thomson created a business management structure that went a step beyond procedures then in use on other railroads and that was virtually unique in the American corporate community. Recognizing that the railroad's business would be so voluminous and so complex that it could not be effectively controlled by managers using more traditional administrative procedures, they divided managerial responsibility. Line officers directed the day-to-day operations of the railroad while staff officers at the General Transportation Office, using information supplied by the line, concentrated on "the broader problems of cost determination, competitive rate making, and strategic expansion, rather than on more routine operating activities."[57] Though the Pennsylvania's organization was unique in the business community, the United States Army had recognized at least thirty years earlier that the division of management functions between line and staff officers yielded improved efficiency and economy. The organizational structure the Pennsylvania adopted was similar to the line/staff division the Army had long practiced. Indeed, it was this structural format that Calhoun used as the basis of his administrative reforms between 1818 and 1825. The structure was new to the business community, but it had a long and successful history in military service.

The Board of Directors adopted the Haupt plan on June 8, 1849. On September 5 the directors approved Thomson's nomination of Herman Haupt to fill the newly created position of superintendent of transportation.[58] In taking these actions, the Board of Directors approved a plan of organization that it seems few of its members fully understood. They vested operational control of their railroad in the hands of Thomson,

---

56. "Report of the Superintendent of Transportation," in 4 AR, PRR, 1850, pp. 43–44, 46–47.
57. Chandler, *The Railroads*, p. 99.
58. See PRR, Board of Directors Minutebook No. 1, pp. 172–204 for information on the board's discussions and decisions on these matters.

Haupt, and their subordinates. The board still had important strategic functions, but in 1849 it could do little more than watch others build and run the railroad because it had already made most of the key decisions allowed under the new system. Operational control of the railroad was now in the hands of professional railroad managers who, as they gained experience, were less inclined to accept advice from the "amateurs" on the Board of Directors. When the board realized the full implications of its actions, a number of directors attempted to reassert their right to control the daily operations of their company. Throughout 1851 the board and the professional managers struggled for control of the Pennsylvania Railroad.

The key battle was waged over a set of regulations the Board of Directors adopted in May 1851 without consulting its subordinates. The new regulations reasserted the right of the company president to supervise and, in some cases, to control directly the engineer and transportation departments. Neither Haupt nor Thomson objected to the organizational structure thus created, but they were critical of the system of reports the regulations demanded from the superintendent of transportation and of a clause that gave the company president broad powers to act without specific board sanction.[59]

When Haupt failed to convince the board that the new regulations were unworkable, he set out to follow them to the letter, hoping by this course of action to demonstrate that the regulations were impractical. After some petty skirmishing during the summer of 1851, a small faction of board members, led by company president William C. Patterson, decided to try to resolve the conflict by removing Haupt from his position.[60]

In September a committee of the Board of Directors accused Haupt of neglect of duty, disobedience of orders, and insubordination and recommended that he be replaced. Before the full board Haupt demonstrated that the evidence used to support the charges against him was fraudulent and based upon altered transcriptions of his official correspondence. The Board vindi-

---

59. The May regulations are in PRR, Board of Directors Minutebook No. 1, pp. 435–40.

60. For an account of this struggle, see Herman Haupt, *Reply of the General Superintendent of the Pennsylvania Railroad, To a Letter from a Large Number of Stockholders of the Company, Requesting Information in Reference to the Management of the Road,* January 20, 1852 (Philadelphia: Collins, 1852); PRR Board of Directors Minutebook No. 2; Herman Haupt, "How J. Edgar Thomson Became President of the Pennsylvania Railroad," manuscript in the Historical Society of Pennsylvania Collection, Philadelphia.

cated Haupt but asked him to resign for the good of the company. Haupt agreed to leave, but a group of the stockholders, aware of the power struggle, got the board to agree not to fill Haupt's position until after a stockholders' meeting scheduled for February 1852. At the meeting a reform ticket headed by Thomson gained control of the Board of Directors. On February 3, 1852, the new board elected J. Edgar Thomson president of the Pennsylvania Railroad Company. Thomson asked Haupt to remain with the company, which he did.[61]

Thomson immediately moved to place his own stamp on the railroad's operations and management structure. On November 23, 1852, the Board of Directors approved yet another set of railroad regulations that were based largely on principles first enunciated in Haupt's 1849 report. Despite all the controversy of the preceding months, the new regulations significantly reinforced the position of the president as the company's chief officer. They gave Thomson all the powers Patterson and his allies claimed, and more.[62]

Haupt's 1849 plan reflected the success of earlier railroads with some of the management procedures established by Calhoun and his staff around 1820. Calhoun's military model stressed a hierarchical, authoritarian organization, demanded strict fiscal and operational accountability, and provided detailed written rules and regulations to guide organizational behavior. By 1849 this was a well-tried, rational system applicable to the management problems Haupt faced, and those three key elements figured prominently in the organizational structure and the administrative procedures Haupt helped to create on the railroad.

Long, McNeill, Whistler, and Swift followed essentially the same model. The procedures these four, but especially Long and McNeill, instituted had a demonstrable impact on managerial developments. Even so, the pattern suggested by Haupt's career might be the most significant in the long run. Of the 1058 graduates of the United States Military Academy at West Point between 1802 and 1866, thirty-five became corporate presidents, forty-eight became chief engineers, forty-one became superintendents and general managers, and eight be-

---

61. The works cited in note 60 also describe the results of the affair.

62. See Pennsylvania Railroad Company, *Organization for Conducting the Business of the Road,* adopted by the Board of Directors, November 23, 1852 (Philadelphia: Crissy & Markley, 1852).

came corporate treasurers of various kinds of businesses. Most of these officers never had the opportunity to make the dramatic contributions Long, McNeill, Whistler, Swift, and even Haupt made. It would be difficult to trace the careers of each of these 132 officers, but if only a few of them used the military model to solve administrative problems in the same way their five colleagues did, it is possible that a significant network of officer/managers helped to shape the development of the business community during the nineteenth century.[63]

For about thirty years United States Army engineers served some of the nation's most important early railroads. During these years the railroads adopted organizational and administrative procedures new to the business community. The officers joined these lines to provide technical assistance, not managerial advice, but their familiarity with some of the nation's most advanced administrative concepts enabled them to exert a great influence on managerial developments at the time when the new industry struggled to cope with organizational problems caused by its own success.

This is not to imply that a mere handful of officers, working sporadically for about thirty years, single-handedly created modern management in the United States. Civilian railroad organizers also played a vital role in the process, especially as some of the more important ideas coalesced and blossomed into complex management structures during the late 1850s and beyond. It would be imprudent, however, to assign too much credit to any one individual, civilian or military, since managerial innovation, like technological advance, "occurs predominantly in the form of small increments contributed by a great

---

63. The figures on the civilian careers of Military Academy graduates are from George W. Cullum, *Biographical Register and Dictionary of the Officers and Graduates of the United States Military Academy at West Point, New York, From Its Establishment in 1802 to 1890, with the Early History of the United States Military Academy*, 3 vols. (New York: Houghton-Mifflin, 1891), Vol. I, p. 7. The notion that a military career and especially an education at a military academy produces a uniquely military "mindset" that can influence nonmilitary behavior is discussed in Samuel P. Huntington, *The Soldier and the State* (Cambridge, MA: Harvard University Press, 1957); Morris Janowitz, *The Professional Soldier* (New York: Free Press, 1971); R. Ernest Dupuy, *Where They Have Trod: The West Point Tradition in American Life* (New York: Frederick A. Stokes Co., 1940). These works suggest that there is a military ethos, but it appears that there has been little work done to show exactly how this ethos shaped individual behavior when the officers moved on to nonmilitary careers. However, if the military personnel discussed in this paper are typical of even a small percentage of these officers, this might be a fruitful field of study.

many talented individuals rather than through the few spectacular innovations or breakthroughs [made] by those we recognize as geniuses."[64] Real progress came only through the efforts of dozens of individuals working for many railroads. As railroad management evolved, the people involved selected practices from many organizations and knit them together into the fabric of managerial procedures that emerged in the 1850s.

The Army's system was not adopted as a complete, fully developed managerial model, applicable to every circumstance or problem encountered in railroad management. Both civilian and military managers adopted and adapted the elements as needed. Military officers were, however, the catalysts of this adaptive process. They knew, or thought they knew, something about the problems the railroads would face, and they thought they knew how to solve them. As a consequence, they helped to lay the organizational and administrative foundation and established precedents that influenced the course of later managerial developments on the railroads and in other industries. The Army management model provided a conceptual and procedural framework that the officers advanced when there were no other equally suitable models available in the business community. Later work would modify the detailed procedures, but no one ever advanced a model that proved more useful in the decades before the Civil War.

64. John B. Rae, "Commentary," in Barbara E. Benson, ed., *Benjamin Henry Latrobe and Moncure Robinson* (Wilmington, DE: Eleutherian Mills-Hagley Foundation, 1975), p. 65.

3

Technological Innovation and Organizational Change: The Navy's Adoption of Radio, 1899–1919
Susan J. Douglas

Radiobroadcasting is a critical ingredient of modern military communications systems. Yet despite the opportunity to employ wireless telegraphy as early as 1899, the Navy shied away from the new technology and had little to do with it until the outbreak of World War I. In this essay Susan Douglas examines why it took the Navy so long to adopt the radio as standard operating equipment. Her account provides an instructive analysis of the tensions that arise when a bureaucracy is confronted with an innovation that threatens to alter the traditional hierarchical arrangements of command and control. Her history of the Navy's adoption of radio underscores the significance of entrepreneurship in innovative activity while illuminating the meaning of technological change as a social process.

In 1919, at the close of World War I, the United States Navy controlled and operated America's radio communications network, which consisted of stations aboard ship, medium-range shore stations, and several high-powered, long-distance stations capable of signaling over thousands of miles. The day after the United States declared war against Germany, President Wilson invoked the authority granted him in the 1912 Radio Act and placed all radio stations under naval control. But the Navy did not merely assume a custodial role; rather, it presided over a "technological revolution" that helped create a "coordinated industry."[1] Having orchestrated such a technological and organizational coda, Navy officials then assumed a central role in the negotiations leading to the formation of an "all-American" communications company, the Radio Corporation of America. By simultaneously promoting technological innovation and envisioning a reorganization of this important industry, the Navy in 1919 displayed considerable entrepreneurial talent and vision. In the RCA-sanctioned history of radio written in the late 1930s, Gleason Archer praised this vision, asserting that from the beginning wireless telegraphy and radio broadcasting owed the Navy "an everlasting debt." He elaborated, adding: "Not only did Navy technicians contribute to the development of the art, but private inventors and private manufacturers with their research departments found in Navy patronage the encouragement and inspiration that led them to persevere in their endeavors."[2]

Archer's portrait of naval patronage has been accepted by historians who have then focused on other important eras and events in the history of broadcasting.[3] Indeed, the extent to

I am indebted to Taylor R. Durham and Hugh G. J. Aitken for their advice and support.

The correspondence and memoranda cited are from two archival collections: The Clark Collection in the Division of Electricity, National Museum of American History, Smithsonian; and the Files of the Bureau of Equipment, National Archives. The Clark Collection is now only partially catalogued and was not catalogued at all when I did my research there; material from the Clark Collection is simply marked "SI." All the material from the Archives is from Record Group 19, File 18301. The box number in which the material was found is indicated after each letter.

1. Erik Barnouw, *A Tower in Babel: A History of Broadcasting in the United States*, Vol. I (New York: Oxford University Press, 1966), p. 52.

2. Gleason Archer, *History of Radio to 1926* (New York: The American Historical Society, Inc., 1938), p. 76.

3. See, for example, Barnouw, *A Tower in Babel*, and Christopher H. Sterling and John M. Kittross, *Stay Tuned: A Concise History of American Broadcasting* (Belmont, CA: Wadsworth Publishing Co., 1978), p. 36; Daniel J. Czitrom, *Media and the American Mind* (Chapel Hill: University of North Carolina Press, 1982).

which radio's technical and institutional fortunes depended almost entirely on military enterprise both during and just after the war and the repeated avowals of the importance of this debt by men such as David Sarnoff and Lee De Forest have suggested a long-standing and harmonious relationship between the Navy and radio unblemished by conflict or change. The progress achieved during the war thus eclipsed the Navy's earlier work with radio, and the events of three very intense years came to overshadow the preceding seventeen-year relationship between one fledgling industry and America's "new Navy." A more detailed examination of those seventeen years, however, suggests there might be much more to this story and that military enterprise in this case did not occur spontaneously nor always willingly, but was the culmination of a protracted and often uneasy process.

The Navy was introduced to radio in 1899, the year before William S. Sims met with resistance over adopting a system of continuous-aim firing on naval ships.[4] Would the same Navy whose members balked at changes in a certain type of weaponry introduced by someone not in Ordnance, but a naval officer nonetheless, embrace a revolutionary new communications technology presented to them by a half-Italian, half-Irish civilian inventor? Even Archer admits that certain officers "opposed with might and main the new agency of communications."[5] Commander Bradley A. Fiske, for example, an officer noted for his technical foresight and expertise, published an article in 1904 which asserted that radio had "no military usefulness whatever."[6] Impatient magazine and newspaper editorials from the early 1900s, such as those in *Electrical World,* strongly criticized the Navy's "moss-backed bureaucrats" and their "procrastination bureau" for taking too long to adopt radio while European navies were eagerly exploiting the invention.[7] As late as 1911 one commanding officer wrote, "We have failed to develop or employ the wireless, as a means of signaling" and urged that "The present systems [of tactical signaling],

4. Elting E. Morison, *Men, Machines and Modern Times* (Cambridge, MA: The MIT Press, 1966), pp. 17–44.

5. Archer, *History of Radio*, p. 73.

6. The article was initially published in the *Proceedings of the United States Naval Institute* and is summarized in Captain L. S. Howeth, USN, *History of Communications-Electronics in the United States Navy* (Washington, D.C.: Government Printing Office, 1963), p. 65.

7. *Electrical World* XXXVI, 5 (August 4, 1900): 157; 8 (August 25, 1900): 273; XL, 10 (September 6, 1902): 354.

so far as they relate to preparations for battle should be blown sky-high."[8]

Clearly, then, the Navy's attitude toward the use of radio changed dramatically between the early 1900s and 1917. Such different stances were separated by nearly twenty years and were bridged by the tortuous process of technical and institutional adaption. What was the nature of this adaptation and how did it occur? If there was initial resistance to radio, as the previous citations suggest, what prompted it and who voiced it? How was such resistance overcome? It is the purpose of this essay to examine how one military organization, the United States Navy, eventually integrated a revolutionary new communications technology, radio, into its operations and how its officers shifted from skeptics to sponsors. Several factors played critical roles in this process. The technical capabilities of radio, which improved dramatically during this twenty-year period, no doubt enhanced military receptivity in certain quarters. But examining the technical changes is not sufficient for understanding how a particular technical system fit into a particular social system over time. Certainly individual officers, with their own attitudes and goals, who served in key positions at critical moments influenced how and when the invention would be used. But possibly the most important factor was the organizational structure of which these men were part and into which this technology had to fit. For organizations can and do, through their traditions, structures, and reward systems, reinforce certain behavior and outlooks among their members while undercutting others. Organizations also consist of hierarchies and of niches, and new technologies can be consigned to the basement or showcased in executive offices.

How was the Navy organized at the turn of the century, and how did this structure affect the introduction of radio into the service? Did radio, in turn, alter the organization? Did the integration of radio mandate innovations in the organizational hierarchy and in managerial roles? Hugh Aitken has suggested that in times of technical uncertainty, and before exchanges of information between the realms of science, technology, and the economy were bureaucratized, individuals he calls "translators" transferred information between differently oriented and

---

8. Capt. W. F. Fullam, USN, Commanding Officer of the U.S.S. *Mississippi*, to R. D. White, Flag Lieutenant, Atlantic Fleet, 2 January 1911, cited in Howeth, *History of Communications-Electronics*, p. 193.

sometimes antagonistic sectors of society.[9] Such people were "bilingual" in that they understood the language and demands of more than one realm, and this facility made them indispensable to the innovation process. In the case of radio and the Navy, by 1899 the Navy was already a bureaucracy; the invention was the product of independent inventors and their fledgling companies. And the Navy, as a hierarchical organization, imposed constraints on the innovation process at the same time that it possessed resources capable of sustaining technical change. Could a translator sensitive to both the constraints and the opportunities emerge *within* this bureaucratic setting and mediate between technical change and institutional realignment? If so, what sort of a man would he be and how would he achieve his goals?

Wireless telegraphy, as radio was initially called, made its debut before a Navy adjusting to the physical transformation of its fleet. After resisting the adoption of steam propulsion and steel hulls for twelve years, the Navy of the 1880s, prodded by Congress, began acquiring bigger, faster, more impervious steel ships.[10] By 1900 the new fleet was nearly complete. The change from canvas to steam and from wood to steel profoundly affected the way a ship was run, what its needs were in port, and how officers thought of their duties and command. The reconstruction, though much needed, was unsettling to the men and to the department, and thus any concomitant alterations in the bureaucracy that might have made this "new Navy" more efficient were not readily forthcoming. The metamorphosis from old to new was initially cosmetic: while the hardware was being modernized, changes in naval administration, organization, and tactics lagged behind.[11] Much as the first steam-powered ships still retained rigging for sails, the naval organization sought to at least preserve a familiar structure during such major, unsettling changes.

In 1899 the Navy Department was comprised of eight

9. Hugh G. J. Aitken, *Syntony and Spark—The Origins of Radio* (New York: John Wiley and Sons, 1976), pp. 330–32.

10. Morison, *Men, Machines and Modern Times*, pp. 98–122; Harold and Margaret Sprout, *The Rise of American Naval Power, 1776–1918* (Princeton: Princeton University Press, 1939), pp. 165–82; Lance C. Buhl, "Mariners and Machines: Resistance to Technological Change in the American Navy, 1865–1869," *Journal of American History* 61, 4 (December 1974): 704; Robert Greenhalgh Albion, *Makers of Naval Policy, 1798–1947* (Annapolis, MD: Naval Institute Press, 1980), pp. 9–10.

11. Sprout, *The Rise of American Naval Power*, illus. facing p. 218 and pp. 270–80.

bureaus, each headed by a bureau chief.[12] The chiefs were responsible to the Secretary of the Navy, a civilian political appointee who usually knew little or nothing about naval affairs and who served at the pleasure of the President.[13] The responsibilities and jurisdiction of the bureaus often overlapped, yet there were no men, committees, or offices facilitating inter-bureau cooperation.[14] Jealously guarding their territory and prerogatives, the bureau chiefs were often embroiled in inter-necine squabbles that generated "friction, circumlocution, and delay."[15] The difficulty of reconciling and coordinating the duties and objectives of the bureaus was a constant source of frustration to the secretaries.[16] The bureaus, on the other hand, could not count on long-term or informed guidance from their chief executive.

This lack of departmental coordination and direction was exacerbated during the first decade of the twentieth century. The Navy was in an organizationally vulnerable position between 1900 and 1912, weakened, ironically, by the activities of its commander-in-chief. President Roosevelt, a staunch Navy booster, took such an active interest in the department that he became its de facto secretary. Between 1902 and 1909, there were six secretaries of the Navy, none of whom had much power or influence.[17] Even someone as energetic as Roosevelt could not provide the department with sustained leadership and continuity while serving as President, and as a result the Navy's top management and public relations position was compromised. This leadership vacuum, coupled with bureau separatism, worsened the organizational isolation of the bureau chiefs. Thus they learned to rely on "precedent and routine,"[18] and the department was guided by the daily grinding of the bureaucracy, which "ruled with an iron hand, usually ignoring,

12. At the turn of the century, these were: Bureau of Navigation, Bureau of Ordnance, Bureau of Equipment, Bureau of Construction and Repairs, Bureau of Steam Engineering, Bureau of Yards and Docks, Bureau of Medicine and Surgery, Bureau of Supplies and Accounts.

13. Charles Oscar Paullin, *Paullin's History of Naval Administration, 1775–1911* (Annapolis: U.S. Naval Institute Press, 1968), p. 438; Sprout, *The Rise of American Naval Power*, p. 274; Albion, *Makers of Naval Policy*, pp. 7, 12.

14. Sprout, *The Rise of American Naval Policy*, p. 193.

15. Ibid.; *Annual Report of the Secretary of the Navy*, 1905, p. 3.

16. *Annual Report of the Secretary of the Navy*, 1900; 1905.

17. Albion, *Makers of Naval Policy*, pp. 212–13; *Annual Report of the Secretary of the Navy*, 1909, p. 6.

18. Sprout, *The Rise of American Naval Policy*, p. 274.

sometimes penalizing, those who attempted to introduce reforms and innovations."[19]

Tactically, the sea-going Navy was equally decentralized until the twentieth century. Before then the fleet had been divided into "small groups of cruising vessels thousands of miles apart," although in reality each ship usually cruised by itself. "Even when in company, the ships rarely engaged in group maneuvers," and the men were more accustomed to thinking of each ship as a "potential solitary raider than as a unit of a fighting fleet."[20] Although the department began mandating periodic exercises and maneuvers in 1894, there was no accompanying "fleet policy," no long-term vision of coordinated activities or strategy within the Bureau of Navigation.[21] Despite Mahan's influence, there was no permanent fleet consisting of ships and commanders trained to operate cooperatively until 1907.[22]

Thus at sea and on shore autonomy and independence at the higher levels of the bureaucracy prevailed. Within each bureau, on each ship, the lines of authority and communication were clear and strong. But between ships or between bureaus, the lines, if there at all, were no more than fragile threads. And once ships were at sea, their lines literally and figuratively cast off, there was no web, either organizational or technical, to connect the ship to shore.

The Navy's rank and reward systems were structured to preserve this luxurious autonomy for ship commanders. After the Civil War, which had enhanced the importance and prestige of the staff departments in general and the engineers in particular, the line officers sought to "humble the engineer corps."[23] Mere "mechanics" could not be allowed to enjoy the same status and perquisites as the line. These efforts included a reduction in 1869 of the relative rank of each grade in the engineer corps as compared to the corresponding grade in the line.[24] This act, which of course outraged the staff, served to devalue the engineers and their work just at the time when the Navy was

19. Ibid., p. 271.

20. Ibid., p. 168; A. T. Mahan, *From Sail to Steam; Recollections on Naval Life* (New York: Harper, 1908), p. 270.

21. Sprout, *The Rise of American Naval Policy*, pp. 277–79.

22. Prior to 1907, squadrons were organized into fleets on a temporary basis, as during the Spanish-American War.

23. Sprout, *The Rise of American Naval Policy*, p. 177.

24. "The History of the Naval Staff Question," *The Nation* 295 (February 23, 1871): 121–22.

modernizing and becoming increasingly dependent on technology and hence on technical competence. The animosity between the two groups smoldered for over thirty years, with predictable repercussions for technical innovation. In addition, the promotion system in the Navy was notoriously slow, so that line officers arrived at command rank late in their careers and "grew grey as lieutenants while their sons caught up with them in the same rank."[25] These older officers were, like most older members of an organization, particularly resistant to new devices and procedures and nostalgic for the old sailing Navy. Their authority and prestige rested on tactical skills and decision making, not on technical mastery, and they were disconcerted by those who challenged this basis of command. But as the secretary noted in 1899, recognition of the engineer's importance had become essential: "A modern ship of war is one of the most complicated machines in existence. It is filled with machinery of various sorts from one end to the other. The finished ship, ready for service is of great cost and enormous value to the government. It is worth nothing unless efficiently handled, cared for, and kept in readiness for immediate service."[26] The Naval Personnel Act of 1899 amalgamated the engineers with the line in an attempt to make "every line officer an engineer, and also every engineer a line officer."[27] The engineers were awarded commensurate rank and salary. In practice, only a few young officers were given the duties of engineers, and it took over ten years for thirty years of tradition to begin changing.[28] Until officers saw engineering expertise as a route to promotion and prestige, few would switch from strategy to technology. Technology would have to advance strategy for total acceptance to occur. Six years after the amalgamation, the secretary complained that "some officers have not yet outgrown the idea that the engines of a ship are, in some sort, an excrescence, and those in charge of them rather auxiliaries to the fighting force than members of it."[29]

The communications systems available to the Navy both served and reinforced its decentralized administration. By 1890 telegraphic or cable communication was available in most

25. Albion, *Makers of Naval Policy*, p. 10; Paullin, *Paullin's History*, pp. 418–19.
26. *Annual Report of the Secretary of the Navy*, 1899, p. 326.
27. *Annual Report of the Secretary of the Navy*, 1905, p. 7.
28. Paullin, *Paullin's History*, p. 463.
29. *Annual Report of the Secretary of the Navy*, 1905, p. 7.

ports and Navy yards. This somewhat eroded autonomy: when squadron commanders were in port, it was possible for them to be more closely in touch with Washington. Although by the turn of the century there was a "growing tendency to make naval strategic decisions at Washington instead of the theater of operations," it was still only a tendency and not a practice that could be enforced when ships were incommunicado.[30] Between ships, flag signaling by day and newly installed light signaling by night were used for intership communications. During rain or fog, or across long distances, intership communication was impossible. Many ships were not equipped with the Ardois lights and could not signal at night.[31] And no ship could communicate with the shore once out at sea.

Responsibility for providing ships with signaling apparatus fell to the Bureau of Equipment, which furnished the vessels with other supplies including coal, rigging, navigation instruments, cordage, and hammocks.[32] Thus the Bureau of Equipment, a procurement and supplies division with no authority over or expertise in engineering, ship construction, and redesign or in maneuvers and fleet tactics, would be responsible for assessing and acquiring wireless telegraphy, which would alter all three.

In addition to these internal organizational constraints, the Navy would find itself in an increasingly delicate political position. President Roosevelt, the Navy's primary lobbyist, began to encounter well-organized Congressional opposition to extending the "New Navy." Senators and congressmen opposed to Roosevelt's brand of imperialism (and allied with others opposed to his liberal reforms) succeeded in reducing several of Roosevelt's requests for large naval appropriations.[33] These budgetary battles increased financial uncertainty and reinforced departmental caution. Thus, even if a bureau chief was technically sophisticated and sought to sponsor a particular innovation, he would confront obstacles above, below, and lateral to him in the organization. On the other hand, an officer's reluctance to make use of a particular technology was protected by the Navy's decentralized structure.

---

30. Howeth, *History of Communications-Electronics*, p. 13.
31. Ibid., p. 11.
32. Paullin, *Paullin's History*, p. 447.
33. Albion, *Makers of Naval Policy*, pp. 212–17.

Elting Morison and others have portrayed the Navy of this era as conservative and tradition-bound, its officers adhering to the time-honored way of doing things and distrustful of new devices or procedures. A review of the Navy's organization at the turn of the century helps explain this state of affairs. A weak and uninformed titular head of the department presided over an organization consisting of jealously guarded bureaucratic fiefdoms. Those with the most power were men who had been promoted slowly and late and were set in their ways. Autonomy was valued as independence, not bemoaned as isolation. The lack of communication between ship and shore was central to the whole spirit and idea of commanding a ship because it required and ensured freedom of action. This was not the sort of organization in which technical sponsorship, especially of an invention that threatened autonomy and decentralization, was either desired or possible. Yet this was the bureaucracy that wireless and its inventors would confront. The inventors were working at the forefront of electrical engineering, tackling both scientific and technical mysteries. But cracking this organizational enigma would elude them.

Guglielmo Marconi, who had first publicly demonstrated wireless telegraphy in England in 1896, brought his apparatus to the United States in the autumn of 1899. As a publicity stunt designed to promote his invention and provide *The New York Herald* with yet another "scoop," Marconi was to report the progress of the America Cup Race by wireless. With his apparatus set up aboard the *Mackay-Bennett*, Marconi "wirelessed" daily developments to *Herald* reporters on shore. Marconi's success was widely celebrated in the press;[34] *The New York Times* declared that if ship-to-shore transmission were the only use for wireless telegraphy, "it would still be one of the greatest and most beneficent discoveries of all time."[35]

The Navy was about to embark upon the first simple but necessary phase of technical adoption, introduction to the invention. Taking advantage of Marconi's American visit, and aware of Marconi's recent contract with the British Admiralty, the Bureau of Equipment sought to inspect his apparatus on

34. See *Electrical World* XXXV, 15 (October 7, 1899) and *The New York Times* and *The New York Herald* for October 1899.
35. *The New York Times*, December 17, 1901, p. 8.

behalf of the Navy. Marconi agreed to allow four officers, all electrical experts, to witness the operation of his equipment throughout the races.[36] In his report to the bureau, Lieutenant J. B. Blish stated that the demonstrations "were most convincing that the system was already excellently adapted for use on board ship; and my investigations since then have strengthened that conviction."[37]

During these observations Marconi was persuaded to allow further naval testing of the apparatus after the yacht races ended. Marconi agreed to the tests only after issuing several disclaimers: he had not expected to give such a demonstration and thus the equipment he had with him was not "sufficient for a government test . . . on a large scale." Nor did he have with him his "devices for preventing interference" from competing transmitters because these devices were not yet "completely patented."[38] Marconi wanted it understood that this was not his standard demonstration for naval vessels, that he did not have all of his state-of-the-art equipment with him, and that consequently he could not guarantee the same success in these tests that he had achieved during the yacht races. Marconi was on the verge of patenting his method of tuning, whereby several wavelengths could be used with a given antenna. Because he was the only one signaling during the yacht races, he had no need for tuning and had not brought the additional apparatus to America. While Marconi may have been seeking only to protect himself from unjust criticism, the Navy eventually came to feel he was trying to cover up a major and unavoidable defect of the system.[39]

Marconi's apparatus was dismantled from the press boats in October 1899 and installed on the armored cruiser *New York* and the battleship *Massachusetts*, both anchored in the New York Harbor. A third set at the Navesink Lighthouse in New Jersey served as the shore station. The members of the "Marconi Board" were to assess the equipment's accuracy, establish maximum operating distance, determine the best location for the instruments, and report on interference. After several days of tests, one of the board members, Lieutenant Commander

36. Howeth, *History of Communications-Electronics,* p. 26.

37. Lieut. J. B. Blish, Report to Bureau of Equipment, November 13, 1899, NA, box 83.

38. Guglielmo Marconi to Wireless Board, 29 October 1899, NA, box 83.

39. George H. Clark, "Radio in War and Peace," unpub. ms., 1940, SI, p. 14.

J. T. Newton, advised the bureau that sending accuracy was not always achieved and that Marconi's temporary set-up aboard the ships would be inadequate for a permanent installation. Transmission speed averaged twelve words per minute. While the two ships exchanged messages over a distance of 36.5 miles, and the *Massachusetts* received the *New York*'s transmissions up to 46.3 miles, this success was overshadowed by a persistent drawback: "In every case, under a great number of varied conditions, the attempted interference was complete,"[40] meaning that whenever they tried to interfere with the messages, they succeeded. Interference had occurred whenever more than one set was signaling, because only one wavelength was being used for all transmissions. Although Marconi had claimed that he could prevent interference, "he never explained how nor made any attempt to demonstrate that it could be done."[41]

Yet despite these failings, Newton and the board recommended that "the system be given a trial in the Navy." Newton pointed out that the system could be adapted for use on all Navy vessels and had the distinct advantage of performing well in "rain, fog, darkness, and motion of ship . . . excessive vibration at high speed apparently produced no bad effect on the instruments." Within the working ranges, "accuracy was good." He noted that the best location for the instruments would be "below, well protected, in easy communication with the commanding officer."[42] Another Board member wrote that "even in its present state the instruments can be made useful in signaling between ships, and ship and shore."[43]

Admiral R. B. Bradford, Chief of the Bureau of Equipment and himself quite knowledgeable about electrical technology, was persuaded by this report and appealed to the Secretary of the Navy on December 1, 1899. "This system is successful and well adapted for Navy use. The chief objection to it is known as 'interference'. . . . Notwithstanding this fact, the Bureau is of the opinion that the system promises to be very useful in the future for the naval service." Citing Marconi as the recognized inventor and noting that no other "makers of electrical instru-

40. Lieut. Comdr. J. T. Newton, Report on the Marconi System, 13 November 1899, NA, box 83.
41. Ibid.
42. Ibid.
43. Lieut. G. W. Denfeld to Chief, Bureau of Equipment, 1 November 1899, NA, box 83.

ments have been able to successfully duplicate Marconi's apparatus," Bradford recommended acquiring sets from Marconi for continued naval experimentation.[44]

Despite this favorable endorsement from the Bureau of Equipment, no such acquisition of Marconi apparatus occurred. The breakdown of relations between the Marconi Company and the U.S. Navy following the preliminary tests and the board's favorable recommendation had far-reaching and long-term repercussions that affected both organizations as well as other fledgling wireless companies.

Why was such an admittedly imperfect, yet extremely promising invention not acquired by the "New Navy"? Not enough evidence exists to answer this question with complete assurance. The reason most often cited explains that the Navy rejected Marconi's contract specifications as too expensive and restrictive. The dispute over the terms of purchase reflected misunderstanding on each side about the needs of and constraints upon the other party to the contractual negotiations.

Marconi would not sell his apparatus to the Navy or to anyone without royalties. Under his terms, the Navy would purchase not less than twenty sets at a total cost of $10,000 and agree to pay a $10,000 annual royalty. The royalty would be reduced if a greater number of sets were purchased.[45] These terms, for Marconi, represented a concession: the Navy got to keep the sets, while other customers could only lease the apparatus. The Marconi Company had decided on this leasing policy after three disappointing and expensive years trying to market wireless in England through outright sales.

When the company was formed in 1897, Marconi had hoped it would become economically viable by selling wireless apparatus, especially to shipping firms. But a customer could not merely buy the equipment: the client would need trained operators; intermittent adjustments, repairs, and improvements; and shore stations with which to communicate. No firm was prepared to make that large an investment in such a new device, and sales were discouraging.[46] Marconi, like Western Union and Bell Telephone, would have to provide an entire

---

44. Adm. R. B. Bradford, Chief, Bureau of Equipment to Secretary of the Navy, 1 December 1899, NA, box 83.

45. Ibid.

46. Susan J. Douglas, "Exploring Pathways in the Ether: The Formative Years of Radio in America, 1896–1912" (Ph.D. dissertation, Brown University, 1979), p. 75.

communications network and then lease access to the system. Leasing encouraged more firms to give wireless a try, and it ensured that the company retained control over both the apparatus and the personnel upon whom its success depended. As Hugh Aitken has observed, "It is hard to conceive of any other market strategy available at that time that would have sustained the growth of the company."[47]

In addition, the Marconi Company had begun to promulgate its nonintercommunication policy. The company continued to invest thousands in shore stations and apparatus. Yet by the very nature of wireless and wave propagation, anyone with a transmitting or receiving set could tap into the system free of charge. Marconi simply could not allow free access to all and realize a return on his investment at the same time. Consequently, all Marconi operators at all Marconi stations were strictly instructed to exchange messages only with other Marconi stations.[48] Without control over a network of wires or cables, to which he could physically have controlled access, what else could Marconi have done to ensure that his system survive financially?

The Navy's reaction to Marconi's terms and policies was influenced by precedent and law, by nationalism, and by an apparent suspicion of inventors and business firms. The Bureau of Equipment did not have enough money to pay Marconi's price and the department was constrained, by law, from obligating funds beyond the current fiscal year.[49] But in addition the Navy viewed the leasing and nonintercommunication policies as unnecessary and grasping monopolistic ploys designed solely for the purpose of granting yet another British company complete control over international communications. As one official noted, "Such a monopoly will be worse than the English submarine cable monopolies which all Europe is groaning under and I hope the Navy Department of the U.S. will not be caught in its meshes."[50] From the Navy's point of view, Marconi was trying to prevent anyone else from gaining access to a resource—"the air"—that had traditionally been free. The

47. Aitken, *Syntony and Spark*, pp. 230–40.

48. Douglas, "Exploring Pathways in the Ether," pp. 77–78.

49. William Moody, Secretary of the Navy to the American Marconi Wireless Telegraph Co., 25 September 1903, NA, box 85.

50. Comdr. F. M. Barber to Chief, Bureau of Equipment, 6 December 1901, NA, box 83.

Navy, however, had not experienced firsthand the financial difficulties surrounding wireless: the research expenses, the patent and legal fees, and the revenue problem. To the Navy, the Marconi financing strategy was not protective but avaricious. As an independent entrepreneur, Marconi, in turn, was often not sympathetic to the financial, legal, and political constraints operating on the Navy.

During the next twelve years, negotiations between the Navy and Marconi rarely transcended this early stalemate. But the Marconi Company was not alone in provoking negative reaction from Navy officials over prices and contract terms. During the acquisition process, every company trying to do business with the Navy encountered an attitude inhospitable to inventors and unappreciative of their technical goals and financial needs.

Naval officers and the wireless inventors were, in fact, approaching each other from two strong but opposite cultural traditions, traditions that influenced self-image and behavior and were laden with prejudice and sterotypes that often affected negotiations between the two. A Navy man and an inventor were very different types of people, differently socialized, with contrary and often conflicting orientations. The Navy officer was an organization man. He spent his life both obeying and giving orders within an institutional context, moving up gradually through the ranks, preserving and identifying with the status quo, honoring tradition, defending the organization that provided him with security and recognition. Except during wartime, "making it" involved diligence and diplomacy, keeping a low profile. Organizational stability surrounded and insulated him, and that was what he came to prize.

The inventor, on the other hand, had no such large organizational affiliation. Often he was a loner, sometimes seeing himself as an outcast who would redeem himself through his inventions.[51] Driven by a desire for fame, money, love, or all three, the inventor sought to make his mark on history by making change possible, by disrupting the status quo. Initially, sometimes continually, plagued by the problems of financing or even of remaining solvent and determined that his contributions remain distinctive, the independent inventor built

---

51. Morison, *Men, Machines and Modern Times*, pp. 9, 27; see also the diaries of Lee De Forest, Manuscript Collection, Library of Congress.

his reputation and career on technical change and improvement. Stability, established ways of doing things, existing schemes—these were what the inventor disrupted, sometimes deliberately, sometimes inadvertently.[52] Because he lived on possibilities, he was of necessity overoptimistic, often given to exaggeration.

Each group acquired and used money differently, an additional and powerful source of mutual distrust. To paraphrase Aitken, inventors responded to market demands, to "signals" they received from the economy, while military men, not usually subject to such outside forces, responded more to "internally generated signals" rarely tied to the marketplace.[53] Their contrasting pecuniary orientations, coupled with widely divergent socializations, induced each group to view the other with suspicion and, occasionally, contempt. While the wireless inventors expressed impatience over what they saw as a constricted and unimaginative bureaucratic outlook, naval officials had to assess the often inflated claims of a range of inventors, some of whom were indeed crackpots. As the Navy continued to investigate wireless over the next ten years, these conflicting traditions, cultures, and attitudes played a salient part in contract disputes. Certainly there were exceptions in both groups to these characterizations; but it is important to remember that membership in a long-standing and tradition-bound institution was a critical factor in the careers and outlooks of one group and that such membership was missing from or anathema to the orientation of the other.

For the next two years, until the autumn of 1901, the Navy took "no active steps . . . with a view of investigating the merits of any particular design of wireless telegraphy."[54] That fall, however, the department decided to explore what European inventors other than Marconi had to offer. By this time the bureau had become concerned that "most naval powers are far in advance of the United States in the installation of wireless telegraphy appliances on board of naval ships."[55] Although Reginald Fessenden had been experimenting with wireless under the auspices of the Weather Bureau and Lee De Forest

52. Morison, *Men, Machines and Modern Times*, p. 9.
53. Aitken, *Syntony and Spark*, p. 322.
54. *Annual Report of the Secretary of the Navy*, 1902, p. 375.
55. Ibid., p. 376.

had recently formed his own company, the Navy believed that there was "no American wireless telegraph company ready to furnish apparatus."[56]

Commander Francis M. Barber, USN, retired, an old classmate and friend of Bradford, was living in Paris at this time. Well connected in diplomatic circles, knowledgeable about electrical engineering, and fluent in French and German, Barber seemed the perfect "translator" between the European inventors and the U.S. Navy. From 1901 until 1908, he monitored the European technical press, solicited information from inventors and naval officers, visited the various companies, and sent extensive and lively reports on all aspects of wireless to the Bureau of Equipment. His thirty-year tenure with the Navy had given him a keen appreciation of how to get technical information from foreign military organizations. He ingratiated himself with the senior officers first, and after he had won them over, he then felt free to go to the real source of information. As he wrote to Bradford, "It's no use commencing with junior officers anyway. They have all the knowledge; but the old busters have to be coddled first."[57]

Barber's correspondence provides a fascinating view of how the bureau's official representative felt about and dealt with the inventors and the still young wireless industry. During 1901 and 1902, he investigated the apparatus of two French inventors, Rochefort and Ducretet, and of two German firms, Slaby-Arco and Braun-Siemens-Halske. He liked Rochefort because "he is a modest gentlemanly little man and not at all captious and prejudiced as inventors usually are."[58] Barber also observed that "An inventor is a visionary, a visionary is a genius, and a genius is a lunatic or next door to it."[59] Some inventors he heard about aroused his interest, but he decided against satisfying his curiosity: "One better have the itch than encounter an impecunious inventor. He never lets up once he makes your acquaintance."[60] One day he would visit Ducretet, who would

56. Comdr. F. M. Barber to Chief, Bureau of Equipment, 2 April 1902, NA, box 85.

57. Comdr. F. M. Barber to Chief, Bureau of Equipment, 31 December 1901, NA, box 84.

58. Comdr. F. M. Barber to Chief, Bureau of Equipment, 30 January 1902, NA, box 84.

59. Comdr. F. M. Barber to Chief, Bureau of Equipment, 15 April 1902, NA, box 85.

60. Comdr. F. M. Barber to Chief, Bureau of Equipment, 19 February 1902, NA, box 84.

call Rochefort a liar and a thief, and the next day hear Rochefort say the same thing about Ducretet.[61] Barber found sorting out the wireless situation in Germany particularly frustrating because he could not get what he felt was reliable information: "These manufacturers are such liars that one often wonders with St. Paul 'What is truth?' "[62] Barber's suspicion of inventors was compounded by his attitudes toward many foreigners. He was unimpressed with the German company Slaby-Arco, which he found "too slippery."[63] This reaction seems to have been reinforced by his impression of Count Arco, whom he described as "a weedy little chap with a great big head—he looks like a tadpole."[64] In his assessment of the British, Barber commented, "You can't hint to an Englishman, you must *kick* him. In my long business experience the English are the most dishonest people I know."[65]

But Barber reserved his most stinging scorn for Marconi. Any information, whether rumor or fact, that reflected badly on Marconi's apparatus or his business was eagerly reported to the Bureau of Equipment. Barber heard—and believed—that in developing wireless, Marconi had "walked off" with others' inventions and therefore was operating with an extremely vulnerable patent structure. Thus he would ultimately fail, but in the meantime his "system" deserved to be circumvented because it was all stolen anyway.[66] Barber doubted the accuracy of press accounts hailing signaling successes, such as Marconi's celebrated transatlantic S in December of 1901.[67] He took particular delight in recounting a conversation he had with Colonel Hozier, the Secretary of Lloyds and a director of the Marconi Company. "He thinks Marconi had never yet got a signal across the Atlantic or 2000 miles at sea either. The whole thing was a stock-jobbing operation worked in the interest of 'a lot of Jews.' This from a director of the company is rather

61. Ibid.
62. Comdr. F. M. Barber to Chief, Bureau of Equipment, 31 December 1901, NA, box 84.
63. Comdr. F. M. Barber to Chief, Bureau of Equipment, 29 July 1902, NA, box 85.
64. Comdr. F. M. Barber to Chief, Bureau of Equipment, 17 June 1902, NA, box 85.
65. Comdr. F. M. Barber to Chief, Bureau of Equipment, 22 May 1908, NA, box 89.
66. Comdr. F. M. Barber to Chief, Bureau of Equipment, 28 November 1901, NA, box 83.
67. Comdr. F. M. Barber to Chief, Bureau of Equipment, 15 January 1902, NA, box 84.

good."[68] He continued to hope and expect that the U.S. Navy would "be able to drive the American Marconi Company out of business."[69]

These are the words and attitudes of the man who was the Navy's primary source of information on the European wireless community. As the bureau's eyes and ears in Europe, he was in a highly influential position. The inventors, no doubt unaware of his true feelings, opened their laboratories and factories to him, advised him, confided in him, boasted to him, and, of course, tried to win him over. While transmitting important technical and business information to the bureau in the United States, Barber was also reflecting, and reinforcing, a particular way of viewing and dealing with inventors. To Barber and the bureau officials, inventors were those eccentric and frequently deceptive people the department was forced to deal with to get the apparatus it needed.[70] The bureau's subsequent business practices were certainly consonant with the overall spirit and outlook of Barber's correspondence.

During the next several years, the Navy experimented with various kinds of wireless, both European and American. These sets were usually tested between the Washington Navy Yard and the Naval Academy at Annapolis as well as between Annapolis and one or more ships. The distance between Annapolis and the Yard was only thirty miles, so, as Barber noted, "almost anything ought to work there."[71] In the spring of 1902, Barber arranged for the Navy to purchase two sets each from Ducretet, Rochefort, Slaby-Arco, and Braun-Siemens-

68. Comdr. F. M. Barber to Chief, Bureau of Equipment, 22 April 1902, NA, box 85.

69. Comdr. F. M. Barber to Chief, Bureau of Equipment, 11 February 1907, NA, box 89.

70. Barber's letters, which comprise the most sustained and detailed record of the Navy's side of naval-civilian wireless correspondence during this period, cannot be dismissed as the idiosyncratic observations and opinions of one officer. In fact, Barber's letters both corroborate and make explicit the views implicit in more bureaucratically constrained correspondence. This is especially true of letters sent from the bureau and the secretary's office to the American Marconi Company and to Reginald Fessenden's company, NESCO. In these letters, naval representatives were wary, brusque, and sometimes antagonistic. This tone was frequently not unprovoked, but it does indicate the level of misunderstanding that existed between these two groups. See, for example, Chief, Bureau of Equipment to Marconi Wireless Telegraphy Company of America, 4 May 1904, NA, box 89; 25 July 1902, NA, box 12; William Moody, Secretary, Navy Department to Marconi Company, 25 September 1903, box 85; Chief, Bureau of Equipment to Fessenden, 14 April 1904; Charles Bonaparte, Secretary, Department of the Navy to Fessenden, 19 April 1906.

71. Comdr. F. M. Barber to Chief, Bureau of Equipment, 20 January 1902, NA, box 84.

Halske.[72] These were tested between August and October 1902. That autumn, two De Forest sets were also purchased and tested. These trials were hindered by a dearth of skilled operators and officers knowledgeable about radio.[73] A Wireless Telegraph Board was established to oversee and report on the tests, but its members had other, conflicting duties and were unable to continue with the board for long. Three of the five members had to be replaced during the course of the tests. The officers ultimately "went their respective ways," leaving three enlisted men to oversee the tests and then notify their superiors of the results.[74]

Admiral Bradford, Chief of the Bureau of Equipment, complained to the Secretary of the Navy about the lack of departmental commitment to the experiments: "The Bureau desires to express its great regret that these important experiments have been interrupted for the want of vessels necessary for the work; also that two members of the Board are under orders for sea. It is feared that no important results can be reached unless a Board can give its uninterrupted attention to the subject."[75]

The early negotiations leading to the purchase and testing of these sets indicated how the Navy would do business with the wireless companies over the next eight years. The Navy enjoyed a buyer's market, and Barber seemed well aware of his advantages. To ensure that their apparatus performed well, the various companies wanted their own engineers to be present at the tests. They expected the Navy to subsidize the travel expenses, especially since the Navy operators and engineers would need the sort of instructions and advice not conveyed in written specifications. From the inventors' perspective, the Navy would be getting the best possible results as well as free training and should therefore feel obliged to cover the travel expenses. The Navy, of course, did not see it that way and refused financial support.[76]

On instructions from Bradford, Barber indicated to the companies that the Navy would not employ the services of any

72. Chief, Bureau of Equipment to Comdr. F. M. Barber, 13 January 1902, box 84; 6 December 1901, box 83; 14 December 1901, box 83.

73. Howeth, *History of Communications-Electronics*, pp. 52, 43.

74. Clark, "Radio in War and Peace," p. 33.

75. Adm. Bradford, Chief, Bureau of Equipment to the Secretary of the Navy, 13 December 1902, NA, box 85.

76. Chief, Bureau of Equipment to Comdr. F. M. Barber, 19 April 1902, NA, box 84.

private specialists, and while it would be helpful to have experts on hand when the tests occurred, the Navy engineers would probably be able to figure out the apparatus.[77] Of course the thought of amateurs tinkering with their instruments drove the inventors wild, especially because proper performance could mean a big contract. Barber knew this and made them quickly see that sending representatives at their own expense was better than sending none at all. By April of 1902 he was able to advise Bradford that "I have them all corralled and they will go at their own expense rather than not at all."[78] Extra expenditures had not been the only consideration: naval pride was operating as well. Barber acknowledged that "it is rather humiliating to be obliged to have 'square heads' come over and show us how to run things, but after all the main idea is to succeed and to get the best apparatus."[79] The Navy spent nearly $12,000 on the eight sets of wireless, with prices ranging between $2250 and $3500 for two sets.[80] When Slaby-Arco, citing recent improvements, tried to raise its prices, Barber notified the company that it could either return to its previous prices or cancel the Navy's order.[81]

During the tests, both the Slaby-Arco and De Forest apparatus out-performed the others.[82] The Slaby-Arco Company, hearing of its success, wrote Barber what he described as a "very cheeky letter." "They wanted to know how soon now they might expect the orders which would repay them for the vast expenditure to which they had been subjected in sending engineers to the U.S. and they wanted me to write and urge that the orders be placed immediately. I replied laconically."[83] Slaby-Arco was eventually awarded a contract. Its prices were low but, more importantly, its apparatus was better suited to the Navy's need for easily adjustable instruments. The receiver used by Marconi and Slaby-Arco, a filings coherer, was con-

77. Ibid.

78. Comdr. F. M. Barber to Chief, Bureau of Equipment, 4 April 1902, NA, box 86.

79. Comdr. F. M. Barber to Chief, Bureau of Equipment, 6 December 1902, NA, box 86.

80. Comdr. F. M. Barber to Chief, Bureau of Equipment, 30 January 1902, box 84; 14 May 1902, box 88; Chief, Bureau of Equipment to Barber, 11 February 1902, box 84.

81. Slaby-Arco to Comdr. F. M. Barber, 19 July 1902, NA, box 84.

82. Report, Wireless Telegraph Board to Chief, Bureau of Equipment, 3 December 1902, NA, box 85.

83. Comdr. F. M. Barber to Chief, Bureau of Equipment, 7 November 1902, NA, box 83.

nected to a recorder which printed signals on a strip of paper. The coherer was insensitive and erratic and would often print static as well as signals, but it provided a written record and required little skill to operate. Marconi quickly found the coherer to be an unsatisfactory detector, and after many futile attempts to improve it, abandoned it in 1903 in favor of his magnetic detector, which relied on a skilled operator wearing headphones to record the signals. De Forest and the other American inventors had also begun to improve on the weakest part of Marconi's system,[84] and De Forest had substituted headphones for the tape as well. He found that the operator could distinguish between true and false signals the way a printer never could and therefore eliminated the coherer by 1902. But the Navy chose Slaby-Arco and the older method: "The De Forest method had the advantage of enabling any speed of reception to be used, depending on the skill of the operator, but the very fact that the Navy did not have even one operator that was skilled militated against the De Forest method."[85]

Thus, the needs of the Navy and the goals of the inventors, particularly the Americans seeking to "invent around" and improve on Marconi, were completely at odds. The inventors were striving for greater distance, greater selectivity, and faster reception. The inventors assumed that any client, including the Navy, would welcome all three. The Navy, on the other hand, preferred apparatus that required little skill to adjust and operate, even if it was less sensitive, far-reaching, or accurate, for the one thing the Navy knew it could not yet provide were skilled, experienced operators. In addition, naval officers may have wanted apparatus that supplied a written record if it were going to be operated by enlisted men. But the inventors' work was guided by the desire for technical improvement, not for organizational accommodation. They assumed the Navy would want the most up-to-date apparatus available. The Navy, however, needed equipment that would compensate for its organizational idiosyncrasies, a factor the inventors were slow to grasp and reluctant to address, especially if it meant supplying components the inventors had already discarded as inadequate.

In March 1903 the Navy ordered twenty sets of Slaby-Arco apparatus. The purchase prompted *Electrical World* to con-

---

84. Douglas, "Exploring Pathways in the Ether," pp. 99–103.
85. Clark, "Radio in War and Peace," p. 35.

demn the Navy for the "cold shoulder it had consistently turned to American workers in the field" and to refer to the system the Navy favored as "that of the German Emperor's court jester."[86] Reginald Fessenden, who in 1902 had formed the National Electric Signaling Company, wrote to the Bureau of Equipment and the Secretary of the Navy suggesting that before buying foreign equipment, the Navy should test his apparatus. Fessenden quoted Bradford a price of $4000 for two sets and offered "to send a couple of men with a pair of sets" to wherever the Navy desired.[87] Bradford agreed, being careful to specify that the entire test would be "at your own expense." The bureau would not, as it had done before, purchase two sets.[88] Why should it? It had already spent approximately $6000 on French apparatus that barely worked. The Americans were here, and if they wanted the business, they would have to take risks.

Fessenden and De Forest continued to brag about their apparatus, particularly to the press. Navy officials, and Barber in particular, believed the claims to be hyperbolic public relations statements, which they sometimes were.[89] Yet if the inventors were going to boast, the Navy was going to hold them to their word in subsequent tests. More frankness on both sides might have better served all concerned. But none of the inventors behaved as if they believed that candor would sell wireless.

Bradford subsequently notified Fessenden of two additional conditions for the tests. The Navy had no rooms available on board ship for Fessenden's apparatus, so he would have to set it up in a hallway. And the specifications for the apparatus included the filings coherer,[90] which Fessenden had replaced with a superior detector in 1901. Fessenden complained that the Slaby-Arco people had not been relegated to a hall and said his company was disinclined to supply the Navy with special and outdated—apparatus at the company's expense.[91] No experiments with Fessenden apparatus were made until August

86. *Electrical World* XLIII, 24 (June 11, 1904).

87. Reginald Fessenden to Chief, Bureau of Equipment, 8 May 1903, NA, box 85.

88. Chief, Bureau of Equipment to Reginald Fessenden, 21 May 1903, NA, box 85.

89. Comdr. F. M. Barber to Chief, Bureau of Equipment, 18 February 1908, NA, box 89.

90. Adm. Bradford, Chief, Bureau of Equipment to Fessenden, 21 May 1903, SI.

91. Reginald Fessenden to Lieut. Hudgins, USS Topeka, 26 May 1903; Fessenden to Adm. Henry N. Manney, June 13, 1903, SI.

and September of 1904, when the department conducted tests of American apparatus between the Brooklyn Navy Yard and the Navesink Highlands.[92]

The provisions for these tests were different from those of the 1902 demonstrations. Now all expenses related to the demonstration (except for supplying the necessary current) were to be assumed by the companies. The companies were allowed to send their specialists to help with the tests, but "the Bureau's operators must be given every opportunity necessary for it to determine whether the system can be successfully operated by them." In addition, the Americans' apparatus had to conform to the technical standards previously set by Slaby-Arco, a system inferior to and different from their own.[93] By qualifying performance specifications with rather specific and continuing technical specifications, the Navy initially eliminated certain types of apparatus from the experiments.

The American Marconi Company scoffed at the terms of the tests. The company said it would participate only if the bureau would guarantee that successful performance would lead to a contract. "In view of the fact that we are working on a commercial basis over greater distances and under varying conditions all over the world . . . no outlay for the purpose of demonstration only commends itself to us."[94] The Bureau would not consider contingent contracts, and the Marconi Company saw no reason to "incur an expense which, in our opinion, would be out of proportion to the value of the result."[95] The chief of the bureau suggested that by not participating in the tests, "you might even be open to suspicion of not desiring to submit to the conditions of interference in the vicinity of New York in competition with the other systems."[96] This warning did not seem to impress the Marconi Company, and the antagonism between the two continued.

Fessenden was anxious about the nearly $1000 he estimated the demonstrations would cost his small company. He tried to

92. Charles Darling, Acting Secretary of the Navy to NESCO, 15 December 1903, SI.
93. Ibid.
94. Marconi Wireless Telegraph Company of America to Chief, Bureau of Equipment, 9 May 1904, NA, box 89.
95. Marconi Wireless Telegraph Co. of America to Chief, Bureau of Equipment, 2 May 1904, NA, box 89.
96. Chief, Bureau of Equipment to Marconi Wireless Telegraph Co. of America, 4 May 1904, NA, box 89.

arrange for an alternative method of testing, preferably at his own stations, but the Navy refused.[97] The Navy had its own needs and requirements: only by testing wireless on its ships and at the Navy yards could it determine suitability. Its men had to be able to operate the equipment. And continued mistrust of inventors' claims reinforced the Navy's desire to test the apparatus on its own turf. As Admiral Manney wrote, the bureau "prefers to conduct the tests in its own way."[98]

The conditions the Navy imposed during these and subsequent tests were, from the inventors' point of view, niggardly and demoralizing. But the Navy demonstrated even less faith in the inventors when negotiating over purchases and contract specifications. The wireless market was still small and the various inventors competed fiercely against each other. Pride as well as money was at stake, and the mutual hostilities provided the Navy with bargaining advantages.

Once the Navy had decided to acquire apparatus, its first goal was to get the price reduced, and its policy was to buy from the lowest bidder. Barber took great pride in his negotiating skills, reporting that Slaby-Arco lost about $7000 on the first twenty sets it sold the Navy. "The company inferred from my letters that they were competing with other people, especially with Braun-Siemens (I *did* mislead them intentionally in that respect) and the result was an impossibly low bid which I accepted by telegraph before they had time to think it over."[99] A year later he got Telefunken (an amalgamation of Slaby-Arco and Braun-Siemens-Halske) to lower its price for the strategically important Nantucket lightship station by threatening to buy from the French at lower prices. Barber exulted: "Evidently they are red hot on the subject and the Bureau can name its own figure—It isn't often that you get a German down on his stomach like that."[100] The Navy paid nothing in advance; in fact, no payment was sent until the apparatus was installed and operating. If the apparatus arrived late, was damaged in transit, or if the enlisted men mishandled the installation, the

---

97. Reginald Fessenden to Adm. Henry N. Manney, March 8, 1904; Manney to Fessenden, 14 April 1902, SI.

98. Chief, Bureau of Equipment to Marconi Wireless Telegraph Co. of America, 10 May 1904, NA, box 89.

99. Comdr. F. M. Barber to Chief, Bureau of Equipment, 25 July 1903, NA, box 90. Barber was able to get Slaby-Arco down to $1077.50 a set.

100. Comdr. F. M. Barber to Chief, Bureau of Equipment, 8 July 1904, NA, box 12.

payment to the supplier was reduced.[101] While it was clearly in the Navy's interest to get the best possible price and not pay until the apparatus was working, its tactics compounded financial uncertainty for the inventors.

The Navy soon added to its contracts other provisions that imposed additional risks and burdens on the fledgling companies. The equipment had to be guaranteed to signal over a certain distance under all conditions, and failure to supply such a guarantee meant elimination from consideration. Once awarded a contract, the wireless company was required to "bond" its apparatus: it paid a security deposit, and if the apparatus failed, the bond was forfeited.[102] Lee De Forest, who was awarded a contract to erect four high-powered stations in the Caribbean, had to guarantee that the stations would be able to maintain communication "at all times and under all atmospheric conditions" over a distance of 1000 miles. He had to put up a bond of over $16,000 and complete all four stations within six months.[103] These were very stringent requirements to impose on a small company erecting radio stations far away from its base of operations and sources of supply. The territory was unknown to De Forest, and there were increasing reports that static was particularly relentless in the tropical regions. Even Barber questioned the Navy's specifications: "When I said some time ago that I did not think that his contract with the Department was legal, I meant that if it came into court, the court would decide against the Department."[104]

If an inventor would not reduce his prices, the Navy got a competitor to copy the invention and supply it at lower cost. Fessenden introduced the Navy to his new receiver, the "electrolytic detector," during the 1904 demonstrations. Fessenden's assistant wrote that naval officials were "highly pleased with the results, we having done very much better than any other system tested by the Navy."[105] Subsequent evidence bears out this re-

101. Chief, Bureau of Equipment to Commandant, Navy Yard, New York, 25 November 1903, NA, box 88; Barber to Chief, Bureau of Equipment, 10 July 1904, NA, box 12.

102. Douglas, "Exploring Pathways in the Ether," p. 159; Chief, Bureau of Equipment to Comdr. F. M. Barber, 9 July 1904, NA, box 12.

103. *The New York Times*, 10 July 1904, pt. 5, p. 26; H. W. Young to Hay Walker, Jr., June 29, 1904; NESCO Memo, "In Regard to the West India Wireless Contract," 7 May 1906, SI.

104. Comdr. F. M. Barber to Chief, Bureau of Equipment, 29 November 1904, NA, box 89.

105. James Boyle to Reginald Fessenden, 17 September 1904, SI.

port: by 1905, the electrolytic detector was the Navy's standard receiver. But Fessenden's prices ($2000 to $5000 per set) were considered too high, so the Navy arranged for De Forest and Telefunken to supply copied receivers at a lower cost.[106] Fessenden was outraged. Yes, his apparatus was more expensive than the Germans', who received government support. Didn't the Navy understand research and development costs? Didn't the Navy respect patents? For over two years he wrote letters of complaint to the bureau and even demanded that the Secretary of the Navy be impeached.[107] The Secretary informed Fessenden that his prices allowed the department to be "relieved of any moral obligation" to honor Fessenden's patents.[108] By 1906 Fessenden refused to have any further dealings with the Navy. "If we do not communicate any more of our inventions to the government, the government cannot steal them."[109] Other inventors complained as well about the bureau's knowingly buying, and even encouraging the manufacture of, pirated goods.[110]

To inventors, patents were central: they established priority in scientific and technical circles, in history books, and in the courtroom. Their strength could ensure one's prestige and fortunes. With so much riding on them, patents were considered inviolate by their owners. The Navy, on the other hand, felt it could not be constrained by patents and, in fact, it had no legal obligation to honor them. Amid all the press releases, claims, charges, and countercharges, how could the Navy really tell who the legitimate patent holder was? The Navy's policy was to

106 George Clark, who became a naval radio technician in 1907, acknowledged that the Navy made "free use of" Fessenden's patent. Full documentation of the dispute between Fessenden and the Navy over the electrolytic detector can be found in the box containing NESCO Navy correspondence in the Clark Collection. Walter Massie, a wireless entrepreneur based in Providence, also warned Fessenden about the Navy's efforts to circumvent his patents. See George H. Clark, "The Life and Creations of John Stone Stone," unpub. ms., 1946, pp. 92–94; Reginald Fessenden to James Hall of *The New York Tribune*, 9 December 1905; Charles Bonaparte to Fessenden, 19 April 1906; and Fessenden to James Hayden, 11 January 1907; all in the Clark Collection; and Comdr. F. M. Barber to Chief, Bureau of Equipment, 20 February 1906, NA, box 89.

107. Reginald Fessenden to Charles Bonaparte, Secretary of the Navy, 5 May 1906; Fessenden to President Theodore Roosevelt, 14 May 1906; NESCO to Adm. Henry N. Manney, 14 June 1905, SI.

108. Charles Bonaparte to Fessenden, 19 April 1906, SI.

109. Fessenden to Lieut. Comdr. Cleland Davis, 30 November 1906, SI.

110. John Firth, "The Story of My Life," unpub. ms., n.d.; John Firth to Cleland Davis, 11 June 1908, SI; H. W. Sullivan to Barber, 30 May 1908, NA, box 89.

acquire apparatus "independently of patents."[111] Barber advised the bureau that he doubted whether anyone really had a defensible patent on a wireless telegraph system. He did not think Fessenden, who was threatening to sue the government for back royalties, should be taken seriously and doubted "if any of the present owners of wireless telegraph patents will ever do anything more than they have done in serving these preliminary notices."[112] Fessenden's threats against the Navy were in fact empty; the government at this time could not be sued for using patents without permission.

Even when priority was established in court, however, the Navy did not acknowledge the rights of the patent holder. Fessenden, after being advised by the Secretary of the Navy that his first successful infringement suit against De Forest was not "conclusive," had to win three more consecutive decisions and file an injunction and contempt of court citation against both De Forest and Telefunken before the Navy would stop purchasing pirated electrolytic detectors from Fessenden's competitors.[113]

Another point of considerable controversy between the inventors and the Navy was the notion that wireless was a "system" and that different wireless systems existed. Since 1900 the Marconi Company's strategy had been to market wireless as a complete system or network. The company erected the shore stations, equipped the ships, and established channels for communication. Other companies followed suit. Although this systems policy was motivated primarily by business considerations, technical considerations played an important part as well.

In each competing wireless set the various components were carefully engineered and adjusted with the efficient operation of the entire system in mind. Not only were the components themselves special, but the adjustments and arrangements between them were also crucial to superior performance. From the number of turns in the induction coil to the type and num-

111. Comdr. F. M. Barber to Chief, Bureau of Equipment, 23 June 1904, NA, box 12.

112. Comdr. F. M. Barber to Chief, Bureau of Equipment, 14 September 1903, NA, box 89.

113. Charles Bonaparte to Reginald Fessenden, 19 April 1906; NESCO v. De Forest Wireless Telegraph Co. et al., Circuit Court of United States, So. District of New York, May, 1906, SI. This was not the only time an agency of the U.S. government tried to circumvent or appropriate patents. See the account of Herman Hollerith's struggle with the Census Bureau in Geoffrey D. Austrian, *Herman Hollerith* (New York: Columbia University Press, 1982), pp. 260–62, 275–78.

ber of condensers and the aerial arrangement, all the inter-
connections were designed to meet the system's special needs.
One could have a very sensitive and reliable detector, but if it
were connected to incompatible or second-rate headphones,
the receiver would appear to be inferior also. Chances were
excellent that rival apparatus would not integrate well into a
competing system and would cause poor performance. No inventor could allow alien and possibly inferior components to
discredit his system or the merits of wireless. While inventors
were trying to protect their business, they took pride in the
distinctiveness of their apparatus and recoiled at the thought of
it being dismantled and recombined with competitors' devices.

To Navy officials, wireless components were individual inventions like telephones or light bulbs. The Navy considered
the inventors' systems rationale nothing more than a justification for monopoly. Was the Navy to buy from only one
company and ignore all others? Naval officials sought to
squelch any maneuverings that resembled Marconi's tactics, including any attempt to treat wireless as a system. Consequently,
the Navy determined to buy only components, to establish its
own "composite" system, and to ignore the inventors' systems
approach. The way the Navy pursued this goal was not to buy
complete sets of transmitters and receivers from several companies, preserving the integrity of those sets, and installing
them in different navy yards or ships. Rather, the Navy began
buying only components and had naval personnel combine the
different devices together on an ad hoc basis. As the chief of
the bureau advised Barber in 1902, "It is proposed to conduct
tests of composite sets, made up of portions supplied by different makers and such a combination may be adopted as standard for the service in case it is found to work better than an
entire set supplied by a single maker."[114]

The bureau, which didn't think civilian "squareheads" were
attuned to the needs of the Navy and which out of pride may
have wanted to develop its own system, no doubt sought to
achieve standardization through the composite route. The
bureau may also have been trying to reduce technical uncertainty; if it mastered the components and controlled its own
system, maybe it could either anticipate or avoid too rapid technological turnover. From the bureau's point of view, acquiesc-

---

114. Chief, Bureau of Equipment to Barber, 15 August 1902, NA, box 85.

ing to the systems notion meant fostering monopolistic goals. Thus the Navy would try to prove that there were no distinct, incompatible systems. But acquiring various components made by competing firms and then successfully combining them into a composite system were two very different processes indeed.

While wireless in the first decade of the twentieth century still had many shortcomings, each year brought improvements in reliability and range that enhanced the invention's military potential. During the 1904 World's Fair, De Forest succeeded in transmitting messages hundreds of miles over land; by 1907, Marconi was providing *The New York Times* with a transatlantic wireless news service. Cunard and the White Star Line had begun installing apparatus on their ocean liners as early as 1900 and found the invention to be highly useful.[115] While the Navy had reason to criticize wireless because of the still unsolved interference and lack of secrecy problems, the invention nonetheless provided a method of intership and ship-to-shore communications superior to anything else then available. While technical uncertainty cannot be discounted as a factor in the Navy's position, neither can it fully explain the lack of harmony between the Navy and the inventors.

The chief of the Bureau of Equipment had little support from the rest of the organization for adopting wireless. To the secretary the first major experiments were of low priority. The other bureau chiefs had their own concerns and had no reason, bureaucratically, to be interested in wireless. Bradford had no organizational allies actively interested in the new technology whose support or influence he could enlist.

Because the Navy was, along with steamship companies and several newspapers, one of the few wireless customers, it was able to impose its uncompromising terms on the fledgling industry. The inventors clearly assumed too much: they thought the Navy would be an early and regular client but did not achieve the patronage they so desperately wanted. In addition, their dealings with the Navy, rather than reducing risks and uncertainty, increased them. The expense of the tests, the bonds, and the efforts to either ignore or subvert patents were, to the inventors, unanticipated and unwelcome costs not of actually getting but of simply trying to get government business. Unlike other innovations in industries such as steel, radio

115. Douglas, "Exploring Pathways in the Ether," pp. 104–5, 236.

was not "shielded" by the Navy "from the rigors of the market-place."[116] The Navy did not assume risks here, it exacerbated them.

During the acquisition phase, inventors and naval officers failed to establish a mutually beneficial relationship. Their different orientations and their subsequent interactions reinforced stereotypes and distrust on both sides. De Forest, who referred to the Caribbean stations as "the hellhole of wireless," became demoralized by the "hostility, open or concealed, on the part of officials, from whom we had every reason to expect cooperation and interest." Revealing his own prejudices in a letter to his attorney, De Forest complained about these "cheap" officers, whom he characterized as men "with more gold tape than brains."[117] These two kinds of men, from two different American subcultures, were simply not communicating. Barber, the link between European inventors and the Navy, failed in his role as translator. He was not truly bilingual. Although he spoke French and German and understood the apparatus, he always maintained the Navy's point of view. He made no effort to understand and then relay to the bureau the inventors' perspectives on research and development costs, prices, marketing or contracts. As a retired officer, he had been a Navy man too long to successfully encode and translate non-military views and needs. And the logistics were wrong. An officer in France could not help a bureau in Washington implement a new invention. The translator the Navy, and the inventors, needed was a different man, probably younger, more comfortable with technical mastery as a basis for power and prestige, more enthusiastic about the possibilities of wireless, and closer to the lines of authority and communication in Washington.

Although the Navy acquired wireless apparatus continually from 1902 onward, the period from 1902 to 1906 marked the initial spurt in purchasing. The panic of 1907, followed by several years of indifference, established the 1907 to 1911 period as a lull in Navy acquisition of equipment. Overlapping with the acquisition phase was the implementation phase, which was also characterized by individual and organizational

116. See Merritt Roe Smith's introductory essay.
117. See Frank E. Butler, "How Wireless Came to Cuba," *Radio Broadcast* (November–April, 1924–25): 916–20; Lee De Forest to Frank Butler, April 20, 1906; De Forest to Francis X. Butler, October 14, 1905, SI.

resistance. The Navy's shore command, represented by the Bureau of Equipment, had tested apparatus and ascertained what the fleet would need, independently of any advice or reaction from ship or squadron commanders. This procedure aggravated the long-standing tension between the officers at sea and those behind desks in Washington. Not all commanders liked coming aboard their ships after a brief stay in port to discover mysterious contraptions they had neither requested or desired. But when the shore command had made few provisions for personnel to set up and use the apparatus, rejection was even less surprising.

When the first twenty Slaby-Arco sets were ordered from Germany, there were no engineers who knew how to install them properly. In the summer of 1903 "the number of men in training capable of taking charge of a station" totaled eight enlisted men.[118] On board ship there were no wireless operators. And few commanders welcomed the apparatus. "No serious effort was made by the various commanders to organize, utilize, or supervise radio communications within the fleet."[119] These men, especially once out at sea, enjoyed complete control of their ships and did not want that authority subverted by wireless, which threatened to render their leadership merely titular. As George Clark observed, "The traditional power of a commanding officer to do as he felt best with his ship or command as soon as he got out of sight of land would have been completely wiped out if someone in the Bureau of Navigation or elsewhere could give him orders. So often the instructions to the wireless room were to shut down the wireless and not acknowledge calls from shore at all."[120]

Flag lieutenants were to supervise wireless on board ship, but they knew nothing about the equipment and had no incentive to learn. With the installation of wireless below decks to protect the apparatus from the rigors of battle, would the flag lieutenant now be consigned to some remote cabin, away from the captain and the action on the bridge? This prospect was hardly appealing and was quite naturally opposed. One flag lieutenant, T. P. Magruder, when inspecting a new installation on his ship, objected to the "unsymmetrical appearance" the antenna

118. Capt. C. H. Arnold, President, Wireless Telegraph Board to Chief, Bureau of Equipment, 10 July 1903, NA, box 85.
119. Howeth, *History of Communications-Electronics*, p. 65.
120. George H. Clark, "Radio in the U.S. Navy," unpub. ms., n.d., SI.

wires and guys produced and ordered the lines and wires realigned to parallel the rest of the ship's rigging. The new arrangement significantly reduced the efficiency of the apparatus. When it was suggested that the new arrangement rendered the sets nearly useless, Magruder said he "didn't give a damn about wireless . . . but he did give a damn for the appearance of the ship."[121]

The Bureau of Equipment had no authority to compel the officers to use the new invention, and the Bureau of Navigation, which oversaw the movement of the ships but had no jurisdiction over wireless, had no incentive to assist implementation. And with no permanent and experienced chief executive to enforce or encourage use, officers saw few organizational inducements countermanding their own recalcitrance. Officers on shore could not compel officers at sea to adopt the invention. The most striking example of the dissonance between the ambitions of the shore command and those of the fleet was the acquisition in 1907 of Lee De Forest's radiotelephones for the "Great White Fleet," which was about to embark on its famous cruise around the world. The bureau ordered twenty-six sets, which transmitted and received speech instead of dots and dashes, so that the commanding officers could talk directly to each other without going through the wireless operators. But Admiral Evans, the commander-in-chief, wanted nothing to do with these devices, and he issued orders to dismantle and stow the apparatus shortly after the fleet set sail.[122]

The performance of wireless, once acquired, was also affected by the ability of the enlisted men and the facilities available for maintenance and repair of the apparatus. Lieutenant J. M. Hudgins, who had helped Barber investigate European apparatus, complained to the secretary in 1904 that "we are not getting one-half the service possible out of the apparatus in use, owing to the lack of skilled operators."[123] He warned that few of the men assigned to take charge of the Navy's new stations were really qualified for such duty, particularly since they had

121. Howeth, *History of Communications-Electronics*, p. 65.
122. G. H. Clark, "Why the De Forest Radiophone Failed," unpub. ms., n.d.; Lieut. Comdr. H. J. Meneratti, "Story of the De Forest Wireless Telephone in the U.S. Navy," unpub. ms., n.d.; Meneratti, "Log of installation and operation of De Forest Radiophones in the U.S. Navy in 1907," SI.
123. Lieut. J. M. Hudgins to Secretary of the Navy, 15 February 1904, NA, box 83.

no experience adjusting or making quick repairs to the sets.[124] Strong criticism of the operators' general incompetence came from both civilian and military quarters and persisted for ten years.[125]

Wireless was installed aboard ships while docked at either the New York or Washington Navy Yards. The apparatus could theoretically be repaired at all the Navy yards. And the yards were also the sites for Navy shore stations. But the nature of the work and supervision at the yards did not promise to provide wireless with a favorable environment. Administration of the Navy yards epitomized the department's decentralized structure and management. Although nominally controlled by the Bureau of Yards and Docks, the yards contained offices and staffs affiliated with and loyal to the other bureaus. Predictably, this led to confusion and waste. For example, several different engineering departments and machine shops, each working for a different bureau, were dispersed throughout the yard. This arrangement militated against concentration of effort and combination of expertise.[126] The inexperienced operators charged with installing and repairing wireless might have carried out their duties more efficiently had they been part of a unified engineering department at the yard. Under the existing arrangement, they had little supervision and often found themselves caught between conflicting orders, one set from the Bureau of Equipment, another from the commandant of the yard.[127] And there was no technical standardization or uniformity from one Navy yard to the next. Disregarding whatever standard plans the bureau may have tried to issue, each Navy yard pursued its own method of wireless installation and repair.[128]

Exacerbating this organizational lack of continuity and fragmentation was the composite system. "Composite" did not mean that the Navy used only one sort of transmitter or one

124. Ibid.

125. Chief, Bureau of Equipment to Commander in Chief, North Atlantic Fleet, 22 March 1905, NA, box 32; F. S. Doane, Master of Light-Vessel no. 85 to Capt. W. G. Cutler, Inspector, July 20, 1909; Cutler to Lighthouse Board, 21 July 1909, NA, box 79; Douglas, "Exploring Pathways in the Ether," pp. 306–12.

126. Paullin, *Paullin's History*, p. 406; *Annual Report of the Secretary of the Navy*, 1883, pp. 17, 107; 1884, pp. 16–19.

127. Commandant, Navy Yard, Washington, D.C. to Chief, Bureau of Equipment, 9 March 1909, NA, box 76.

128. Clark, "Radio in War and Peace," pp. 316–17.

sort of receiver connected according to standard specifications. The Navy concocted these systems from whichever components were available at the time at the lowest price and left it to the operators to piece them together. This encouraged untrained and inexperienced men to tinker with the apparatus and to conduct their own trial and error experiments. The use of the "composite system" also meant that an operator transferred from one yard to another or from one ship to another "had to learn an entirely different run of wiring and placement of apparatus in many cases, which not infrequently resulted in his total ignorance of the status of his new assignment."[129] The composite system and the independence of each Navy yard and of each wireless station led to a proliferation of many different types of wireless sets throughout the service. The chief of the Bureau of Equipment in 1907 described the costs imposed by lack of supervision and standardization:

Certain operators when first ordered to a station, and who were perhaps familiar with other systems, would not use that provided but improvised systems of their own. The original instruments would thus fall into disuse and deteriorate, and when these operators were detached they would take away the improvised instruments. The stations would thus remain inefficient for a considerable period and in some cases could hardly be operated at all until new instruments were provided.[130]

Wireless entrepreneurs were not pleased with the situation, which caused their apparatus to be "abused frightfully."[131] One company claimed that after loaning some apparatus to the Navy "it was in such a condition that we had to throw it aside as a lot of junk."[132]

Some Navy yards, particularly those on the west coast, complained of hand-me-down equipment and unsuitable facilities. And once a ship or station was equipped, little effort was made to update the apparatus. The commandant of the Mare Island

129. Ibid., p. 317.

130. Chief, Bureau of Equipment to Commandant, Navy Yard, Washington, D.C., 5 June 1907, NA, box 13.

131. S. M. Kintner, General Manager, NESCO to General Storekeeper, Navy Yard, Brooklyn, 5 October 1911, NA, box 13.

132. John Firth, WSA, to Chief, Bureau of Equipment, 9 November 1909, NA, box 82; see also complaints from William Walker, Massachusetts Wireless Equipment Co. to Chief, Bureau of Equipment, 31 December 1908, NA, box 75; and Douglas, "Exploring Pathways in the Ether," pp. 159–60.

Navy Yard suggested in 1904 that the yard's wireless station be moved from the deteriorating pigeon coop in which it was first installed.[133] Six years later, the wireless building was so decrepit and leaky that it was too dangerous for the operators to work there.[134] The commandant of the Philadelphia Navy Yard was informed in 1910 that the apparatus at his station was all jerry-built and obsolete. In fact, the files of the Bureau of Equipment for the years 1909 and 1910 are filled with reports from Navy yards around the country criticizing the obsolete, poorly maintained, and barely functioning wireless sets at the shore stations.[135] By 1910 wireless telegraphy, now more frequently referred to as radio, was hardly being used to its full advantage in the Navy. Radio had reached an organizational dead end.

Several changes, both within and outside the Navy, began to pave the way for improvement. For decades the various secretaries had recommended that the number of bureaus be reduced and their duties consolidated. While this much-needed reform was not enacted until World War II, in 1910 the Bureau of Equipment was abolished and its duties distributed among the remaining bureaus.[136] The Bureau of Steam Engineering, in existence since 1862, had long been the department's center for steam and then for electrical engineering, and it assumed control of radio in 1910. The Bureau of Steam Engineering was responsible for designing, constructing, maintaining, and repairing the machinery on board naval vessels. It was not just a procurement bureau but one actively involved in the building and successful mechanical operation of the ships. With its strong engineering tradition and greater influence within the fleet, it provided a more propitious organizational niche for radio's deployment.

In 1909 Secretary Newberry began reorganizing the Navy yards. All the previously dispersed mechanical departments and their personnel were placed under the direction of a man-

133. Commandant, Navy Yard, Mare Island to Chief, Bureau of Equipment, 16 April 1904, NA, box 83.
134. Lieut. E. H. Dodd, Mare Island to Chief, Bureau of Equipment, 6 April 1910, NA, box 81.
135. W. L. Howard, Inspection Officer to Commandant, U.S. Navy Yard, Philadelphia, 16 June 1910, box 80; C. D. Mills, Chief Electrician, Tatoosh Island, Washington to Inspector, Navy Yard, Puget Sound, 19 January 1910, box 80; Inspector of Equipment to Commandant, Navy Yard, Norfolk, 29 June 1909, box 80; Chief Electrician, Navy Yard to Inspector of Equipment, Navy Yard, Charleston, SC, 10 June 1909, box 78.
136. *Annual Report of the Secretary of the Navy,* 1910.

ager who consolidated both manufacture and repair work. Eventually the manager was replaced by a line officer.[137] Secretary Meyer, Newberry's successor, appointed four aides with specialized expertise to advise him on operations, personnel, materiel, and inspections and to coordinate the work of the bureaus.[138] Meyer also sought to improve the business methods of the department.[139] This effort may have been propelled by a law enacted in 1910 (partly as a result of Fessenden's lobbying) which authorized the owners of patents that were used by the government without permission to sue in the Court of Claims.[140]

Between 1907 and 1912 significant improvements in both transmitters and receivers were introduced by De Forest, Fessenden, Marconi, and Telefunken. Radio signals now had more power behind them and were higher pitched and easier to read. Reliability, durability, and transmission distance had been greatly enhanced. The Navy had not helped foster these changes, but for the department technical uncertainty was being reduced dramatically.

Most importantly, new legislation required the Navy to increase its radio activities. Prior to 1912 radio in America was unregulated. Anyone with a transmitter could send messages whenever he wanted, and as a result the airwaves were frequently congested. This anarchy in the spectrum had drastic repercussions. On April 16, 1912, America learned that the *Titanic* had sunk, and hundreds of radio stations along the northeast coast of North America clogged the airwaves with inquiries and messages. The resulting interference prevented speedy, efficient communication and produced misinformation as well. Within four months of the *Titanic* disaster, Congress enacted the 1912 Radio Act, which prohibited independent "amateur" operators from transmitting in the preferred portion of the spectrum.[141] The Act also sought to ensure that ships' passengers would always have access to wireless services, even if they were not near a commercial wireless station. Thus Navy radio stations were now required to transmit and receive commercial messages if there was no commercial station within

137. Paullin, *Paullin's History*, p. 479.
138. Ibid., p. 442.
139. Ibid.
140. *Annual Report of the Secretary of the Navy*, 1911.
141. Douglas, "Exploring Pathways in the Ether," pp. 337–48.

a 100-mile radius.[142] As Secretary Meyer observed in his Annual Report for 1912, "The radio work and expenses of the department will be largely increased. It will be necessary to modernize and improve the apparatus of coast stations so that the commercial work may be successfully handled . . . the added work will undoubtedly prove an incentive to increased efficiency."[143]

By mid-1912 this new constellation of technical, legal, and organizational changes confronted the department. But the changes did not guarantee that radio would be efficiently integrated into naval operations. Radio, with the potential to establish new and strong channels of communication in the Navy, had been forced to operate within a nineteenth-century organizational structure. Only the efforts of a very enterprising translator, adept at exploiting unusual external pressures, would compel this structure to yield to and be realigned by this technology.

Stanford C. Hooper has been called the "Father of Naval Radio."[144] It is a title he enjoyed and believed he had earned. His version of the Navy's ultimate adoption of radio has the self-aggrandizing tone not uncommon to the autobiographies of many people who were pioneers in their field,[145] yet the record does support Hooper's story of his efforts to integrate radio into naval operations. He achieved this integration at a propitious moment in naval history, but this does not detract from the adroitness and ultimate success of his strategy or methods. He was a man who read his organization—and the times—very shrewdly indeed.

The son of a banker and named after Leland Stanford, Hooper grew up in an entrepreneurial environment. When he was eight, his father built him a telegraph sender and key and taught Hooper the Morse Code. By the age of ten he was working part time for the railroad as a relief ticket agent and then

---

142. *Annual Report of the Secretary of the Navy*, 1912, pp. 38–39.

143. Ibid.

144. Howeth, *History of Communications-Electronics*, p. 114.

145. See, for example, Lee De Forest, *Father of Radio* (Chicago: Wilcox & Follet Co., 1950) or Helen Fessenden, *Fessenden—Builder of Tomorrows* (New York: Coward-McCann, Inc., 1940). Material on Hooper from George H. Clark, "Radio in War and Peace." In 1940, Clark, who had worked as a radio inspector for the Navy between 1907 and 1919, persuaded Hooper to dictate his memoirs to him. Some of the material is based on Clark's own recollections of this period and some on Hooper's reminiscences.

relief telegraph operator. His seven years experience with telegraphy provided a necessary foundation for his later radio work. He saw at a young age how a transportation and communication network were integrated and operated cooperatively. In 1901 his father arranged for Hooper to attend the Naval Academy, and at the age of fifteen he entered Annapolis and was embarked upon his career. One of his earliest challenges was studying to become proficient in Navy signaling. Because Hooper had grown up with the Morse Code, learning another code, also based on dots and dashes but completely different, proved confusing and difficult. Yet by mastering both, Hooper began to build the "bilingualism" that would later prove so important. To get Navy signalmen and officers to switch ultimately from one code to another, Hooper had to understand and be proficient in both.[146]

After graduating from the Academy in 1905, Hooper served on various ships of the Pacific Fleet. He began to read about and tinker with wireless. Sometime between 1907 and 1908 Hooper put in a request for postgraduate training at the Naval Academy, specializing in wireless. This request was denied by Lieutenant Commander S. S. Robison, who believed "wireless would never be enough to warrant an officer giving it his full attention."[147] Hooper continued to pursue his goal, trying various tactics and routes, and finally was sent to the Academy in 1910 as an instructor in electrical engineering, with wireless instruction added to his regular duties.[148] Thus, not unlike many teachers, Hooper was to learn his subject matter shortly before teaching it to a class. But his assignment, as George Clark noted, marked a turning point in his career: "From then on he was in charge of a 'radio division' of the Navy, be it of the Department of Electrical Engineering at the Academy, or the Bureau of Steam Engineering in Washington, or of the Fleet."[149]

Now a lieutenant, Hooper was one of the few officers in the Navy who had experience as an operator. By 1911 several commanders working in the Bureau of Steam Engineering had begun to consider more seriously the use of radio for com-

146. Ibid., p. 74.
147. Ibid., p. 70.
148. Ibid., p. 71.
149. Ibid.

municating between vessels when in battle formation but could not adequately implement this plan without firsthand knowledge of radio communication. Consequently Hooper was assigned to develop and write up instructions for tactical signaling between battleships. His plan would be tested during Spring Target Practice of 1912, when radio signals would accompany all visual signals.[150]

Meanwhile another young officer, who also had childhood experience with telegraphy, was assigned to report on the use of wireless during the Autumn Battle Practice of 1911. Ensign C. H. Maddox assessed the technical merits of the apparatus and analyzed wireless's potential for tactical signaling. Much of his information was intended for Dr. Louis Austin at the Naval Radio Lab and not for those with authority over the fleet. Yet Maddox became less concerned with technical problems than with organizational matters. In his first report he urged that wireless have its own set of tests rather than be tested in conjunction with target practice. Only then would wireless "get the full consideration that it deserves." During target practice, "a wireless test is too liable to be relegated to the list of those things that can be slighted for the sake of possible increase of 'hits per gun per minute.' "[151]

Maddox saw as the most immediate and pressing need "officers in the fleet who possess a thorough practical and theoretical knowledge of wireless and who are themselves expert operators. At present the real head of the fleet's wireless system is the enlisted operating force of the flagship." He found these operators to be "mediocre," in part because no specialization existed in the electrical force aboard ship. Wireless operators and dynamo tenders were often rotated between these two jobs and thus there was "small chance for improvement." He also recommended that the Atlantic Fleet have an officer in charge of the fleet's wireless "who will systematize and control this important factor in naval efficiency." He looked forward to the day when the Navy would possess enough officers proficient in wireless that "one might be assigned to each division of the fleet, and eventually one to each battleship."[152]

150. Howeth, *History of Communications-Electronics*, p. 181.

151. Ensign C. H. Maddox, "Report of Battle Practice, Autumn 1911," USN, Washington, GPO: 1911, SI, p. 230.

152. Ibid.

Reportedly, Hooper did not see Maddox's report or recommendations. In the spring of 1912, Hooper would reiterate the same suggestions. But Hooper, now twenty-six, was about four years older than Maddox, more experienced, higher in rank, and better connected. One important ally was Lieutenant Commander D. W. Todd, head of the Radio Division of the Bureau of Steam Engineering. Hooper's observations and report would have more sway when passed through him.

Prior to the April 1912 tests, Hooper devised a tactical signaling code and made several other specific recommendations. From the earliest demonstrations in 1899, Navy observers had suggested protecting the wireless apparatus by placing it below decks. For tactical signaling the distance between the captain on the bridge and the apparatus below caused an unacceptable delay between orders given and orders sent. Others had tried speeding up the communication between the bridge and the radio room with voice tubes or telephones. Hooper's suggestion was characteristic of his unconstrained view of naval organization: move the operator up to the bridge. Portable equipment would be installed quickly and the transmitting key on the bridge could be connected to the main transmitter below decks. This move, while eliminating any delay between the captain's message and transmission of that message, symbolically and actually demonstrated the importance of tying radio directly to the chain of command.

Hooper's tactical signals and instructions on general signaling procedure for the maneuvers were printed up and included in a booklet of general instructions written by Commander Craven, in charge of fleet training in the Division of Operations, Bureau of Navigation. When the tests began, Hooper and Craven went to the flagship to observe results. Hooper began to monitor the radio signals and heard nothing all day. All the signaling was done by flag. At the end of the day, he visited several of the ships in the fleet to investigate what had happened. The only encouraging discovery Hooper made was that the Bureau of Steam Engineering had set up radio apparatus on the bridge. But no other steps in Hooper's plan were followed:

The Navy, as usual up to that date, did not take radio seriously. The commanding officers had handed the instruction pamphlet to the officer in charge of communications, but he, in every case not at all familiar with radio, did nothing more with it, probably not having faith in the ability of radio to do the work. In a few cases, the booklet had found its way down to the radio cabin, but the Chief had not had time to read and understand the scheme, nor did he have it explained to him.[153]

Thus, responsibility for testing radio signaling had been passed down to those in the organization with no authority or accountability, particularly in the sphere of tactics or strategy. It was 1912, and radio was still being treated as an afterthought.

Hooper reported his discoveries to Craven, who in turn took the issue up with the commander-in-chief. The commander ordered that Hooper's instructions be followed the next day. Hooper tried to ensure better performance by personally instructing some of the officers and men in the use of his plan. The next two days proved disappointing and on the fourth day the flagship's transmitter died, bringing the experiment to a close. Hooper's report reflected his disappointment. He criticized the officers for not incorporating the operators into the tactical signaling process aboard ship. "The operators did not understand what they were to do," he stated, adding "there was no inter-ship teamwork." He also wanted the skills of the operators upgraded. "About one-third of the operators are not operators and delay the general business about one-half." He urged that the operators increase their transmission speed from their current ten to eighteen words per minute as "our standard is about half the commercial standard." Noting that "the wireless is running away from us in certain regards," Hooper also recommended that "there should be an officer in charge of radio matters in the Fleet who is an expert operator."[154]

Again, the spur for improvement came from the officers on shore, in Washington, not from ships' commanders. Craven persuaded the chief of the Bureau of Navigation to add to the staff of the fleet commander-in-chief the position of Fleet Radio Officer. Craven and his friend Todd in Steam Engineering both recommended Hooper, who, in August of 1912, became the Navy's first officer in charge of coordinating the use

153. Clark, "Radio in War and Peace," pp. 84–85.
154. Lieut. S. C. Hooper to Chief, Bureau of Steam Engineering, "Report on Radio-Telegraphy in the Atlantic Fleet, Spring Target Practice of 1912," Hooper Papers, Library of Congress.

of radio at sea.[155] Rear Admiral Hugo Osterhaus "objected strenuously" to having such an officer on his staff and then arranged for Hooper's duties to include tactics and athletics as well.[156] Having to organize and supervise boat races and boxing matches helped to delay Hooper's work and to undermine the importance and prestige of his main task. Only by working extra hours was Hooper able to perform his radio duties adequately.[157] Thus, despite the efforts of the Navy's equivalent of "mid-management," recalcitrant top executives could still preserve the status quo. No fleet officer could do it alone. He needed allies not only on shore, but also within the fleet. Hooper and Maddox had recognized the need for a new organizational tier in the fleet: officers, and not enlisted men, had to have control over radio. In the fall of 1912 Hooper recommended to Osterhaus that commanding officers of all battleships, flagships, or cruiser and gunboat divisions, and flotilla flagships of destroyers designate an ensign as radio officer and require him to become a proficient operator. Osterhaus followed Hooper's suggestion and issued the order.[158]

This was Hooper's first shrewd strategic move as Fleet Radio Officer. Young officers would be less bound by naval tradition; unhindered by years of flag-signaling, they would be more open to the new technology. They had, at this juncture in their careers, little to lose and much to gain by becoming proficient in radio. And Hooper would shortly have dozens of officers just junior to him in rank, bound to his authority, competent with and sold on the new technology. He was arranging the beginnings of his network by creating an organizational cadre which would both permit and benefit from full integration of radio use within Navy operations.

In 1913 Rear Admiral Charles J. Badger, who had served on the 1902 Wireless Telegraph Board, became Commander-in-Chief of the Atlantic Fleet. His chief of staff, Commander Charles F. Hughes, was more sympathetic to Hooper's goals and recommended that Hooper be relieved of his duties as fleet athletic and tactical officer. Hooper was now free to concentrate completely on radio, and his attempts to incorporate radio

155. Clark, "Radio in War and Peace," p. 118.
156. Howeth, *History of Communications-Electronics*, p. 195.
157. Ibid.
158. Ibid.

more fully into the daily functioning of the fleet would be more sympathetically and seriously considered.[159]

While the ensigns were learning about radio, Hooper had another, more difficult task. He had to upgrade the performance of the operators while simultaneously wresting control of fleet radio from them. Because wireless had been kicked down the naval hierarchy from its earliest introduction, it was now controlled almost entirely by the enlisted men.

From rear admiral to ensign, few officers had considered wireless important and therefore few sought to learn how to use it. But the bureau kept sending the instruments to the ships and someone had to oversee them. The "technological buck" stopped with the enlisted men who worked under the broad heading of electrical engineer. At this level of the Navy, wireless did not mean subversion of autonomy or tradition. On the contrary, an enlisted man who knew about radio gained some small distinction. He enjoyed a certain degree of autonomy— he could transmit whatever he wanted—and he often possessed privileged information. Most of the messages sent out by naval operators before 1913 were personal messages such as this one cited by Clark: "Longing for you darling, and waiting for the fog to lift. Lieutenant _____."[160] The ship's operator also conversed with other naval and commercial operators in the vicinity and could eavesdrop on various conversations.

As radio apparatus began to proliferate in the fleet, control of wireless was often maintained by the chief engineer of the flagship. "Friends of his among the Fleet operators could use their sets anytime they wished to do so; those who were not 'in' with the Chiefs had to wait until the 'Flag' was good and ready to let them open up." Use of the airwaves came to be dispensed by the chief as a privilege, a perquisite.[161] Radio technology provided the chief and the operators with control and diversion, two things often denied aboard ship, and they were not about to relinquish these easily.

Hooper's goal was to compel the operators to "use their station for nothing but official business, to make use of a military routine whose first requirement was obedience to orders, and

159. Ibid., p. 196; Clark, "Radio in War and Peace," p. 118.

160. Clark, "Radio in War and Peace," p. 115. This is one of many examples cited by Clark.

161. Ibid., p. 121.

to improve their operating ability."[162] The Fleet Radio Officer ordered that operators were to send and receive only official messages using official Navy forms and that personal conversations were to stop. Operators responded to this reform both by ignoring the orders and denigrating Hooper over the air. Commanding officers had shown little interest in radio before and certainly were not monitoring sending and receiving, so the operators were apparently not convinced that the new officer could truly enforce change. Hooper listened in to this initial reaction, which evidently included some Darwinian speculation on his true origins.[163]

Hooper spent his evenings learning to distinguish the sound of each ship's spark and the "fist" of each operator in order to determine which operators were violating his new regulations. He then devised a scheme he hoped would end all resistance. Admiral Badger authorized Hooper to send the following message to the commanding officer of any ship guilty of disobeying Hooper's rules: "Your attention is invited to Fleet Regulations. . . . Your radio operator is disregarding instructions and is using unofficial language. Badger, C-in-C."[164] By transmitting this message, Hooper set up the operator, who had to deliver himself to his commander for disciplinary action. The first operator who refused to acknowledge receipt of such a message was court-martialed.[165] Early control of the technology was no match for access to the lines of authority. Hooper gradually began to enforce the discipline and obedience he needed. His next task was to build from the bottom up an efficient operating network that commanding officers at the top would eventually be convinced was indispensable.

While the enlisted men had enjoyed control over radio until Hooper's reforms, they had had no need to be fast or efficient operators. Commanding officers had placed no premium on speed or accuracy, so why should the operators? They had been given no compelling reason to make this technology work well for the organization. They were not rewarded for doing so nor chastised for failing. Hooper had to structure an incentive system that would give the operators a continuing interest in good signaling.

162. Ibid., p. 120.
163. Ibid., p. 121.
164. Ibid., p. 122.
165. Howeth, *History of Communications-Electronics*, p. 195.

The most frequently given excuse for poor reception was static.[166] This excuse worked well for preserving the autonomy of the operators. It cited an outside, uncontrollable element for which the operator could not be blamed and thus protected the operator against personal rebuke. In addition, it served to perpetuate the notion among the officers that radio was not very reliable or flexible and thus was not a signaling system worth taking seriously. Such an attitude would naturally keep ultimate control of radio in the operators' hands.

Though static was indeed a problem, an experienced and attentive operator could "weed it out." Hooper began making surprise visits to radio rooms in the fleet, and if static was listed on the log, Hooper would show up the operator by putting on the headphones and reading the message himself. He then issued an order that was posted in every radio room of the fleet: "Henceforth static disturbance will not be considered as an excuse for non-reception of a message."[167] In addition, Hooper introduced a rating system whereby every operator would be labeled according to his level of proficiency. Linking their performance to organizational rewards, Hooper also initiated sending and receiving competitions among the operators with "promotions as prizes." He began drilling the operators in learning and using his battle signal code. He standardized the number of operators on each ship, thus eliminating the practice of one ship having only one operator while another had six.[168]

By 1913 Hooper had succeeded in having an officer on every ship in the fleet assigned to oversee the radio room. Removing control over radio from the enlisted men, who had no role in strategy and planning, and assigning control instead to a new managerial tier of young officers was probably Hooper's most important reorganizational ploy as Fleet Radio Officer.

Hooper also sought to physically reorganize radio aboard ship. The permanent radio installations were in cabins on the main deck, a spot most vulnerable to enemy shells.[169] This exposed position did solve a technical problem, however: the apparatus was not far from the antennas. Hooper wanted to move the radio rooms below the protective deck and below the water-

---

166. Clark, "Radio in War and Peace," p. 125.
167. Ibid., p. 126.
168. Ibid., p. 127.
169. Ibid., p. 161.

line. In doing this he would sacrifice some efficiency because of losses along the great length of under-deck wiring connecting the set to the aerial. Yet Hooper preferred to have a less efficient set that was less likely to be knocked out during battle.

Space aboard ship was jealously protected, yet Hooper managed to get different territory for radio. He ultimately moved the radio rooms to coal bunkers and similar protected places. In 1913 these first reinstallations were "very crude indeed," an improvisation which Hooper would correct later. He also added permanent installations on the bridge, so the commander would always have radio at his side for tactical signaling.[170]

By 1913 Hooper had realigned and disciplined the lower levels of the fleet hierarchy. He needed this tier of enlisted men, ensigns, and lieutenants to be efficient, coordinated, and obedient to his authority in order to impress the men at the top levels that the new technology could be an invaluable tool for commanders. Hooper had recognized that men and "machines" had to be fully integrated at the lower levels first for an organizational resource to exist and for it to be perceived as such. Only then could the top brass legitimize the system through successful and continued use of this new communications and personnel network.

Hooper got his critical opportunity in 1913. Admiral Osterhaus "would not permit his ships to be maneuvered by radio and would only execute his signals by flaghoist."[171] Admiral Badger, a younger officer, was more inclined to try radio. We have no exact dates, but sometime in 1913 Badger ordered that during one day's exercises all maneuvering would be handled by radio. For the first time, no flags were to be used at all. All of the commander-in-chief's instructions were accurately relayed and carried out, and the maneuvers were, for radio, a complete success. The next week a similar but unexpected test occurred. While en route in Chesapeake Bay, the fleet hit a sudden squall, and visibility was reduced to zero: the flags were of no use. Radio had to transmit all instructions. The storm lasted for half an hour, and when it cleared all the ships could be seen in formation, exactly as they had been ordered. Tactical signaling by radio, done previously only on an experimental basis, now

---

170. Ibid., pp. 131, 161.
171. Howeth, *History of Communications-Electronics*, p. 196.

became regular practice in the fleet.[172] In this transition period, as during others in the Navy, the old and the new would operate side by side: flags and radio would transmit tactical signals simultaneously. Hooper had achieved an important breakthrough: commanders saw firsthand how radio could avert disaster. This was a major step toward gaining full acceptance. Another incident illustrates how Hooper, with both experience and luck, succeeded in convincing commanding officers that radio could give them a decided advantage in their area of expertise and challenge, strategy. During the war games of January 1914, Badger asked Hooper if it was possible to locate the position of the "enemy" with radio. At this time Hooper had no direction finder, so he had to determine location by the strength of the enemy destroyer's radio signals. After monitoring their transmissions for several hours, Hooper predicted that the "enemy" was moving closer and would strike around 2:00 a.m. His prediction was uncannily close: shortly after two, the destroyers sent up rockets, signaling that they had "torpedoed" the fleet. Badger was extremely impressed, as were the other officers, who began to view radio as more instrumental to their own victories and advancement.[173] By 1913 the chief of the Bureau of Steam Engineering was able to report that "more careful control and the extended use of radio for signaling, especially in the Atlantic Fleet, have resulted in the development of clearer ideas as to the ultimate value of radiotelegraphy for military purposes." The annual report for 1914 praised "the personal interest of the fleet radio officer, Lieut. S. C. Hooper, who has assiduously labored to bring about this high state of efficiency."[174]

Hooper's reorganization of radio operations within the fleet took place between August 1912 and August 1914. This was also the period when the Navy had to respond to the provisions of the Radio Act of 1912. The Act became effective in December and designated naval radio stations were required to handle commercial business as of February 1913. This increased participation in the spectrum had less to do with engineering and more to do with the movement of ships at sea. Consequently, in late 1912 the department established the Naval

---

172. Ibid., pp. 196–97; Clark, "Radio in War and Peace," pp. 141A–141D.
173. Ibid.
174. *Annual Report of the Secretary of the Navy*, 1913, p. 225; 1914, p. 215.

Radio Service in the Bureau of Navigation. The Bureau of Steam Engineering concentrated on the technical development of radio,[175] while the Radio Service handled the administrative and accounting chores generated by commercial operation and worked to establish "the adoption of standard practices in the matter of operating."[176] The separation of these two very different but highly interdependent duties was important to improved management of naval radio. The traditionally most influential bureau, Navigation, now supervised radio operations. Steam Engineering was free to "modernize and improve the apparatus of coast stations so that the commercial work may be successfully handled."[177] The two bureaus central to the successful deployment of the "New Navy" were now organizationally allied by their clearly delineated and reciprocal radio duties. While this arrangement produced more efficient and technically upgraded stations, the Navy's shore station network still lacked systematic coordination.

In August 1914 Hooper was ordered to Europe to observe the use of radio during the early months of the war. In early 1915 he returned to Washington, where he served for two weeks in February with three other officers, all with radio experience, on a radio reorganization committee. The committee concentrated on the need to strengthen and better coordinate the Navy's "coastal chain." At that time each shore station communicated with ships at sea, listened for distress calls, and worked with the two stations adjacent to it along the chain.[178] Messages were relayed from one station to the next along the north-south linkage. Thus a message from Boston to Pensacola would be relayed many times.[179] The stations were set up in series: if one link broke down, no transmissions were relayed beyond that point. In addition, most of the stations were at the Navy yards, under the control of the commandant, an officer with multiple responsibilities and concerns and little reason or incentive to seek improvement in the use of radio. The commandant's influence was confined to the yard; he had no jurisdiction over the radio waves beyond it. Under this structure, coastal radio could not be coordinated.

175. *Annual Report of the Secretary of the Navy*, 1914, p. 226.
176. Ibid., pp. 213–14.
177. *Annual Report of the Secretary of the Navy*, 1912, p. 38.
178. Clark, "Radio in War and Peace," p. 158.
179. Ibid., p. 212.

Hooper found such a situation antiquated and potentially dangerous. Again he proposed a marriage of technical improvement and organization building. No one oversaw the coordination of the shore stations because no slot existed in the organization for this purpose and because no one had conceived of the "airwaves" as an appropriate jurisdictional "turf" for an officer. Through the committee, Hooper proposed a series of high-powered stations, preferably with a range of a thousand miles or more; these large areas would be called Naval Communication Districts and each would be supervised by a District Communications Officer. The existing coastal stations would have their power and apparatus upgraded and serve as a secondary signaling tier. Again Hooper devised a highly centralized network, with clearly defined and articulated lines of authority leading from the bottom to the top of the hierarchy and from the "field units" to the "central office" and with a specific scope of responsibility in the organization's adoption of radio.

When the secretary approved these recommendations in February 1915, some elements of the plan were already in place: the government took over two high-powered stations under German control, one at Tuckerton, New Jersey in 1914 and the other in Sayville, New York in 1915. In April Hooper became head of the Radio Division of the Bureau of Steam Engineering. He held this position until July 1917. Under his leadership, "a coordination plan" was instituted, "whereby each yard is kept informed of the experimental work of every other yard, by interchange of information of work at various yards and at the bureau, with consequent prevention of duplication and increase in efficiency and economy."[180]

In 1915 Congress enacted a bill creating the post of Chief of Naval Operations. This officer would serve as the much needed liaison between the Secretary of the Navy and the bureau chiefs, gathering information on material, operations, and personnel, which would help the secretary develop more informed and long-range strategy. Although the Chief of Naval Operations was the ranking active officer of the Navy, he was not empowered with direct authority over the bureaus. Nonetheless, the creation of this influential advisory position just above the bureaus and just below the secretary provided the de-

180. *Annual Report of the Secretary of the Navy*, 1915, p. 270.

partment with "professional coordination and operational direction."[181] During the same year, the Radio Service was reorganized and became the Office of Communications, which supervised telegraph, telephone, cable, and radio communications. The service was moved out of the Bureau of Navigation and became an important department office, its director reporting not to a bureau chief but to the Chief of Naval Operations.[182] These elevations in title, location, and organizational niche indicated how far up the hierarchy this technology had come. The centralization and consolidation of radio operations and their placement much closer to the center of power ensured radio's progress under naval auspices. Radio's portent at the turn of the century had been fulfilled: it had brought about a more centralized structure at sea and on shore and had become central to naval strategy.

While Hooper's influence and prestige had increased considerably, his role as translator became less critical between 1915 and 1919. Hooper had been instrumental in laying the structural groundwork for change, and his contributions continued to be important. But as the outside pressures generated by the European war to modernize and reorganize increased, the need for a translator was eclipsed. More and more officers in the Navy were persuaded by the war that whatever technical improvements the Navy could adopt, should be adopted.

Alfred Chandler cites rapid growth and increased competition as two external factors which prompt changes in strategy and structure. For the military, such rapid growth and competition occurred during wartime: for the Navy, war was competition and of the fiercest sort. Under these circumstances, the Navy behaved not like the resistant bureaucracy Elting Morison found, but like a modern industrial enterprise.

On April 6, 1917, the United States entered the Great War. Under section three of the 1912 Radio Act, President Wilson was authorized "in time of war or public peril or disaster" either to close down private radio stations or to place these stations under the control of "any department of the government, upon just compensation to the owners."[183] Consequently, on April 7 all radio stations in the United States, except those already

181. Albion, *Makers of Naval Policy,* p. 13, 218–19.

182. *Annual Report of the Secretary of the Navy,* 1915, pp. 10, 263; 1916, p. 27.

183. Frank J. Kahn, ed., *Documents of American Broadcasting* (New York: Appleton-Century-Crofts, 1972), p. 9.

under Army control, were taken over by the Navy. The progressive and gradual growth Navy radio had been undergoing turned into sudden, very rapid expansion. The Navy suddenly had five high-powered stations (two Marconi, two German, and its own at Arlington) plus the entire network of private stations, most of them American Marconi, at its disposal. In addition, it required increased numbers of efficient and sturdy transmitters and receivers as well as portable sets. Rapid production and rapid integration, with centralized coordination, were essential. The Navy could not just focus on day-to-day operations: now there was a critical need for "coordinating, appraising, and planning."[184]

During the war the Navy controlled the design, purchase, installation, and upkeep of all governmental radio except the Army's. This centralization led to standardization of apparatus, the Navy's long-sought goal, and better control over suppliers, rate of production and delivery, and "competition" from other agencies needing radio. For example, while there had always been several suppliers of most radio components, the Crocker-Wheeler Company enjoyed a near monopoly in the production of motor-generators. When demand increased during the war, Crocker-Wheeler was deluged with orders from various companies all demanding to be supplied immediately because of war contracts. Hooper intervened and set up a schedule for production and delivery based on the Navy's needs.[185] This incident prompted Hooper to view production more like a corporate executive: he wanted to ensure that he had at least two sources of supply for whatever he might need. He demanded that Crocker-Wheeler turn their blueprints over to General Electric. Failure to do so could mean government takeover of Crocker-Wheeler. In the face of this threat, Crocker-Wheeler naturally compromised and suggested the Triumph Electric Company as second supplier.[186]

Hooper also contrived to eliminate competition over the limited output of radio equipment. He learned of a plan to build a new merchant fleet, called the Emergency Fleet Corporation. Hooper saw this fleet as a potential competing buyer and feared that civilian (and in his view inexperienced) purchasers

---

184. Alfred D. Chandler, Jr., *Strategy and Structure: Chapters in the History of the American Industrial Enterprise* (Cambridge, MA: The MIT Press, 1962), p. 4.
185. Clark, "Radio in War and Peace," p. 301.
186. Ibid., p. 302.

would be willing to pay more for radio apparatus and thus both raise prices and deprive the Navy of needed equipment. He intervened even before construction began and persuaded the chief engineer of the Emergency Fleet Corporation to let the Navy supply it with radio.[187]

Because of the great demand for radio, American companies producing radio apparatus, such as General Electric, Western Electric, De Forest, and AT & T, now began to enjoy Navy patronage. One of the American companies' major competitors, the German company Telefunken, would obviously no longer be supplying the Navy. The government's patent moratorium instructed all suppliers to make use of the best components, no matter who owned the patent. The government guaranteed to protect all suppliers against infringement claims and encouraged the inventors not to be oversensitive to relatively free use of their apparatus during the national emergency.[188] Under this arrangement, with the inventors and radio companies concentrating less on marketing strategies and litigation and more on research and development, significant advances in continuous wave technology were achieved. Civilian-military cooperation produced apparatus more ideally suited to the Navy's special needs.

At the end of the war, the American radio companies, technically strong and confident, were ready to embark on new commercial ventures Navy sponsorship had made possible. During the teens, the General Electric Company had supplied the Navy with a powerful, long-distance transmitter, the high-frequency alternator. When the war ended, GE was eager to find a customer for this machine and entered into negotiations with the Marconi Company, which was trying to bargain for exclusive rights. The Marconi Company, never popular with the Navy, now symbolized British domination over international communications. When Hooper learned of the GE-Marconi negotiations, he warned Secretary Daniels that Marconi's acquisition of exclusive rights to the alternator would ensure foreign control of radio communications in America. On behalf of the Navy, and to their minds, the country's national interest, Hooper and his colleagues approached GE executives and persuaded them to suspend the negotiations. In a

---

187. Ibid.
188. Archer, *History of Radio*, p. 138.

subsequent meeting, the Navy's officials went one step further: they suggested that GE itself take advantage of the alternator and form an international communications company. If GE would buy out American Marconi and form such a corporation, the Navy would help the new company negotiate for other necessary and related patents and licenses. These talks, which led in October 1919 to the formation of an American-controlled company, the Radio Corporation of America, demonstrated how far the Navy's vision of radio's value and potential had come. Through a long and unsettling process, shaped by extraordinary external events and shrewd individuals, this military organization had indeed evolved from resistant skeptic to farsighted entrepreneur.

The most critical factor in this twenty-year process of technical adaptation was organizational realignment. Technical improvements, legislative mandates, and the European war all pushed the Navy closer to implementation. But the Navy would not have been able to exploit the invention properly or expeditiously without restructuring how and where the technology fit into the bureaucracy. The Navy was a relatively decentralized organization in 1899, and the absence of communications links between ships, and between ships and the shore, reinforced autonomous action and the jealous protection of institutional turf. Radio, with the potential to establish invisible yet powerful links between previously poorly connected or unconnected segments of the service, portended nothing less than structural revolution. Lines of authority both on ship and shore were disrupted, redefined, or strengthened as the invention was deployed, a change some welcomed and others deplored. Command and control relationships in any military service are sacrosanct, being simultaneously delicate and firmly enforced, and radio, a command and control technology, got to the heart of these relationships. Certainly the fact that radio was a new *communications* technology goes a long way towards explaining why it was not immediately embraced by many naval officers. For radio directly affected organizational interactions, redefining who had to communicate with whom and under what circumstances. It transformed how activities central to naval strategy were coordinated. Such profound transformations required institutional realignment, for the decentralized Navy of 1899 could not coordinate or facilitate such change.

Elting Morison's analysis of military resistance to technical change was a landmark study. But it is time for historians to extend his work. Individuals who shunned innovations because of their personal, psychological outlooks were also members of organizations. To understand their resistance more fully, we must understand the organizations in which they operated. Only then can we appreciate how personal interests and organizational direction reverberate, often furthering insularity and reaction. Both the acquisition and the implementation of radio were frustrated not only by tradition-minded officers, but also by a bureaucratic structure which, through its procedures and rewards, preserved stasis and retarded innovation. We must also examine when particular individuals became members of their organization, for organizations emphasize and value different types of activities and skills in different eras and often exacerbate tensions between older members, invested in one ethos, and younger members encouraged to embrace quite another.

Because organizations themselves experience change through growth, elaboration, or decline, we must assess what stage of development an organization has reached and how that stage has been shaped by larger political and socio-economic forces to determine how, or whether, technical adaptation will occur. In this particular case, it is critical to note that the reorganization within the Navy, which occurred in the early teens and the war years, was part of a much larger push towards centralization and the consolidation of management then transforming both the government and the corporate sphere. Business executives came to appreciate the importance to managerial power of technical control, a realization manifested during this time in the rise of automation, the assembly line, and Taylorism. Although the Navy lagged behind the corporations in recognizing the advantages of technical control from the top, when it did so during the war, the Navy became in fact another powerful agent in the institutional movement towards centralized control over both technology and people.

Hugh Aitken's concept of the translator is also compelling and instructive. The function and role of such an individual remained unexplored prior to Aitken's work. Yet the term "translator" suggests a middleman who passively conveys information from one sector of society to another. There is not sufficient emphasis on the importance of pushing, of marketing. This is what Hooper did: he served as a broker. In 1912 he confronted an Atlantic Fleet which had not integrated radio into its strategic operations. Through persistent and astute management and salesmanship, he redefined the technology's relationship to the organization. He sold the methods and management of commercial radio to the Navy. By developing radio as an organizational resource, with a distinct tier of operations and outlets on both ship and shore, he created and then marketed an in-house communications system. By 1915 the jealous protection of autonomy at sea had been replaced by a desire to ensure that the commander-in-chief, fleet commanders, and the department be "in communication at all times."[189] Ironically, Hooper did for the Navy what Marconi, the inventor he distrusted, did for the commerical market. They each built their networks, got them operating, and then showed the important buyers how the system would serve their special needs. Hooper did not do all this in a vacuum: technically and organizationally, the Navy was beginning to change. Yet Hooper must be credited with discerning the change in the organizational tide and both harnessing and accelerating that change to radio's—and his own—advantage. Hooper's role as broker was especially crucial in the case of radio, because this invention initially came from a civilian and a foreigner, two factors which would compromise its ready acceptance. The invention also required a certain level of technical mastery and learning a code. Hooper had to sell a new, technically based method of coordinating the fleet and the service to men who had, they firmly believed, run the Navy just fine for decades without such contraptions. Hooper thus mediated between a technical system and a social system and also between the innovations of civilians and the ongoing needs of the military. He successfully challenged institutional constraints while exploiting opportunities and resources others had neglected.

What does Hooper's success and the naval experience with radio teach us? This account offers several lessons about what organizations need to do to integrate a new communications technology successfully into their structure. Management of

---

189. *Annual Report of the Secretary of the Navy,* 1915, p. 274.

the technology must be sufficiently high up in the organization's hierarchy to ensure maximum exploitation of the technology's potential. Integration into strategic operations could not occur when radio was managed by enlisted men. The organization may need to create a position of relative independence and authority—such as Fleet Radio Officer—to oversee and evaluate the performance of the technology, both technically and as part of the organization. The person in this position must be familiar with and sympathetic to the technology and knowledgeable about his or her organization's structure and requirements. Finding such a person, who is both organizationally and technically sophisticated, may not be easy. The technology's implementation must be directly linked to the organization's long-term goals and strategies. And once the organization adopts the technology, it must participate in and support innovations. If the technology is going to disrupt or realign the organization's structure, then top management must be convinced that this realignment will better serve them and their careers. Older and higher ranking executives may be especially reluctant to adopt a technology that was first used only by members of the organization's lower tiers. Thus, integration may have to be handled by newer, younger executives who have no hard and fast associations between specific duties and power. These lessons appear to be as important today as they were seventy years ago. Researchers have recently determined that companies wishing to successfully integrate computer systems into their management and long-term planning need to follow the strategies just listed.[190]

As historians continue to study the innovation process and to explore what circumstances contribute to military enterprise and technological change, a more detailed analysis of organizational structure and mission may help our understanding of when and how such enterprise occurs. Institutions, both military and civilian, have assumed greater importance in American society, and by examining their dynamics and the personalities that guide them, we may enrich our appreciation of the complex interactions between technology and culture.

190. Study completed by Robert Mautz, Alan Merten, and Dennis Severance of the University of Michigan Business School described in *The New York Times*, November 28. 1982, Section F, p. 21

4

Ford Eagle Boats and Mass Production during World War I

David A. Hounshell

There are many instances in which inventions made in the private sector find their way into military use. A prominent example, discussed in the previous essay, is radio, although one can also point to various types of aircraft, internal combustion engines, tracked vehicles, and telephones, to mention only a few. In this chapter David Hounshell examines the Ford Motor Company's ill-fated attempt to mass-produce Eagle boats for the U.S. Navy during World War I. The effort foundered for a number of reasons, but among the most prominent were the company's unbridled confidence in the wide applicability of its assembly-line methods as well as its failure to recognize that marine engineering involved design problems and construction techniques different from auto making. Hounshell provides valuable insight into the limits of technological systems by placing these issues in historical perspective.

Writing in 1934 after "the war to end all wars," Lewis Mumford identified the military as the source of—and indeed the salvation of—mass production. Mumford saw that mass production had grown out of the standardization and quantity production of military weapons in the late eighteenth and early nineteenth centuries; mass production was at once the means of support for and an extension of the regimentation and quantity consumption of soldiers outfitted in standardized uniforms consuming standardized goods. The army was, Mumford argued, the perfect "pattern not only of [mass] production but of ideal consumption under the machine system. . . . Quantity production must rely for its success upon quantity consumption; and nothing ensures replacement like organized destruction." As he concluded, "War. . . is the health of the machine."[1]

Mumford arrived at this view not only through his own study of the history of modern technology but also by considering the arguments of the noted American anthropologist Otis T. Mason and especially the overarching and penetrating analysis of Werner Sombart in his *Krieg und Kapitalismus*. Mason concluded as early as 1895 that "war. . .stands forth pre-eminently as an incitement to the genius of invention and discovery," while less than a generation later Sombart argued that the rise of modern industrial capitalism had resulted in large measure from Western military enterprise.[2]

Since the appearance of Mumford's classic work *Technics and Civilization,* a number of scholars have published analyses that have supported, in broad outline, Mumford's conclusions about the general effect of warfare and military enterprise on technological or industrial development.[3] Most recently, William H. McNeill, in his sweeping survey *The Pursuit of Power: Technology, Armed Force, and Society since A.D. 1000,* has focused on the relationship between military enterprise and technologi-

1. Lewis Mumford, *Technics and Civilization* (New York: Harcourt, Brace & World, 1934), pp. 93–94.

2. Otis T. Mason, *The Origins of Invention* (1895; reprint ed., Cambridge, MA: The MIT Press, 1966), p. 409; Werner Sombart, *Krieg und Kapitalismus* (München: Dunker and Humblot, 1913).

3. See, for example, J. A. A. van Doorn, *The Soldier and Social Change: Comparative Studies in the History and Sociology of the Military* (Beverly Hills, CA: Sage Publications, 1975) and Albert G. Mumma, "Technology's Motivating Force," *Mechanical Engineering* 89 (April 1967): 18–21. In many respects, the present volume is but a continuation of this tradition. It should be noted that there has been one major study arguing against the Sombart/Mumford thesis: John U. Nef, *War and Human Progress* (Cambridge, MA: Harvard University Press, 1950).

cal development. Like Mumford, McNeill specifically addresses the subject of mass production and identifies warfare as the principal force in the creation and diffusion of mass production in the economies of Western Europe. World War I was especially critical, McNeill argues, for stimulating the adoption of mass production technology in Europe for turning out war materiel. Subsequent to the war, the technology of mass production was applied to domestic purposes. Concludes McNeill:

As so often before, military demand thus blazed the way for new techniques, and on a very broad front, from shell fuses and telephones to trench mortars and wrist watches. The subsequent industrial and social history of the world turned very largely on the continuing applications of mass production whose scope widened so remarkably during the emergency of World War I. Anyone looking at the equipment installed in a modern house will readily recognize how much we in the late twentieth century are indebted to industrial changes pioneered in near-panic circumstances when more and more shells, gunpowder, and machine guns suddenly became the price of survival as a sovereign state.[4]

McNeill's analysis, like Mumford's, deserves careful attention, and a case study of military enterprise and mass production in the World War I era will provide an additional perspective on their overarching views. Henry Ford is often and rightfully associated with the rise of mass production in America. But in the context of World War I, he is perhaps better remembered for his support of the much-ridiculed Peace Ship, dreamed up in late 1915 by Rosika Schwimmer to bring about a negotiated peace among the European belligerents, than for any significant involvement with the production of war materiel.[5] Yet two days after President Woodrow Wilson ended American diplomacy with Germany, the once pacifist Ford declared that he would place the productive resources of the Ford Motor Company "at the disposal of the United States government and will operate it without one cent of profit."[6]

Ford's entry into the preparedness movement was almost as newsworthy as Wilson's diplomacy. For many Americans it was far more concrete and, therefore, more understandable. The

4. William H. McNeill, *The Pursuit of Power: Technology, Armed Force, and Society since A.D. 1000* (Chicago: The University of Chicago Press, 1982), p. 331.

5. On Ford's Peace Ship venture, see Allan Nevins and Frank Ernest Hill, *Ford: Expansion and Challenge, 1915–1933* (New York: Charles Scribner's Sons, 1957), pp. 26–54.

6. *New York Herald,* February 6, 1917; *New York Times,* February 6, 1917.

productive resources of the Ford Motor Company were enormous and growing still greater. Ford had sought to manufacture as many automobiles as he could, while lowering the price of the Model T and paying unprecedented wages to his employees. His industrial empire had turned out almost 600,000 Model T automobiles in 1916 alone, using methods that had been widely publicized in both the technical literature and the popular press. These production techniques and the philosophy that lay behind them would soon become widely known as "Fordism." Not until a decade later, when an article on mass production attributed to Henry Ford appeared in the Sunday *New York Times* and the *Encyclopaedia Britannica* did "mass production" supplant "Fordism" and become the generic description of the methods and philosophy the Ford Motor Company had pioneered.[7] "Mass production is not merely quantity production," wrote Ford, "for this may be had with none of the requisites of mass production. Nor is it merely machine production, which also may exist without resemblance to mass production. Mass production is the focussing upon a manufacturing project of the principles of power, accuracy, economy, system, continuity, and speed."[8] Although vague, this definition pointed to the assembly line, first developed by Ford in 1913, as the critical ingredient of mass production. And the mere existence of a moving assembly line presupposed that behind these assembly techniques lay a manufacturing system where abundant, accurately machined parts flowed to the right place at precisely the correct time. Although the techniques of mass production would be refined by the Ford Motor Company in the years after World War I, they certainly had been mastered by the company by the time the United States entered the war. The company had achieved levels of output and productivity considered highly impressive, if not incredible.

Given Ford Motor Company's achievement and technological virtuosity, it is interesting to examine the company's efforts in World War I. Although Ford's war work ranged from the manufacture of twelve-cylinder aircraft engines, soldiers' helmets, and tanks to the production of gun caissons and armor

---

7. On Ford and the rise of mass production in America, see David A. Hounshell, *From the American System to Mass Production, 1800–1932* (Baltimore: Johns Hopkins University Press, 1984), pp. 1, 217–61.

8. Henry Ford, "Mass Production," *Encyclopaedia Britannica*, 13th ed., Suppl. Vol. 2 (1926), p. 821 and *New York Times*, September 19, 1926, Sec. 10, p. 1.

plate, one project in particular is noteworthy: the construction of submarine patrol boats, called Eagles. The history of this project provides an excellent opportunity to study how the company that pioneered mass production responded to the war emergency, how the military viewed the promising new technology of mass production, how the public imagination was captured by Ford's work, and what the limitations of the technology actually were.

Henry Ford's account of the Eagle boat project, which appeared in his widely read *My Life and Work* (1922), is rather simple. He pointed out that the Eagle had been designed by the Navy and noted that "on December 22, 1917, I offered to build the boats for the Navy." A contract was let on January 15, 1918, and Ford casually recorded the ensuing events:

On July 11th, the first completed boat was launched. We made both the hulls and the engines, and not a forging or a rolled beam entered into the construction of other than the engine. We stamped the hulls entirely out of sheet steel. They were built indoors. . . . These boats were not built by marine engineers. They were built simply by applying our production principles to a new product.[9]

Ford's tone suggests that in his view the new product could just as easily have been a vacuum sweeper as a 200-foot long patrol boat; his firm's mass production technology was universally applicable.

The postwar account of Secretary of the Navy Josephus Daniels is equally matter-of-fact. Discussing the German submarine menace during World War I, Daniels noted that the Navy had worked very hard and effectively on developing detection devices but that they fell short of protecting American shipping lanes. What was needed, Daniels argued, was the design and construction of submarine chasers: "I sent an SOS to Henry Ford and at his plant were built a number of the ideal type of submarine-hunting vessels. They were called Eagle boats and were of high speed and noiseless operation."[10]

Both accounts suggest in a most perfunctory way that the Ford Motor Company agreed to manufacture submarine chasers using the same techniques of mass production as employed

9. Henry Ford and Samuel Crowther, *My Life and Work* (Garden City, NY: Doubleday, Page, & Co., 1922), pp. 246–47.
10. Josephus Daniels, *The Wilson Era: Years of War and After, 1917–1923* (Chapel Hill: The University of North Carolina Press, 1946), p. 136.

for the Model T and then readily carried out this task, but such was not the case. And although contemporary technical and nontechnical journalists hailed Ford's submarine chaser venture as "one of the most extraordinary manufacturing activities of the war,"[11] the Eagle boat venture cannot be considered successful either by the standards of the war era or of the present. Rather, Ford's attempt to mass-produce ships using the latest and best production techniques points up some of the inherent limitations of mass production and, more fundamentally, the importance of deep-seated, traditional "craft" knowledge within the institution of the Navy. This latter point is particularly noteworthy because the Navy of World War I was the "new" Navy that had been carefully built over a thirty-year period by a group of scientifically and technologically trained naval officers. In pushing the Navy into the twentieth century, these men had brought about a virtual technological revolution in ship design, construction, and armament. The "craft" knowledge they possessed in 1917 was that obtained through years of experimenting with designs, testing, and creating performance and production standards. Ford and his engineers believed that, with their mass production technology, they could dispense entirely with such experience. The Eagle boat venture is important if for no other reason than it identifies both a certain "vanity" in Henry Ford and his company's leading production engineers in 1917 and 1918 and the failure of the Navy—or more particularly, the Secretary of the Navy—to appreciate the role tradition played in equipping its fleet.

The Eagle boat project had roots in a meeting in February 1917 between Ford and Secretary Daniels in which Ford offered to turn over the full productive resources of the Ford Motor Company to the Navy. Following this meeting, Ford informed an eager press, "I can build 1000 small submarines and 3000 motors a day."[12] Although Ford's claims may not

11. Fred E. Rogers, "Ford Methods in Ship Manufacture—VII," *Industrial Management* 58 (1919): 11.

12. *New York Times*, February 6, 1917. See also Ford's ideas on a small "U-Flivver" in *New York Times*, February 10, 1917. Why Ford devoted his company's resources principally to the Navy rather than the Army is an interesting question. I suspect that Thomas A. Edison's relationship with Josephus Daniels was the determining factor. Edison, whom Ford literally worshipped, served as head of Daniels's Naval Consulting Board, a body of well-known, independent inventors and well-established professional engineers and scientists intended to screen military inventions coming to the Navy from Yankee-type inventors throughout the United States. Scholarship on Daniels and the Naval Consulting Board suggests that Daniels had a nineteenth-century mentality

have inspired or excited U.S. naval officers, who declined to comment on them,[13] both the public and Daniels were captivated by Ford's proposal. President Wilson was, too. During the summer of 1917, Wilson called Ford to Washington to recruit the Flivver King for membership on the United States Shipping Board, the body established to direct America's wartime merchant shipping efforts. Ford's chief administrative assistant, E. G. Liebold, advised him against accepting the appointment because, he contended, the Shipping Board had drawn negative reaction from the public, particularly about the performance of the board's Emergency Fleet Corporation. But Ford and Liebold met with the chairperson of the Shipping Board, Edward N. Hurley, and discussed how Ford could contribute to the war effort. According to Liebold, Hurley showed Ford data on the tonnage of vessels sunk by German submarines and said that unless something changed, the Germans would shortly be sinking ships faster than they could be built. Liebold suggested that rather than merely building more ships to be sunk, the United States should devise some type of submarine chaser to root out and eliminate the menacing submarines. The suggestion stuck, recalled Liebold, and soon the Navy and the Ford Motor Company were on their way to the Eagle venture.[14]

In fact, the Navy had already taken bold steps toward eliminating preying U-boats. Without fanfare, the Navy's Bureau of Construction and Repair, in cooperation with a number of yacht builders, designed a 110-foot, 75-ton submarine chaser. To minimize conflict with other (both merchant and naval) shipbuilding efforts, the Navy settled upon a wooden boat powered by a gasoline engine, which could be built by a large number of maritime firms on the Atlantic,

---

about how technological research and development were done. Specifically, he believed that all American technology had been the product of independent inventors—of the Thomas Edisons of America—and failed utterly to see that modern technology had become the province of formal, bureaucratic research and development organizations. Daniels's views led to the complete failure of the Naval Consulting Board, and they probably were the reason why Daniels pushed hard—perhaps over the objections of seasoned naval officers—for Ford's manufacture of the Eagle boats. On Daniels and the Naval Consulting Board, see Thomas P. Hughes, *Elmer Sperry: Inventor and Engineer* (Baltimore: Johns Hopkins University Press, 1971), pp. 243–74 and Daniel J. Kevles, *The Physicists* (New York: Knopf, 1978), pp. 102–38.

13. *New York Times*, February 10, 1917.

14. E. G. Liebold, Reminiscences, Ford Archives, Edison Institute, Dearborn, Michigan.

Pacific, and Gulf coasts as well as on the Great Lakes. Some 440 such submarine chasers were rapidly built. Able to cruise at almost seventeen knots and equipped with a depth charge launcher, two machine guns, and a 3-inch, 23-caliber boat gun, these submarine chasers nevertheless appeared to be something more than a military vessel. Their traditional, clean lines and their mahogany construction, complete with brass fittings, spoke as much to the yachtsman as to the naval commander. Although these boats achieved several of the Navy's objectives, the submarine menace continued into the late summer and fall of 1917, and the Navy determined that it needed to design a steel ship midway between the wooden chasers and the modern destroyers that distinguished the "new" Navy. These vessels would be designed to have the same speed as the wooden chasers but would have a far greater range and would be more heavily armed.[15]

Henry Ford's decision in November 1917 to accept Wilson's plea that he join the United States Shipping Board dramatically influenced the Navy's planning of these new submarine chasers. Officially, Ford served as a special assistant to the vice president of the Shipping Board's Emergency Fleet Corporation. The *New York Times* reported that Ford's "particular task will be to introduce into shipbuilding the multiple production methods he has used with such marked success in producing automobiles." The same article made clear that the Emergency Fleet Corporation was undergoing a "complete reorganization," with businessmen displacing naval officers in the major departments of the corporation.[16] In a prepared statement delivered the day after accepting his new duties, Ford said that more thought had to be given to machinery in the war effort. He argued that the war would be won by the nation "that knows the secret of quantity production through standardization on one model," and he expressed his disappointment that the U.S. government had not yet standardized its instruments of war.[17]

By December of 1917 Ford had offered the services of his company to the Shipping Board for the purpose of ship construction. Ford had two of his top engineers, William B. Mayo

15. Benedict Crowell and Robert Forrest Wilson, *The Armies of Industry* (New Haven: Yale University Press, 1921), 2: 467–69. See also J. A. Furer, "The 110-Foot Submarine Chasers and Eagle Boats," *U.S. Naval Institute Proceedings* 45 (1919): 743–52.

16. *New York Times*, November 17, 1917.

17. *New York Times*, November 18, 1917.

and William S. Knudsen, study the shipbuilding situation and visit leading shipyards in Newport News, Virginia; Camden, New Jersey; and Philadelphia. In their report to Ford, Mayo and Knudsen made several important recommendations that would soon be hallmarks of the Eagle boat venture. Above all, they recommended standardization of ship design, including their power plants. Mayo and Knudsen also argued for much of the "rough angle bending and smithing work" to be done in factories away from shipyards, for the invention of a new type of rivet heating furnace, for the adoption of a "continuous travelling system" for conveying steel parts to their point of use, for the adoption of electric welding, and for the construction of naval destroyers on "new temporary ways leaving the present ways to be utilized for merchant shipping."[18]

Ford's promotion of these recommendations in Washington intersected with the work of the Navy's Bureau of Construction and Repair on their projected steel submarine chaser. On December 24, 1917, the automobile manufacturer formally offered to build for the Navy a new ship called a "submarine detector-destroyer" with the same quantity production techniques that had been so successful for the Model T. Ford claimed that his company could build six such boats with mass production methods for every one built by conventional methods and at savings of more than 80 percent.[19] At once, Secretary Daniels requested Ford to dispatch a team of his best production engineers to Washington to confer with the Navy's Bureau of Construction and Repair.[20] Through a series of conferences, the Ford experts and the Navy's chief designers arrived at basic design principles for the new submarine chaser. By mid-January the chief of the Bureau of Construction and Repair, Admiral David W. Taylor, and his assistant, Captain Robert Stocker, had carried the design and specifications sufficiently far to allow Ford Motor Company to make a formal offer to manufacture the boats. By this time they had become known as Eagle boats thanks to a *Washington Post* editorial

18. W. B. Mayo and W. S. Knudsen, "Report of Engineers to Mr. Henry Ford Relative to Shipbuilding," December 17, 1917, Accession 572, Box 26, Ford Archives.

19. As quoted above, Henry Ford claimed that his offer was made on December 22, 1917. But Crowell and Wilson, *Armies of Industry*, 2, p. 469, noted the date as December 24. I have not found evidence for either date in the Ford Archives, but I tend to trust Crowell and Wilson rather than Ford and his ghost writer Crowther. Ford's quantity and price claims appear in Crowell and Wilson, *Armies of Industry*, 2, p. 471.

20. Ibid., pp. 469–70.

Ford's assistant Liebold had written for Ira Bennett to convince Congress of the necessity of building these vessels.[21]

Ford's offer reiterated the bureau's specifications and estimated a cost of \$275,000 per boat. More important, it laid out a production schedule for the boats to be built at a new factory on Ford's property along the River Rouge near Detroit: one Eagle by mid-July, ten by mid-August, twenty by mid-September, and twenty-five each month thereafter (or roughly one Eagle per working day).[22] This schedule became part of the formal contract between Ford and the Navy executed on March 1, 1918.[23] A few days before submitting the formal offer, Henry Ford had actually told his engineer W. B. Mayo that his company could build one Eagle within "four months then one per day for two months, then two per day for three months."[24] But the schedule in the final proposal was more modest. Three days after Ford's formal proposal went out, Daniels telegraphed Ford, "PROCEED WITH ONE HUNDRED SUBMARINE PATROL VESSELS . . . DETAILS OF CONTRACT TO BE ARRANGED AS SOON AS POSSIBLE."[25] The Eagle project had taken flight.

When the Bureau of Construction and Repair had finished its rough design of the Eagle, it projected a ship that rivaled destroyers built a decade earlier. In fact, at 200 feet in length and 500 tons displacement, the Eagle boat exceeded the size of any U.S. destroyer built prior to 1903. Its 2000-horsepower, steam-driven turbine also testified to its size.[26] But the Eagle represented a radical departure for the Navy. Assisted by

21. E. G. Liebold, Reminiscences, Ford Archives. See also Crowell and Wilson, *Armies of Industry*, 2, p. 470. The names of Taylor and Stocker appear in "Submarine Chasers Made by Quantity-Production Methods," *American Machinist* 49 (1918): 842 and Fred E. Rogers, "Ford Methods in Ship Manufacture—I," *Industrial Management* 57 (1919): 3. However, the *New York Times*'s account of the Eagle's design downplays the active role of Taylor and Stocker (February 1, 1918). The *Washington Post* editorial, "A Challenge to America," appeared December 23, 1917.

22. Ford Motor Company to Josephus Daniels, January 14, 1918, Accession 572, Box 26, Ford Archives.

23. A copy of the contract is in Accession 38, Box 42, Ford Archives. Details appear in Nevins's and Hill's notes in Accession 572, Box 26, Ford Archives.

24. Telegram, W. B. Mayo to E. G. Liebold, in Accession 62, Box 58, Ford Archives, quoted in Nevins's and Hill's notes, Accession 572, Box 26, Ford Archives.

25. Telegram, Josephus Daniels to Ford Motor Company, January 17, 1918, Accession 572, Box 26, Ford Archives.

26. The comparative size of the Eagle boat was pointed out in J. A. Furer, "The 110-Foot Submarine Chasers and Eagle Boats," p. 766 and Crowell and Wilson, *Armies of Industry*, 2, p. 470. The Eagle was designed to cruise at 18.3 knots and have a radius of 3500 miles.

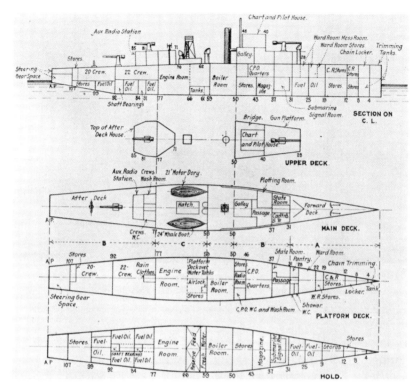

Figure 1
Section and plan views of the Eagle Submarine Chaser. The top view is
the section along the center line. The view below shows the plans of the
hold, the platform deck, the main deck, and the upper deck. The numbers
that appear are frame numbers rather than dimensions. *Source: Industrial
Management* 57 (1919): 293.

Ford engineers, the Navy designed the Eagle for quantity production by an automobile manufacturer—for mass production—rather than for construction by a shipbuilding firm.
Consequently, in the place of smooth curves and rolled members were straight lines, flanged plates, and structural angles. A
New York Times reporter was not wrong when he compared the
Eagle boat with the already famous Ford Flivver (figures 1, 2).[27]

Because the Eagle boat was to be mass-produced, the Ford
Motor Company immediately moved in two important ways. It
began construction of a full-scale model or pattern of the Eagle
in the main craneway at Ford's Highland Park factory (figure 3),
where the assembly line had been born five years earlier, and it

27. *New York Times*, July 7, 1918. See also *New York Times*, February 1, 1918.

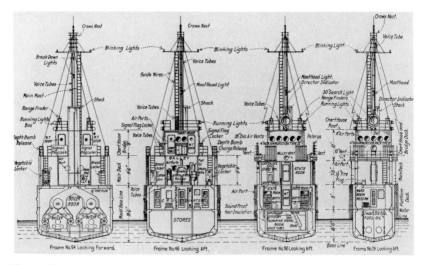

Figure 2
Cross sections of the Eagle at frames 54, 46, 36, and 26. *Source: Industrial Management* 57 (1919): 291.

initiated design and construction of the B-Building or boat assembly plant and other facilities at Ford's River Rouge site. Building the pattern accomplished several important things. First, construction of a model gave Ford engineers and naval officers assigned to the task a chance to revise their initial, rough design. Second, as with musket manufacture at Springfield Armory in the antebellum period, the model served as the basis for the design and construction of special jigs, fixtures, templates, and other patterns for parts of the Eagle boat.[28] The model also enabled the designers and production men to decide the position of all rivet holes. Finally, erecting the model provided Ford's production experts with the opportunity to write operations sheets—minutely detailed specifications for the order or sequence of steps in the manufacturing process.[29] Operations sheets had been brought to a high degree of perfection in the production of Model T's and were essential to the

28. On the use of a model-based jig, fixture, and gauging system in the antebellum federal armories, see Hounshell, *From the American System to Mass Production*, pp. 6, 28–29, 34, 41–42 and Merritt Roe Smith, *Harpers Ferry Armory and the New Technology: The Challenge of Change* (Ithaca, NY: Cornell University Press, 1977), pp. 92–93, 109–10, 225–27.

29. Ford's use of the model is discussed in Furer, "The 110-Foot Submarine Chasers and Eagle Boats," p. 766; Rogers, "Ford Methods in Ship Manufacture—I," p. 3; "Submarine Chasers Made by Quantity-Production Methods," p. 843; and William C. Klann, Reminiscences, Ford Archives.

Figure 3
Full-scale Eagle model being constructed in main craneway of Ford's Highland Park Factory. *Source:* Ford Archives, Edison Institute, Dearborn, Michigan.

production methods of the Ford organization.[30] By the time
the model had been finished and fully utilized, the plant at the
River Rouge site had been partially erected, and hulls were
beginning to be fabricated there.

The construction of the River Rouge facility in about four
months was a remarkable achievement. Henry Ford had purchased the 200-acre site in 1915 with an ill-formed and (in
retrospect) ill-conceived idea of building a fully integrated
manufacturing establishment there, of bringing in raw materials and shipping out finished automobiles and tractors.[31]
Though the war interrupted Ford's plans, it gave him an opportunity to build and to build big. For Eagle boat manufacture, the company hastily designed and constructed (with
public funds) the fabricating shop and tool room, or A-
Building (150 by 600 feet) where the steel sheets were formed
and where many other Eagle parts were fabricated; the main
assembly, or B-Building (350 by 1700 feet), which accommodated three parallel assembly lines, two large outfitting
buildings, and an elaborate transfer and launching table that
moved the Eagle boats out of the assembly building and into
the water (figure 4).[32] (After the war, Ford would quickly convert the fabricating shop and the assembly building into a tractor and automobile factory.) While the buildings awed visitors
with their size, the transfer and launching tables captivated
observers for what they did and how they operated. Designed
by Ford engineer W. B. Mayo, the transfer table, 202 feet long
and supported on eleven railroad tracks, carried Eagle boats
out of the assembly building and then moved them perpen-
dicularly 300 to 600 feet away from this building into position
with the launching table. Once moved onto the launching table,
Eagle boats were then lowered gently into the water by means
of hydraulic jacks (figures 4–7).[33]

Although the size of the Rouge structures and the speed with

30. On the importance of operations sheets at Ford, see Hounshell, *From the American System to Mass Production*, pp. 224, 270–71. For an example of operations sheets for the Eagle boat, see Fred E. Rogers, "Ford Methods in Ship Manufacture—III," *Industrial Management* 67 (1919): 190–97.

31. On Ford's purchase and early plans for the Rouge site, see Nevins and Hill, *Ford: Expansion and Challenge*, pp. 200–216.

32. Rogers, "Ford Methods in Ship Manufacture—I," pp. 2–6. See also "Building the 'Eagle's' Nest," *The Ford Man* 2 (May 17, 1918): 1.

33. Ibid. See also "Submarine Chasers Made by Quantity-Production Methods," pp. 841–43 and Crowell and Wilson, *Armies of Industry*, 2, pp. 471–72.

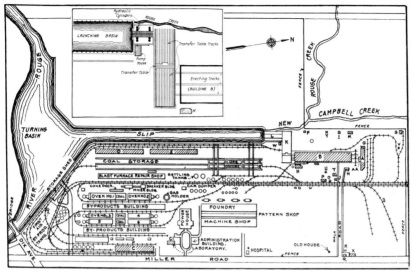

Figure 4
Plan of River Rouge factory. Note the structures related directly to Eagle boat production: A. Fabricating building, B. Assembly building, D. Outfitting shop, E. Outfitting shed, K. Transfer table and tracks (see inset for greater detail), L. Launching table. The length of the B-building (1700 feet) provides an idea of the scale of the Rouge's operations. *Source: Industrial Management* 57 (1919): 6, 456.

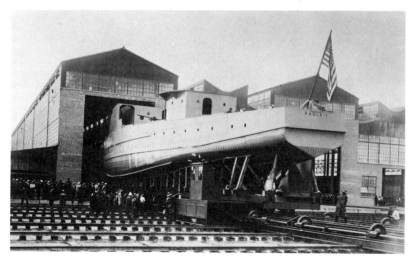

Figure 5
Eagle 1 moving out of assembly building onto transfer table. *Source:* Ford Archives, Edison Institute, Dearborn, Michigan.

Figure 6
Eagle moving off transfer table onto launching platform. *Source:* Ford Archives, Edison Institute, Dearborn, Michigan.

Figure 7
Eagle on launching platform. Note the large hydraulic jacks to lower the Eagle gently into the launching basin. *Source:* Ford Archives, Edison Institute, Dearborn, Michigan.

which they were erected were impressive, the haste in their construction as well as excessive confidence on the part of Ford's production engineers immediately led to problems. No doubt when Henry Ford first broached the idea of mass-producing ships for the Navy, he, his engineers, and everyone else envisioned the Eagle boats being swiftly assembled from parts as the keel (not unlike the frame of the Model T) moved smoothly and deliberately down an assembly line. That image was soon shattered. Although assembly lines would be built, the boats would not be moved along them continuously. Rather, they would be stopped and worked on at seven distinct assembly stations (figure 8). After plant construction had begun, Ford's engineers found that seven stations would be inadequate, and they were forced to add another 200 feet onto the B-Building to house an initial preassembly operation.[34] Moreover, the engineers rapidly concluded that a ship was such a complex and heavy piece of machinery that it would be impossible to complete it while on the assembly line. Unlike the Model T when it came off the assembly line, the Eagle boat would be fitted out after being launched and with techniques that bore little relationship to Ford's much-hailed "work in motion" or "flow production" methods (figure 9).[35] Moreover, major unanticipated problems arose once production got underway.

In a show of technological virtuosity, the Ford Motor Company moved quickly to lay the keel of the first Eagle. Sixteen weeks after Josephus Daniels gave Henry Ford the go-ahead to build Eagle boats, the keel-laying event took place. As Liebold wired Daniels, "Keel for the first boat has been laid on blocks and side frames are now ready to go up."[36] Two months later, on July 11, 1918, with carefully orchestrated press coverage, the complete hull was launched via the transfer table. To the awestruck American public, Ford's announcement of this event implied that Eagle 1 was ready to chase menacing submarines. Moreover, Ford promised that twelve more boats would be launched within two or three days and that soon there would be

34. "Submarine Chasers Made by Quantity-Production Methods," pp. 841–44.

35. Fred E. Rogers, "Ford Methods in Ship Manufacture—VII," *Industrial Management* 58 (1919): 8–9.

36. Telegram, E. G. Liebold to Josephus Daniels, May 7, 1918, Accession 572, Box 26, Ford Archives.

ORDER OF OPERATIONS

Progressive erection of the fabricated parts and unit assemblies is carried through in the following order:

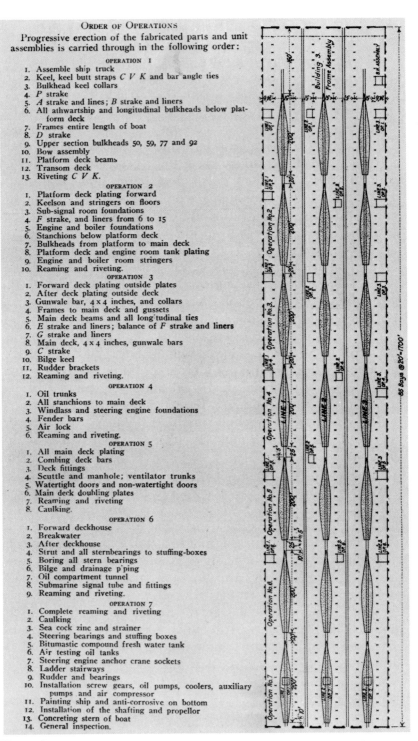

OPERATION 1
1. Assemble ship truck
2. Keel, keel butt straps $C V K$ and bar angle ties
3. Bulkhead keel collars
4. P strake
5. A strake and lines; B strake and liners
6. All athwartship and longitudinal bulkheads below platform deck
7. Frames entire length of boat
8. D strake
9. Upper section bulkheads 50, 59, 77 and 92
10. Bow assembly
11. Platform deck beams
12. Transom deck
13. Riveting $C V K$.

OPERATION 2
1. Platform deck plating forward
2. Keelson and stringers on floors
3. Sub-signal room foundations
4. F strake, and liners from 6 to 15
5. Engine and boiler foundations
6. Stanchions below platform deck
7. Bulkheads from platform to main deck
8. Platform deck and engine room tank plating
9. Engine and boiler room stringers
10. Reaming and riveting.

OPERATION 3
1. Forward deck plating outside plates
2. After deck plating outside deck
3. Gunwale bar, 4 x 4 inches, and collars
4. Frames to main deck and gussets
5. Main deck beams and all longitudinal ties
6. E strake and liners; balance of F strake and liners
7. G strake and liners
8. Main deck, 4 x 4 inches, gunwale bars
9. C strake
10. Bilge keel
11. Rudder brackets
12. Reaming and riveting.

OPERATION 4
1. Oil trunks
2. All stanchions to main deck
3. Windlass and steering engine foundations
4. Fender bars
5. Air lock
6. Reaming and riveting.

OPERATION 5
1. All main deck plating
2. Combing deck bars
3. Deck fittings
4. Scuttle and manhole; ventilator trunks
5. Watertight doors and non-watertight doors
6. Main deck doubling plates
7. Reaming and riveting
8. Caulking.

OPERATION 6
1. Forward deckhouse
2. Breakwater
3. After deckhouse
4. Strut and all sternbearings to stuffing-boxes
5. Boring all stern bearings
6. Bilge and drainage p'ping
7. Oil compartment tunnel
8. Submarine signal tube and fittings
9. Reaming and riveting.

OPERATION 7
1. Complete reaming and riveting
2. Caulking
3. Sea cock zinc and strainer
4. Steering bearings and stuffing boxes
5. Bitumastic compound fresh water tank
6. Air testing oil tanks
7. Steering engine anchor crane sockets
8. Ladder stairways
9. Rudder and bearings
10. Installation screw gears, oil pumps, coolers, auxiliary pumps and air compressor
11. Painting ship and anti-corrosive on bottom
12. Installation of the shafting and propeller
13. Concreting stern of boat
14. General inspection.

Figure 8

Operations sheet for Eagle boat manufacture. *Source: Industrial Management* 57 (1919): 192.

Figure 9
Eagle boats being outfitted in launching basin. *Source:* Ford Archives,
Edison Institute, Dearborn, Michigan.

one Eagle per day.[37] But those notions quickly vanished. Not
until October 30, 1918, was Eagle 1 commissioned. The last of
the next dozen promised was not finished until mid-April 1919
(figure 10).[38]

Problems plagued these first Eagles, many of them growing
out of inferior production techniques. By far the worst problem centered around riveting. Although Ford proudly claimed
that the Eagle boats had been built without the help of seasoned
marine engineers, such was not entirely the case. The Navy
hired Charles C. West of the Manitowoc Shipbuilding Company as its superintending constructor. West tried to help the
Ford engineers work out hull manufacturing methods but met
with pronounced resistance.[39] His powerful position allowed
him to force Ford to hire additional marine engineers. But he
noted in September 1918 that these men were being ignored.[40]

37. *New York Times,* July 12, 1918. Eagle 1 was launched on July 11, as indicated in a
telegram from E. G. Liebold to Josephus Daniels, July 11, 1918, Accession 572, Box 26,
Ford Archives, and "Launching of Eagle No. 1," *The Ford Man* 2 (July 17, 1918): 1.

38. "Eagle Boat Record," Accession 572, Box 26, Ford Archives.

39. The article "Submarine Chasers Made by Quantity-Production Methods" (see note
21), which was first printed in *Engineering News-Record,* called West "one of the most
energetic shipbuilders on the Great Lakes" (p. 842), while Furer regarded him as "a
skilful and experienced shipbuilder, [who] made it possible for a company with no
previous shipbuilding experience to build successfully seven ships of 500 tons displacement in nine months." "The 110-Foot Submarine Chasers and Eagle Boats," p. 766.

40. Charles C. West to Ford Motor Company, September 9, 1918, Accession 572, Box
26, Ford Archives.

Figure 10
Keel for Eagle boat. *Source:* Ford Archives, Edison Institute, Dearborn, Michigan.

Ford engineers and workers found riveting to be more difficult than they had anticipated. In particular, West and his inspectors discovered that because of poor riveting, the Eagle boats were not water-tight or oil-tight (the Eagle was designed with ten separate fuel oil compartments). West reported that the compartments of the first four Eagle boats were leaky and that workmen had "experienc[ed] considerable difficulty" in getting them sealed. In an official complaint, he argued that these problems had arisen because the advice of skilled ship-builders had not been followed. For example, West had observed workers trying to bolt steel plates together prior to riveting. Using short-handled wrenches while standing on ladders, they were unable to bring the plates together tightly. Metal shavings lodged between the plates and made it virtually impossible for the rivets to pull the plates together for a tight seal.[41] One of the riveters complained independently to Henry Ford about the ladders and urged him to use scaffolding in-

41. Ibid.

stead. This was done eventually, but ladders were not entirely abandoned (figures 11, 12).[42] West reported other problems as well, most notably the poor quality of electric arc welding, which was a technique not used in Model T production. The superintending constructor specifically requested Ford to "do as little electric welding on oil and water tight bulkheads as possible as your welders are so inexperienced that the welds are both defective and porous."[43]

West continued to issue formal complaints about the poor quality of the work on Eagle boats. Moreover, he and other naval officials also began to worry about *quantity*. As late as May 18, 1918, Josephus Daniels continued to believe that Ford would produce all 100 Eagles by January 1, 1919.[44] Ford had, after all, promised to do so. Soon after the initial agreement between Ford and Daniels, Ford Motor Company's engineers had concluded that a plant almost identical to the River Rouge factory would be necessary to achieve the 100 Eagle boats as projected. Quickly approved by the Navy, this plant was to be built near Newark, New Jersey, and would cost another $2.5 million.[45] After giving the go-ahead for this plant, Daniels must have been confident of Ford's meeting the production schedule. But by late August 1918, when only two Eagle boats had been launched and none completely outfitted, the Navy officers in the Bureau of Construction and Repair asked Ford for an accurate revision of its Eagle production schedule. The company projected that only 26 boats would be finished by the end of December rather than the promised 112. (The Italian government had ordered 12 Eagles in addition to the 100 ordered by the Navy.) The remainder would be finished by the end of

42. Riveter to Dear Sir [Henry Ford], September 23, 1918, Accession 572, box 26, Ford Archives. Lt. Julius A. Furer, who observed the Eagle venture, wrote that Ford's policy of not consulting sufficiently with shipbuilding experts "is considered by some to have been a mistake, as the work could have been very materially accelerated had [Ford] employed a sufficient number of men skilled in the various shipbuilding trades to act as instructors for his men." "The 110-Foot Submarine Chasers and Eagle Boats," p. 765.

43. West to Ford, September 9, 1918.

44. Josephus Daniels to Ford Motor Company, May 18, 1918, Accession 62, Box 8, as noted in Nevins's and Hill's notes, Accession 572, Box 26, Ford Archives.

45. The first mention of the New Jersey plant is in a letter from W. B. Mayo to Josephus Daniels, June 27, 1918, which states that Ford Motor Company will proceed with the construction of a plant at Newark Meadows which "will be very nearly a duplicate of that at our Detroit plant." Accession 572, Box 26, Ford Archives. See also Ford Motor Company to Graham Egerton, Solicitor, Navy Department, October 14, 1918, Accession 572, Box 26, Ford Archives.

Figure 11
Working on the Eagle from ladders. *Source: Industrial Management* 57
(1919): 290.

Figure 12
Working on the Eagle from scaffolding. *Source:* Ford Archives, Edison
Institute, Dearborn, Michigan.

April 1919.[46] As with Ford's initial schedule, this proved to be unrealistic.

At the end of 1918—more than a month after the armistice—Ford had delivered only seven boats,[47] and all of them had problems. West reported that all seven Eagles had "oil and machinery troubles." Apparently, the oil problem stemmed from the leakage of oil out of the holding tanks, a problem all too familiar to West. In a letter to Ford, he quoted a communication from Navy headquarters saying that the Bureau of Construction and Repair had gotten what it had expected: "Considering all the circumstances surrounding the initiation of the Eagle program, the Bureau was prepared to accept a quality of workmanship below standard in some particulars and on the earlier vessels delivered."[48] (The riveter who had written to Henry Ford complaining about having to use ladders put the matter more bluntly: the boats had been only "half-built" and then only at a rate of one a month rather than one a day.)[49] The Bureau of Construction and Repair informed Ford that work on the remaining Eagles would have to be "brought up to standard." Especially important was rectifying the problem with leaky fuel oil compartments.[50]

Soon *Detroit Saturday Night* came out with an article that pronounced Ford's Eagle boat venture a failure—one that the Navy would not openly admit. The article noted that the "boasted one-a-day program dwindled and shrank to about one a month" and labeled the reasons: "Ford workmen found shipbuilding a more difficult art than 'flivver' building. Most serious of all was their inability to rivet." Although the Navy and Ford had agreed to reduce the final output of Eagles to sixty, the

46. David W. Taylor to Ford Motor Company, n.d. [received August 26, 1918], Accession 572, Box 26, Ford Archives. Ford's reply came in a letter to Admiral D. W. Taylor, September 10, 1918. The first draft of this letter is dated September 4, 1918. Both are in Accession 572, Box 26, Ford Archives. As late as September 27, 1918, Ford Motor Company was claiming that it would turn out one Eagle per day and it argued that "this is an accomplishment unequalled in the industrial annals of the world, and is a fair illustration of Ford efficiency when carried out by Ford initiative, unobstructed or impeded by outside influences, suggestions and orders." "Patrol Boats," in "Summary of War Activities of the Ford Motor Company," September 27, 1918, Accession 572, Box 26, Ford Archives.

47. "Eagle Boat Record."

48. Charles C. West to Ford Motor Company, December 30, 1918, Accession 572, Box 26, Ford Archives.

49. Riveter to Dear Sir, September 23, 1918.

50. West to Ford Motor Company, December 30, 1918.

Navy, according to the article, showed "no sign of abandoning the reduced program."[51] The Navy's refusal to terminate the Eagle venture after the armistice in spite of increasing Congressional criticism stemmed from Josephus Daniels's enthusiasm for Ford's work—a factor that probably had been decisive in Ford being given the contract in the first place. Daniels argued that the Ford plant was ideal for manufacturing the Eagle and "had the war continued, the Navy would have obtained 500 more of them."[52]

It is difficult to imagine the Rouge factory and its twin plant in New Jersey turning out that many. On May 1, 1919, when Ford had delivered only seventeen Eagles, naval officials wrote the company that it was highly unlikely that Ford would complete the sixty Eagles by November 15, 1919, the deadline Daniels had established when he reduced the Navy's order from one hundred to sixty Eagles. The letter called for "energetic action" on Ford's part lest the company fall short by some twenty boats.[53] Less than a month later, Whitford Drake, who had succeeded West as superintending constructor, wrote the Bureau of Construction and Repair that "unless a very radical improvement is made immediately the program will not be completed this year, and it is very doubtful if it can be completed in any event."[54] Ford Motor Company did indeed marshal the necessary energy and manpower to meet the deadline. The sixtieth boat was delivered on October 15, 1919, slightly less than a year after Henry Ford had initially promised it. As a publicity ploy, the company put all of its riveters to work on the hull of the last keel laid (which turned out to be Eagle 59) in April 1919. Only ten days were required from the laying of the

---

51. "Why Congress is Investigating the Building of Ford Eagles," *Detroit Saturday Night* 13 (January 11, 1919): 3, as noted in Nevins's and Hill's notes, Accession 572, Box 26, Ford Archives. It is unclear exactly when Daniels cut back the Eagle program to sixty boats. A report signed by Whitford Drake, May 29, 1919, alludes to a telegram of November 1918 from the Navy Department to Ford Motor Company saying that "the sixty boats" were to be completed by November 15, 1919. Accession 38, Box 42, Ford Archives.

52. *New York Herald,* January 5, 1919. The congressional controversy about the Eagle boat program in light of the declared peace began immediately after the armistice. See *New York Times,* November 20, 1918, for the first of the salvos.

53. Carlos Bean [Inspector of Machinery] and Whitford Drake [Superintending Constructor] to Edsel Ford, May 1, 1919, Accession 572, Box 26, Ford Archives. Drake had succeeded West sometime earlier in 1919.

54. Inspector of Machinery and Superintending Constructor to Bureau of Steam Engineering and Bureau of Construction and Repair, May 29, 1919, Accession 38, Box 42, Ford Archives. Drake signed this document.

keel to the launching of the hull. But it took until the end of August to finish the boat. By contrast, the keel of Eagle 60 was laid earlier, on March 30, 1919; the boat was launched August 13 and, as noted, commissioned October 15, 1919.[55]

Without question, the outfitting operations for Eagle boats— installing the turbine, boilers, piping, wiring, armament, and all other equipment—required more time on average than the construction of the hulls. Though the Eagle had been standard-ized, its very nature and its cramped interior spaces prevented a massed assault on the finishing operations. The detailed oper-ations sheets used for hull fabrication and assembly were not applicable here. Rather, specifications became general, such as "set boilers, turbine, piping flanges, [and] voice tube flanges."[56] These operations, involving a good deal of fitting, mediated against the rapid manufacture of Eagle boats. Certainly the boats were not "mass produced," if one adopts Henry Ford's own criterion for mass production: "In mass production," he later wrote, "there are no fitters."[57] In terms of quantity manu-factured, time required, and means of production, Ford's Eagle boat venture did not live up to the promise that Henry Ford had made during the height of the submarine menace.

How did the Eagle boats perform after they were commis-sioned and what became of them? As indicated, the Navy re-ported that the initial seven Eagles had had oil and water leakage problems. But at the same time it noted that "Eagles 1, 2, and 3 encountered considerable heavy weather [on their maiden voyages] and their behavior in a seaway has been pro-nounced excellent."[58] After the six-month guarantee on Eagle 1 expired, the boat's commanding officer reported that "there were no defects in the hull, hull fitting, and equipment of this vessel, which were due to material or workmanship before the commissioning of this vessel." He pointed out that the vessel had been used heavily, particularly on maneuvers in the icy waters off the Russian coast: "It speaks very well for the con-struction of this vessel that in bucking heavy ice, no damage was

55. "Eagle Boat Record." For an example of the publicity surrounding the big push on Eagle 59's hull, see Fred E. Rogers, "Ford Methods in Ship Manufacture—VI: Launch-ing and the Record of Eagle No. 59 Erected in Ten Days," *Industrial Management* 57 (1919): 456–64.

56. Rogers, "Ford Methods in Ship Manufacture—VII," p. 8.

57. Henry Ford, "Mass Production," p. 822.

58. West to Ford Motor Company, December 30, 1918.

experienced to hull, frames or bulkheads."[59] By contrast, in May 1919 West's successor Drake complained about the quality of Eagles being produced. He noted "the very extensive poor work done in the early stages of the program, and the considerable amount still being done." Reliance upon unskilled supervisors and unskilled mechanics had led to work that continually had to be "patched up or done again."[60] Later, in 1924, Theodore Roosevelt, Jr., reported for the Navy Department that Eagles 1, 2 and 3 had been used until 1920 and had operated "without serious incident . . . although they were very uncomfortable at sea."[61]

Exactly what being "uncomfortable at sea" meant is a matter of definite import. Some charged that the Eagle boats were unseaworthy.[62] Frank A. Cianflone reported that the Eagles handled "like the four-stack destroyers in high seas," and he noted that Eagle 25 capsized in a squall on the Delaware River below New Castle, Delaware, on June 11, 1920. Since nine men died in the accident, the Eagle's reputation of unseaworthiness may have begun here. By 1924 two other Eagles had been lost at sea, enhancing the negative reputation. Thirty of the Eagles had been decommissioned by this date, twenty-two were being used in training naval reservists, and five had been turned over to the Coast Guard. By World War II only eight Eagles remained, and these were used only for duty within United States continental waters. Seven were decommissioned after the war; the eighth, Eagle 56, was the victim of a German torpedo.[63]

The Ford Motor Company cannot be held solely responsible for the Eagle's lack of seaworthiness, if indeed the Eagle had such a problem. The Navy's engineers, David W. Taylor and Robert Stocker, took credit for the Eagle's design, though

59. Statement of Commanding Officer, U.S.S. Eagle One, Brest, France, August 17, 1919, Accession 6, Box 258, as quoted in Nevins's and Hill's notes, Accession 572, Box 26, Ford Archives.

60. Inspector of Machinery and Superintending Constructor to Bureau of Steam Engineering and Bureau of Construction and Repair, May 29, 1919.

61. T. Roosevelt to Gentlemen [Ford Motor Company], April 3, 1924, Accession 572, Box 26, Ford Archives.

62. The first published claim about the instability of the Eagle appeared in "Why Congress is Investigating the Building of Ford Eagles." Josephus Daniels chalked up this and other criticism of the Eagle program to politics related to the senatorial election of Truman H. Newberry. See E. David Cronon, ed., *The Cabinet Diaries of Josephus Daniels, 1913–1921* (Lincoln: University of Nebraska Press, 1963), entries of January 5 and 21, 1919, pp. 364, 369.

63. Frank A. Cianflone, "The Eagle Boats of World War I," *U.S. Naval Institute Proceedings* 99 (June 1973): 76–80.

Ford's production engineers most certainly contributed their ideas to make production easier. Tests of the model conducted by Taylor and Stocker in the tank at the Washington Navy Yard obviously seemed satisfactory to them.[64] But the full-size ship may have had problems. If so, these clearly grew out of Eagle's being designed for mass production by an automobile maker. The tacit knowledge of what constituted a good hull design was consciously disregarded in favor of production priorities. Given the poor performance by the Ford Motor Company, which had been exacerbated by the company's failure to heed the advice of experienced shipbuilders, this tradeoff was probably not worth it in the end.

The Eagle boat venture was, therefore, an ill-fated attempt to bring the supposed power of mass production technology to bear on the problem of procuring ships for warfare. Ford's venture failed to revolutionize shipbuilding techniques. Rather, it pointed up the difficulty of transferring the methods used to manufacture a high-volume consumer durable like an automobile into an area like shipbuilding, which had its own tradition of knowledge and skills and which produced a comparatively small number of units. Moreover, because in many respects the Eagle project was foisted upon the Navy by Josephus Daniels and the wartime shipping crisis, its failure and the "success" of other, non-Ford programs suggests that a close coupling of and careful interplay between military project managers and civilian contractors is essential to military enterprise. The relationship between Ford and the Navy was shaky at best, held together only by the bond between Daniels and Ford.

When in World War II Henry J. Kaiser undertook to "mass-produce" the infamous Liberty ships, he faced many of the same problems as the Ford Motor Company had in the First World War. Like Ford's Eagle venture, Kaiser's project captured the imagination of Americans, but his techniques did not radically change shipbuilding practice either. The Liberty ship, like the Eagle, was an expedient, a stopgap measure born of a wartime emergency. Although allowing for rapid production, Kaiser's techniques had their costs: many, many of the Liberty ships cracked up during normal sea operation, and at least a

---

64. On February 19, 1918, Daniels recorded in his diary that he had been to the Navy Yard to see the model tests for the Eagle. Cronon, ed., *Cabinet Diaries*, p. 281. See also *New York Times*, July 7, 1918.

third of those that did not "suffered major fractures."[65] Like the history of Kaiser and the Liberty ship venture, the case of Eagle boat manufacture does not represent a failure of mass production but rather the failure of Henry Ford and his engineers and Josephus Daniels and the Navy's engineers to realize the inherent limitations of mass production technology. The technology was not, indeed, universally applicable. Shipbuilding remains an industry where craft knowledge and traditional design and construction continue to play important roles that have long since disappeared in the mass production of consumer durables such as the automobile.

65. Unlike the Eagle boat, the Liberty ship was based on a design that had been well-tested by the British. But the construction techniques worked out by Kaiser, who could speak only of the "front end" and "back end" of a ship, were the key to the Liberty program. Most important among them was the all-welded hull. See L. A. Sawyer and W. H. Mitchell, *The Liberty Ships: The History of the "Emergency" Type Cargo Ships Constructed in the United States during World War II* (Newton Abbot: David & Charles, 1970). Quotation appears on p. 19.

5

Adjusting to Military Life: The Social Sciences Go to War, 1941–1950
Peter Buck

Discipline and training are essential to the smooth functioning of military organizations. They are particularly important to armies that employ increasingly sophisticated technologies and rely primarily on draftees to operate them. In this essay Peter Buck examines sociological research mounted during World War II to help the Army ease the adjustment of civilian soldiers to military life and the tools of war. He notes that, during and immediately after the conflict, social scientists lost interest in studying the impact of technological change on either the Army or the enlisted men serving in it. They turned instead to questions that were more directly related to their own wartime experiences of working, often for the first time in their careers, in a large organization. Buck shows how this exposure to bureaucratic life and politics shaped important developments in American social science after 1945: the growth of small group sociology and the renewed appeal of social engineering; the dominance of social theory and the preoccupation with values; the lack of attention to racial, class, and religious divisions; and, finally, the near disappearance of an entire field—the study of technological change and its implications for social policy—from the social sciences' research agenda.

Familiar television commercials show lower-middle-class high-school graduates enlisting in the armed services in order to acquire complex skills and master sophisticated technologies, usually computers. In promotional spots for the Army, Navy, and Air Force reserves, rather more upper-middle-class weekend warriors are seen spending what looks like a day in the country with tank and helicopter. Every so often there are worried reports in the news media about the military's problems training and then keeping enough competent, low-level personnel to handle its increasingly complex weapons. On a slightly more scholarly level, social scientists inside and outside the Department of Defense engage in learned discussions of how the Pentagon is or is not making good use of the advanced technologies that have transformed modern warfare.

As other essays in this volume show, Americans interested in military affairs have long thought that promising, problematic, or just simply important interactions between men and machines go on in the armed forces. But there are other features of military life besides the "man-machine interface," to use the current jargon, that have also seemed worth emphasizing. For perhaps obvious reasons, recruiting posters in World War II held out no promises of technically demanding trades to be learned in the Army. Insofar as military service was advertised as being good for the enlistee, the benefits were presented as more behavioral and personal. The Marine Corps made men, not technicians, and so did the other branches of the War Department. In the movies juvenile delinquents, jaded college students, naive farm boys, and so on all became responsible adults as they survived the rigors of basic training; enjoyed the mixed pains and pleasures of leaving sisters, mothers, sweethearts, and wives behind when they shipped out; and then went through the searing experiences of combat that forever changed, or ended, their lives.

The academic view of what serving in the World War II armed forces involved for ordinary soldiers and sailors was pretty much the same as the movie view. Sociologists and psychologists contributing to the war effort saw themselves as helping to solve problems of organization, morale, and personal adjustment that were only incidentally, if at all, related to anything distinctive about the military technologies of the day. In a conscript army where draftees raised in a free society had to adapt to an unusually rigid institutional order and become accustomed to doing what they were told, how they also accommodated themselves to the unfamiliar tools of war did not rate much attention from experts interested in why men fight, not what they fight with.

Or so it eventually appeared. At the start of the war, prominent American social scientists assumed that the armed forces' main organizational and personnel problems would prove to be closely tied to the peculiarities of technological warfare. This was a reasonable extrapolation, they believed, from their knowledge of how many problems in prewar American life could be traced to the difficulties that individuals and societies necessarily had in adapting to advances in technology. The most thoroughly mechanized conflict in history changed their minds, however, and the onset of a Cold War waged in the shadow of nuclear weapons cemented a new consensus: too much talk about technology as a variable in human affairs interfered with efforts to get down to the more urgent business of social engineering.

This somewhat paradoxical outcome was part of a broad, World War II induced transformation in how American social scientists thought about their work and its practical uses. They lost interest in the study of technology and its social relations more or less in proportion to their understanding that the doing of social science under military auspices had placed them in some awkward positions. World War II gave many academics their first taste of working within rather than merely consulting for large operating agencies. The realities of bureaucratic life and politics taught them that power and influence went together with the right to define problems for others to study. That insight shaped one of the most familiar features of the postwar social scientific landscape: the almost single-minded determination of sociologists and social psychologists to ground their empirical and applied research on basic theoretical principles.

In stressing the dependence of applied and empirical studies on social theory, social scientists were claiming a measure of autonomy for themselves that they had failed to achieve while serving the War Department. There was no reason in principle why either the quest for independence or the commitment to theorizing as a means to that end should have deflected attention away from the role of technology in human affairs. But World War II also presented sociologists, psychologists, and anthropologists with a new problem set, as it were, one that looked so rich in possibilities for social engineering, broadly conceived, as to leave little room, time, or need for worrying in any detail about whatever real engineers might have been doing recently.

Here again, the fact that social scientists were themselves working in unfamiliar settings during the war affected their grasp of the novel terrain before them. It helped them appreciate the significance of one particular feature of the situations they were studying and trying to influence. American soldiers, enemy soldiers and civilians, Japanese-Americans interned in relocation camps, and prospective recruits to the Office of Strategic Services—to cite the examples that figured in the more prominent accounts of World War II social research—all had something striking in common with the sociologists, psychologists, and anthropologists interested in their behavior: they too had been uprooted, usually far more drastically than the academics, from their accustomed social environments. Yet initial attempts to specify, much less make sense out of, the consequences of those displacements produced results whose contradictory character seemed only partly explicable by obvious contextual disparities. The failure of observant social scientists to agree on the implications of what they had seen appeared to underscore the need for theoretical stocktaking. The axes along which their findings diverged indicated that the main difficulties lay in an area of study known as "culture and personality."

The most influential postwar effort to develop new theoretical resources for research in that field was made at Harvard in the university's new Department of Social Relations. Organized in 1946, Social Relations brought together sociologists, anthropologists, and psychologists who were convinced, first, that there were "theoretical foundations underlying the synthesis" that they had "worked out on the organizational level" in their new department; and second, that a "careful analysis" of those foundations would contribute to a "major movement" of considerable "significance to the future of social science."[1] Of the key figures in the department—Talcott Parsons, Gordon Allport, Clyde Kluckhohn, Henry Murray, and Samuel Stouffer—

1. Talcott Parsons and Edward A Shils, eds., *Toward a General Theory of Action* (Cambridge, MA: 1951), pp. v, viii.

three had played important roles in wartime social science. Stouffer had directed the Research Branch of the Information and Education Division of the War Department. Murray had headed the assessment staff of the Office of Strategic Services. Kluckhohn had been co-chief, with Alexander Leighton, of the Foreign Morale Analysis Division of the Office of War Information. Lessons drawn from those organizations made it seem obvious that the great need in postwar social science was for a "general theory of action" capable of integrating "three different interdependent and interpenetrating, but not mutually reducible, kinds of systems": "personalities, social systems, and cultural systems."[2]

Three books suggest the military background for that abstract theoretical prospectus. The first is the 1948 monograph *Assessment of Men* that Murray and members of his staff wrote describing their work for the OSS. The other two are by Kluckhohn's associate Alexander Leighton: his 1945 *The Governing of Men*, which deals with the Japanese Relocation Center at Poston, Arizona; and his 1949 account of the Foreign Morale Analysis Division, *Human Relations in a Changing World*.[3]

As *Assessment of Men* describes, Murray and his staff were responsible for evaluating men and women recruited for the OSS. Candidates were taken in groups to a country estate, known as "S," some forty minutes by truck outside Washington, D.C. There they were dressed in army fatigues, given false names and false biographies, and put through a three-day routine involving "all sorts of stressful situations, indoors and outdoors," designed to test their "intelligence and stamina." The staff viewed S as "a society like a ship's crew organized by a temporary necessity which separated them from the rest of the world." With "formal differences among candidates" eliminated by the use of aliases and army clothing, "a uniform anonymity" was achieved that obliterated "irrelevant and misleading factors," and thus allowed the staff to see each man "whole and to see him real." If a candidate "achieved a desirable niche in the society of S he did so on his own merit," the argument ran.

2. Ibid., p. 47.

3. The OSS Assessment Staff, *Assessment of Men: Selection of Personnel for the Office of Strategic Services* (New York: 1948); Alexander H. Leighton, *The Governing of Men: General Principles and Recommendations Based on Experience at a Japanese Relocation Camp* (Princeton: 1945); Alexander H. Leighton, *Human Relations in a Changing World: Observations on the Use of the Social Sciences* (New York: 1949).

If he carried distinction, it was not the distinction bought at Brooks Brothers or bestowed by his grandfather's shrewd speculations in railroad stock . . . All started on the same level at S, for they could not boast of a high status if they had it, and any status they mentioned was known to be false . . . [They] came naked. So far as past history and achievement were concerned, all men met on the same terms . . . S was the truly classless society.[4]

This is a remarkable statement in its assertion that past histories and achievements, no less than status and class distinctions, are irrelevant and misleading when it comes to understanding how men and women behave; they are mere formal differences that individuals can shed as easily as their clothes. Matters seemed quite different to social scientists who spent World War II trying to understand and influence Japanese and Japanese-American behavior. Consider first the relocation camp where Alexander Leighton worked before joining Kluckhohn at the Foreign Morale Analysis Division. Poston, Arizona was no less artificial and isolated an environment than S. It too "broke apart and reshuffled" established social arrangements, leaving its inhabitants strewn about like parts of "a dissected watch," and it had more than its share of stressful situations. But the result was not a stripping away of the sorts of historically and socially sedimented factors that Murray and his staff thought they were removing at S in the name of seeing people whole and real. Instead, Poston sharpened distinctions "between rural and urban people, between rich and poor, between high social prestige level and low, and between the educated and the uneducated." Far from exposing a kind of pure personality to view, it provided a nearly textbook example of how "in times of stress" individuals behave almost exactly as if they were just agents of the "belief systems" deeply ingrained in them by their pasts.[5]

When Leighton moved from Poston to Washington to join Kluckhohn as co-chief of the Foreign Morale Analysis Division, he carried along the main lesson of the relocation camp: "Never, under any circumstances, attempt to eliminate a system of belief by force; there is no surer way to give it life and permanence."[6] This maxim served him and Kluckhohn well as

---

4. *Assessment of Men*, pp. 5, 221–22.

5. Leighton, *Governing of Men*, pp. 233, 234, 288, 292.

6. Leighton, *Human Relations*, p. 321.

they and their staff attempted to assess the state of Japanese morale, both civilian and military, and offer advice about how propaganda materials might be fashioned to take advantage of opportunities for psychological warfare revealed by their research. Once again the problem was to understand the behavior of people under stress and cut off from accustomed social environments. Once again the salient finding concerned the increased dependence of individuals on systems of belief deeply rooted in their culture, the beliefs in this case being ones centered on the Japanese emperor and expressive of "faith, devotion, loyalty, and conviction as to his importance." And once again the policy implications were clear: in planning and executing a psychological warfare campaign, "it was not profitable to attack the deep-seated beliefs of the enemy"; the preferred approach was "to pay attention to the current and changeable factors" about which the enemy would be sure to have "major misgivings when suffering reverses."[7]

There are several obvious points to be made about the OSS assessment staff, the Foreign Morale Analysis Division, and their conflicting views of which is more basic, culture or personality. The jobs the two organizations were being asked to perform, and hence the data that were valuable to them and their employers, were different. It is hard to see, for example, what use the OSS might have made in selecting its personnel of the observation that "most of us have symbols in which we believe with a firmness which resembles that displayed by the Japanese," or the remark that "The American Way of Life" was like the Japanese emperor in the sense of being a symbol "having cultural prevalence together with strong personal-emotional ties."[8] Equally different were the backgrounds Murray, Kluckhohn, and Leighton brought to their work. Murray was a psychologist who had made his reputation studying Harvard undergraduates. Kluckhohn was an anthropologist who had spent most of his career studying the Ramah Navaho in New Mexico. Leighton, while trained as a psychiatrist, was in effect also an anthropologist: he owed his appointment to the Poston staff to research he had done previously in connection with Kluckhohn's Ramah project on Navaho "healing institutions." That work had brought him to the attention of the Office of

7. Ibid., pp. 56, 61, 89.
8. Ibid., p. 94.

Indian Affairs, which shared responsibility for the camp with the War Relocation Authority. Having studied Indian communities whose cultures remained intact despite years of persecution and exploitation by "white civilization," he and Kluckhohn were not apt to imagine that Japanese soldiers or Japanese civilians would suddenly find themselves culturally naked. What was true of the Navaho could be expected to be true of them: "by the time a person is an adult he is so conditioned to the patterns and attitudes of his culture that subsequent change is unlikely."[9]

Although these variations in job specifications, disciplinary backgrounds, and prior research experiences are important and interesting, they do not help explain the main reaction provoked by the findings of Murray's OSS assessment staff and the Kluckhohn-Leighton Foreign Morale Analysis Division when they became widely known after the war. Instead of observing how different disciplinary perspectives combine with different practical needs to elicit different estimates of the determinants of behavior, American social scientists drew a far more striking conclusion: Americans were different from other people. They were much less inclined to locate their personal experiences in relation to widely held beliefs and values that might give meaning and purpose to their lives. They were therefore either much more open to having their behavior directly managed by social engineers or much more in need of having their consciousnesses raised by social scientists versed in cultural studies.

By the time these inferences were being drawn, the onset of the Cold War had given special urgency to questions about how far Americans understood or were even aware of their cultural traditions. This was a context that Kluckhohn in particular was well positioned to appreciate. In addition to being the chief anthropologist in Social Relations, he became the director of Harvard's Russian Research Center when it was founded in 1948. Yet neither he nor Murray provided the main bridge between wartime and postwar social science. At Harvard and elsewhere their World War II achievements were overshadowed by the much larger project that Samuel Stouffer had supervised as director of the Research Branch of the Informa-

---

9. Alexander H. Leighton and Dorethea C. Leighton, *The Navaho Door: An Introduction to Navaho Life* (Cambridge, MA: 1945), pp. 8, 137.

tion and Education Division of the War Department. Although the essential tension informing research on "culture and personality" was evident in the contrasting conclusions to be found in *Assessment of Men, Human Relations in a Changing World,* and *The Governing of Men,* the primary empirical reference point for subsequent debate was the four-volume *Studies in Social Psychology in World War II* that Stouffer and his staff produced.[10] Published in 1949–50 and usually referred to by the titles of volumes I and II, *The American Soldier,* these books were regarded even by their critics as defining the grounds for all subsequent arguments about what it would mean "to direct human behavior on a basis of scientific evidence."[11]

The American Soldier was enormously influential because it did what neither Murray nor Kluckhohn and Leighton were in a position to do: show that the central problem for postwar social science was to take something like the psychologist's insistence on the power of artificially created social situations to separate individuals from their own and their society's pasts and to reconcile that thesis with something like the two anthropologists' convictions about the irreducibly cultural character of the ultimate determinants of behavior. How that was done, and how it helped to shape the postwar consensus concerning the dependence of applied social research and social progress on social theory, is the subject of the rest of this paper.

The argument that follows moves from the wartime experiences of social scientists working in the bureaucratic context of the Army to a discussion of the constraints and opportunities that the Cold War subsequently imposed on social research. It explores how, against that double background, the theoretical difficulties involved in "culture and personality" studies,

---

10. *Studies in Social Psychology in World War II* (Princeton, NJ: Princeton University Press). Vol. I: S. A. Stouffer, E. A. Suchman, L. C. DeVinney, S. A. Star, and R. M. Williams, Jr., *The American Soldier: Adjustment During Army Life* (1949). Vol. II: S. A. Stouffer, A. A. Lumsdaine, M. H. Lumsdaine, R. M. Williams, Jr., M. B. Smith, I. L. Janis, S. A. Star, and L. S. Cottrell, Jr., *The American Soldier: Combat and Its Aftermath* (1949). Vol. III: Carl I. Hovland, Arthur A. Lumsdaine, and Fred D. Sheffield, *Experiments on Mass Communication* (1949). Vol. IV: S. A. Stouffer, L. Guttman, E. A. Suchman, P. F. Lazarsfeld, S. A. Star, and J. A. Clausen, *Measurement and Prediction* (1950). Cited hereafter as *TAS* I, II, III, and IV, respectively. Although not properly part of the series, Robert K. Merton and Paul F. Lazarsfeld, eds., *Continuities in Social Research: Studies in the Shape and Method of "The American Soldier"* (1950) is usually cited along with the other four volumes. It will be referred to hereafter as *TAS* V.

11. The phrase is from the dust jacket of volume I.

broadly conceived, came to be linked to other salient features of postwar American social science: the development of small group sociology and the renewed appeal of social engineering; the dominance of social theory and the preoccupation with values; the obscuring of racial, class, and religious divisions; and the near disappearance of a whole subject from the social scientific field of vision—technological change and its implications for social policy.

Competent Research and Statesmanlike Policy

On the eve of World War II, social research had an ambiguous relationship to the national government. The New Deal had moved beyond the Progressive tradition of fact-finding commissions staffed by nonpartisan professionals and supported by organized private philanthropy rather than the state. But it had followed two different strategies in doing so. One led simply to an expanded role for advisory commissions: their charge was broadened to include planning and the recommending of courses of action, but they were still left free from responsibility for carrying out their proposals. The other approach pointed toward a more thorough blurring of distinctions between research, policy, and program: federal agencies were encouraged to employ their own research staffs, and social scientists were brought into the actual management and control of the programs they were designing.[12]

Originally posed in relation to natural resource policymaking, these alternatives were carried over into a larger arena as the focus of New Deal planning widened to cover national, including human, resources. In that guise they framed the initial prospectus for the studies of "adjustment to military life" that Samuel Stouffer's Research Branch conducted during the war. The proposal was contained in an August 1941 "Memorandum on Prediction and National Defense" that Stouffer and several colleagues prepared for a three-member Committee on Social Adjustment of the Social Science Research Council (SSRC). Using the language of resource planning, Stouffer and his associates outlined what they called a

---

12. See Barry Karl, *Charles E. Merriam and the Study of Politics* (Chicago: 1974).

task of "overwhelming magnitude."[13] "The basic problem of efficient use of human resources" was one of devising "an arrangement which permits each citizen to function in a situation where his particular characteristics enable him to be most useful, both to himself and to the nation." In the past that had been left to a "wasteful process in which a person had to try and fail in a number of different activities before finding the one for which he was best fitted." But at a time of impending crisis, it was no longer reasonable to "hope that some way, somehow, by incidental procedures and methods, the right person will find the right job." Fortunately, the social sciences were in a position to help: if the previous century had seen "great technological advances and . . . tremendous improvement in the utilization of material resources," the last twenty-five years had produced "real advances . . . in the improved use of human resources" through the development of "improved scientific methods" for educational and vocational guidance and personnel selection. The problem was that those methods were not being fully exploited. American industry, "the happy hunting ground of charlatans and quacks," had been slow to adopt them, and the federal government's efforts were scattered, uncoordinated, and not always under the direction of "competent professionally trained" people. It followed that the urgent need was organizational. There was no question of introducing "centralized administrative control of scientific predictive methods for personnel"; anyone acquainted with Washington knew that would be "impracticable and undesirable." Instead, an "agency under governmental or quasi-governmental auspices" should be created to "coordinate" existing activities ("without exercising any control") and arrange for their "logical extension" into an area "very close to the center of the most important problems in American life today"—research on "social adjustment, especially adjustment to military life."[14]

The substantive burden of this proposal, the emphasis on matching individuals to situations, rested on a combination of logic and empirical evidence that the SSRC Committee on So-

13. E. W. Burgess, Leonard S. Cottrell Jr., Paul E. Horst, E. Lowell Kelly, M. W. Richardson, and Samuel A. Stouffer, "Memorandum on Prediction and National Defense," in Paul Horst, et al., *The Prediction of Personal Adjustment* (New York: 1941), p. 167.

14. Ibid., pp. 159, 160, 162, 168–71.

cial Adjustment accepted without question. On the assumption that the good society is one where the matches between people and positions are close, the observed fact that the fits are always imperfect led almost inevitably to the conclusion that "adjustment" was the central social process or problem in the modern world. The committee was equally familiar with the more procedural thrusts of the proposal—the tacit conviction that distinctions between governmental and quasi-governmental are unimportant and the related belief that coordination can be separated from control, placed in the hands of the professionally trained, and usefully linked to a new area of research. These propositions were all consistent with the expanded advisory commission model of applied social research that the SSRC and its interlocking relation with the National Resources Planning Board stood for by 1941. The "Memorandum on Prediction and National Defense" also reflected the experiences of the two men most directly involved in implementing its call for studies on adjustment to military life: Stouffer and his future chief at the Information and Education Division, Fredrick Osborn, one of the members of the Committee on Social Adjustment.

Described in 1941, on becoming head of the Information and Education Division (or Morale Branch as it was initially called), as a Hudson River squire, friend and neighbor of Franklin Roosevelt, eugenist, population expert, conservationist, business researcher, corporation executive, banker, art connoisseur, and traffic expert, Fredrick Osborn was almost typecast for the role, well established by the start of the New Deal, of the prominent citizen serving voluntarily on government advisory boards. He first went to Washington in 1940 as a consultant on population for the division of statistical standards in the Bureau of the Budget. He soon became head of a civilian advisory committee for the Selective Service Program and then chairman of the Joint Army and Navy Committee for Welfare and Recreation. He only missed fitting exactly the stereotype of the dollar-a-year man by virtue of having written or edited three scholarly books on eugenics and population.

Stouffer represented a different social type: the academic specialist and professional researcher who did much of the work of the advisory bodies on which men like Osborn served. By 1942 when he became director of the Research Branch of Osborn's division, he was firmly established as a University of

Chicago professor who knew how to use his position as a base for forays into the world of professional public service. This meant that, while he shared Osborn's interests in population problems, for example, and was familiar with eugenics through his studies with Karl Pearson and Ronald Fisher, he did not place his demographic concerns in the tradition evoked by the titles of Osborn's books: *Heredity and Environment, The Dynamics of Population, Preface to Eugenics.*[15] Instead, his preferred titles—"Trends in the Fertility of Catholics and Non-Catholics," *Research Memorandum on the Family in the Depression*—suggest the intellectual, as well as the social, style of the technical expert.[16] He had served as staff director of an SSRC committee on the social and psychological effects of unemployment and collaborated with Paul Lazarsfeld in writing its report. In addition to participating in Lazarsfeld's radio research project, sponsored by the Rockefeller Foundation, he had worked with Gunnar Myrdal on the Carnegie Commission studies that led to *An American Dilemma.* He had also served as a consultant to the Census Bureau, preparing a survey of its problems for the 1938 report of the National Resources Planning Board, *Research—A National Resource.*

Stouffer drew on these experiences when he assembled the staff of the Research Branch. Of the eleven senior social scientists who were formally appointed as consultants, for example, three had been involved with the SSRC committee on the effects of the depression: Lazarsfeld, Philip Hauser, and Donald Young. A fourth, Quinn McNemar, was on the SSRC staff at the start of the war, and three others—Hadley Cantril, Robert Merton, and Frank Stanton—had been connected with Lazarsfeld's radio research work. McNemar had been at the University of Chicago with Stouffer, as had John Dollard, who had written his dissertation in one of Stouffer's areas of interest, the sociology of the family. Hauser, Rensis Likert, and Kimball Young all had ties to the Bureau of the Census: Hauser had been chief statistician there since 1938; Likert, as director of the Division of Program Services in the Bureau of Agricultural

15. Fredrick H. Osborn and Gladys Schwesinger, eds., *Heredity and Environment* (New York: 1933); Osborn and Frank Lorimer, eds., *The Dynamics of Population* (New York: 1954); and Osborn, ed., *Preface to Eugenics* (New York: 1940).

16. Samuel A. Stouffer, "Trends in the Fertility of Catholics and Non-Catholics," *American Journal of Sociology* 41 (1935): 143–66; Stouffer and Paul Lazarsfeld, *Research Memorandum on the Family in the Depression* (New York: 1937).

Economics at the Department of Agriculture, had collaborated with the Census Bureau on the development of sampling procedures; and Young had spent a year on Likert's staff as a senior sociologist. The remaining senior consultant, Edwin Guthrie, did not fit this pattern and was on the list for a simple reason; as chief consultant to the overseas branch of the general staff of the War Department in 1941, he was already present when Stouffer arrived. As for the professional and administrative personnel who did the actual work, leaving aside thirty-six men assigned to the Research Branch by the Army and thirty-seven women who appear to have had mainly clerical duties, twenty-nine of the remaining forty-six had some graduate-level training in one or another social science. Of those twenty-nine, all but three had some connection with the University of Chicago, the Bureau of the Census, the Carnegie Commission, the SSRC, or some combination thereof.

Despite the experiences of Stouffer and his staff and despite the prominent role assigned to academic consultants like Lazarsfeld, Dollard, and Merton, the Research Branch never assumed the advisory commission form that the SSRC Committee on Social Adjustment had anticipated for it. Instead of fashioning a vehicle for university-based professionals to use to influence policy, Stouffer and Osborn promptly and self-consciously stepped over the boundary separating research and planning from operations. In Osborn's view the crucial decision was to have all work related to military morale placed firmly under government supervision. This seemed to be the obvious lesson of World War I. Describing the origins of the Information and Education Division in 1942, he looked back to 1919 when Raymond Fosdick, fresh from overseeing a "gigantic civilian effort . . . to provide for the welfare and recreation of the American soldier," had recommended that in the future the government should assume responsibility for much of that activity, which had been "hitherto left to private initiative."[17] For his part, Stouffer was perfectly prepared to see himself as in charge of the research staff of an operating agency, provided that certain conditions were met. His work for the National Resources Planning Board, he wrote to Osborn in mid-1942, had shown him "why research organizations too often fail in

---

17. Fredrick H. Osborn, "Recreation, Welfare, and Morale of the American Soldier," *The Annals of the American Academy of Political and Social Science* (1942), p. 50.

Washington": they had too little contact with the administrators they were supposed to be serving, and they were given too little opportunity to plan for future administrative problems. But once recognized, those difficulties looked manageable, and Stouffer pronounced himself satisfied that the Research Branch could become "a model in Washington for its marriage of honest, competent research to statesmanlike policy."[18] The question was whether the union could be extended to include programs as well as policies, and on that point Stouffer was equally optimistic and ambitious. He saw that insofar as social scientists were being asked to move beyond the problem of finding the right person for the right job and, under the rubric of dealing with morale, consider how to improve adjustment to military life, they were being invited to make their influence felt at an operational level. Morale was "a function of command," something for which "every commanding officer down to the company commander" was responsible; while this meant that "in many respects" social scientists could still only "advise but not execute," it also meant that in other respects they would have a "direct influence upon the soldier."[19]

With one large exception, nothing that the Research Branch did had any broad effect on military policy during the war. The exception concerned how soldiers were discharged from the Army at the end of hostilities. The famous "point system" for demobilization was the branch's invention. It offered a procedure for returning men to civilian life in an order, although not at a rate, determined by "what the soldiers themselves wanted." Properly proportioned samples of enlisted men stationed around the world were asked a series of questions designed to show the importance they attached to how long a soldier had been in the Army, whether he had been overseas or seen combat, how many dependents he had, whether he had skills needed in the civilian economy, and how old he was. Length of service, overseas duty, combat experience, and having children turned out to be significant factors, and "weights" were assigned to those variables that "yielded point scores which had a close correspondence with the wishes of the maximum number

18. *TAS* I, pp. 14, 18.
19. Samuel A. Stouffer, "Social Science and the Soldier," in William Fielding Ogburn, ed., *American Society in Wartime* (Chicago: 1943), pp. 110–11.

of soldiers, even if they did not exactly reproduce those wishes."[20]

This program was the Research Branch's greatest accomplishment. By providing an order of demobilization that could be defended, as Franklin Roosevelt successfully did, on the politically potent ground of being "based on the wishes of the soldiers themselves," it carried the War Department through a potentially explosive process without serious incident.[21] With hostility toward the Army and cynicism about its promises mounting at the end of the war, the point system had the particular virtue of being proof against "charges of favoritism" and the inevitable exercise of "political pressure to discharge certain individuals or certain categories of individuals" out of turn.[22] In addition to minimizing possibilities for scandal, the system also helped divert the military from a genuinely foolish plan of its own that would have kept combat veterans in the armed forces long after "men who made the least sacrifice"—service troops, soldiers with limited periods of service, and men not yet fully trained—had been discharged, this in the name of "preserving intact, at all cost, the combat fighting teams." That the Army did not abandon its aims in this regard except after protracted argument, and then only grudgingly, is reflected in Stouffer's subsequent comment about how "incredible . . . it seemed at the time" that after VJ Day there was still "strong sentiment within the War Department for eliminating combat credit entirely" in determining the order of discharge. The idea might have made sense, he wrote, "if men were robots," but they are not, and in any event "even robots wear out." "The efficiency as well as the morale of the men exposed to long, protracted ground combat was at very least questionable" in 1945, and getting the military to behave as if it accepted that fact, even while it actually did not, was a signal achievement that contrasted conspicuously with the conditions under which the Research Branch usually functioned.[23]

The point system aside, the work of the branch was relentlessly routine. Stouffer and his staff were primarily in the business of survey research, and during the war they distributed

20. *TAS* I, p. 7.
21. *New York Times*, 24 September 1944, quoted in *TAS* II, p. 525.
22. *TAS* II, p. 521.
23. Ibid., pp. 520, 539, 548.

over 200 different self-administered questionnaires to a total of more than half a million soldiers, asking them everything from what they wanted for Christmas to how they felt about Army discipline, their officers, venereal diseases, and combat. The heterogeneity of these questionnaires resulted from the branch's nearly complete lack of autonomy in deciding which issues to investigate. Stouffer tried to make a case for "planning surveys" to identify problems before they became sufficiently troublesome to attract attention, but he got nowhere with that idea. From the beginning of World War II to its end, the branch could only respond to specific requests from the military for information; it was not allowed to initiate studies on its own.

This situation brought a largely unsuspected difficulty to light. Sympathetic critics of the later New Deal's more expansive visions of social scientists as managers would have doubted that Stouffer had quite hit the mark in his explanation of why government research organizations so often failed. He worried about a "vicious circle" developing if research and administration were not kept "in close and continual touch":

I have seen it happen several times in Washington. The research men, frustrated because their stuff is not being sought or used, became more and more "academic," satisfying their desires for expression by doing what may be good work from the scientific standpoint, but useless from the standpoint of policy determination.[24]

The more usual prewar fear was almost exactly the opposite: social scientists brought into government would become so entangled in the day-to-day problems of administration that they would lose sight of long-term considerations of scientific planning. In fact, the main trouble that the Research Branch actually encountered in serving its military masters lay neither in the putative allure of bureaucratic politics for social scientists nor in the attractions of good science for the bureaucratically impotent. It was to be found instead in the sheer inability of staff sociologists and social psychologists to establish scientific connections between their research and their policy recommendations. They could not even convince themselves that their empirical results, much less the courses of action they were proposing, had any scientific grounding. As Stouffer ac-

24. *TAS* I, p. 15.

knowledged, in late 1944 he and his staff were still "trying to blueprint the basic psychological concepts implicit in [their] research." They were still looking for ways to design their studies "so that the basis for inference as to corrective action rests more securely in the data themselves." They were also not succeeding:[25]

> If the war were to end today and if the Army should ask us what single practice [our] million-dollar research operation has *proved* to be helpful to morale, we honestly could not cite a scrap of scientific evidence. The curtain would go up on the stage and there we would stand—stark naked.[26]

The depth of those failures was obscured after the war by the self-deprecating but nonetheless confident references to "social engineering," as distinct from social science, that were used to describe the mission of the Research Branch when *The American Soldier* was published in 1949–50. The message then was that had the Army not been so insistent on getting the branch to do "a fast, practical job," the staff would have been able to proceed with the more interesting and ultimately more useful business of contributing "to the verification of scientific hypothesis of some generality."[27] But in 1944 the situation had seemed quite the reverse. As some of the younger social scientists saw it, the branch had no choice but to stick to the questions the military was concerned with, "because it had no larger hypotheses to test." In a situation where the "two undefined variables" had all along been "(a) what *ought* to be the subject of our research and (b) what the army thinks it ought to be," little imagination was required to appreciate the importance of constructing "a master plan of research" and getting it accepted as the basis for evaluating "any single study as to importance and urgency." Unfortunately, the requisite "general design of research either was beyond us or our power to carry it out," which meant that (a) had turned out to be "strictly a function of (b)."[28]

25. Samuel A. Stouffer, "Memorandum for Major Williams, 31 July 1944," in Samuel A. Stouffer, "Papers Relating to Wartime Research for the United States Army 1942–45," University Archives, Harvard University.

26. Samuel A. Stouffer, "Some Notes on Research Methods," 13 October 1944, Stouffer Papers.

27. *TAS* I, p. 30.

28. Hausknecht "Guess Sheet," n.d., in file marked "Estimates of priorities which should be given to research projects which have been proposed February 1942," Stouffer Papers.

The physical or biological sciences might be able to defend "apparently useless" studies on the grounds that they were serving "as a crucial test, as a way of determining the nature of a keystone in an elaborate theoretical structure," but "we cannot claim that for any study we might carry out."[29] Stouffer put the point in a slightly more positive context, but the effect was the same. He and his colleagues "might have helped channel [their] research much more effectively," he wrote in 1944, had they managed "to redefine the basic variables involved . . . two years ago," but since they had not, they were still unable "to determine critical points at which minimum immediate research resources would make a maximum contribution."[30]

Victory over Germany and Japan changed these acknowledgements of theoretical poverty into a program for future work. Four years with the military had shown that prewar applied social science was insufficiently scientific to withstand the demands made on it by an army inevitably interested in getting a practical job done fast. Having been thus brought to grasp the potential utility of general theories, should any present themselves, as weapons of bureaucratic warfare, the leaders of the Research Branch were quick to grasp the significance of the Manhattan Project and its demonstration that pure physics was the best foundation of good engineering. As they returned to or assumed prominent positions in the social science community (by 1967 seven had served as presidents of the American Sociological Association),[31] they helped to refocus discussions of applied social research away from the prewar contrast of advising and influencing versus controlling and managing and toward a new distinction between addressing practical problems directly and dealing with them indirectly by contributing to the development of fundamental theories and concepts.

In Stouffer's view this involved a thoroughgoing redefinition of the social sciences themselves and the abandonment of older compartmentalizations of such fields as sociology according to categories that reflected the dominance of practical rather than theoretical concerns—"rural sociology, urban sociology, social

29. Arnold Rose to S. A. Stouffer, "The Methodological Study," 27 March 1944, Stouffer Papers.

30. Stouffer, "Memorandum for Major Williams."

31. Raymond V. Bowers, "The Military Establishment," in Paul F. Lazarsfeld, William H. Sewell, and Harold L. Wilensky, eds., *The Uses of Sociology* (New York: 1967), pp. 250–51.

pathology, and the like." He acknowledged that there would still be an important role for people devoting themselves primarily "to developing research data, if not research ideas, for a single community, and doing a great service indirectly to sociology in general, as well as directly to the Chamber of Commerce, business firms, health agencies, and others in their home community." But this approach, characteristic of the Chicago sociologists who had trained him, no longer seemed "the best way to make sociology useful." Instead, it appeared "more profitable" for sociologists to view urban communities, for example, as sites

for the exploration of general sociological ideas, just as government is such a locus, the factory, or the school, or China, or the consuming public. In other words, I would argue that it is a process which should be the main object of study, and the process should be studied in whatever setting it is most easily available for examination.[32]

Why Men Fight: Small Groups and Large Principles

As the early debates about the place of the social sciences in the National Science Foundation indicate, the virtues of general sociological ideas were not immediately obvious to the natural scientists whose achievements were being cited as a warrant for the pursuit of basic knowledge.[33] What eventually made the idea of basing social policy on pure social science attractive was the promise of fundamental research to fill a void left by the major substantive discovery of the Research Branch: American soldiers and, by implication Americans in general, lacked the cultural resources necessary to place even so significant an event as World War II in a meaningful moral framework. Try as they might, Stouffer and his staff had been unable to find any principled statement of the national interest involved in World War II that had any operative significance for the citizen-soldier. It was not just that soldiers disagreed among themselves or with their commanding officers about the interests at stake. More damaging was the apparently clear evidence of a simple "absence of thinking about the meaning of war." As the Research Branch concluded, there was "little support of

32. Samuel Stouffer to Paul Lazarsfeld, 30 November 1945, Stouffer Papers.
33. See Talcott Parsons, "The Sciences Legislation and the Role of the Social Sciences," *American Sociological Review* II (1946): 653–66.

attempts to give the war meaning in terms of principles and causes involved, and little apparent desire for such formulations."[34]

This was effectively the same discovery that the OSS assessment staff made: separated from civilian life, American soldiers shed their civilian culture along with their civilian clothes. But no one in the Research Branch concluded that what had been stripped away were merely "irrelevant and misleading factors" or that the result was a chance, in Henry Murray's striking phrase, to see the American soldier "whole and to see him real." These contrasting reactions had in part to do with differences in the populations under review. Murray and his associates were studying people as often as not very much like themselves, while Stouffer and his staff were dealing with something closer to a cross section of American men under thirty-five years of age. Tables 1 to 4, which appear as tables 41 to 44 in *Assessment of Men,* overstate the case because they compare OSS candidates at S with the total population of the United States, not with the subset who served in the Army. Yet there is no reason to doubt that the general contrasts were as the assessment staff saw them:

The assessed groups were far more cosmopolitan than the average American, as their language facility and travel experiences indicate. . . . [It] was a young, male, native-born white group. . . . The most striking disparity between the assessed group and the general American population was in educational attainment. . . . Professional fields [represented] a much larger proportion of the assessed group than of the whole American population. . . . The relatively high proportion of professional men and women, business executives, proprietors, and

34. *TAS* I, pp. 431, 433. There may also have been little support among enlisted men for attempts to find such things out a quite different issue that the Research Branch had little interest in pursuing. One reviewer of this essay commented in his referee's report that he recalled "as a soldier in World War II being greatly put off by the 'social science' questionnaires and interviewers inflicted on us with the consequence that many of us dug in our heels and declined to play." The refusal "to play" followed from quite conscious motives: "If they insisted on asking us silly questions, we would give them silly answers. And many of their questions were indeed silly, so transparently contrived to be blind or hidden ways to extract our 'real' values and beliefs, that we reacted very negatively. If this was a typical pattern amongst those queried, it may well account for the failure of Stouffer and his associates to identify 'any principled statement of the national interest.' Many of us, including the less well educated among the enlisted men, had such principles and concepts of the role we were playing, and these we revealed in bull sessions with one another though not to probing questioners from outside who often came across as nosey, intrusive, and sometimes draft-delayers even when in uniform since they were doing seemingly 'non-military' tasks not contributing directly to the war effort."

Table 1
Comparison, by age distribution, of the assessed populations with the total population of the United States

Age (in years)	U.S. (18 and over) %	S (N = 2,344) %
18–20	7.9	1.1
20–24	12.4	19.9
25–29	11.8	26.9
30–34	10.9	20.5
35–39	10.2	13.4
40–44	9.4	7.9
45–54	16.5	8.1
55 and over	20.9	2.2
Total	100.0	100.0

Table 2
Comparison, by educational attainment, of the assessed populations with the total population of the United States

Years of school completed	U.S. %	S (N = 2,343) %
Elementary school or less	59.5	1.6
High school:		
1 to 3 years	15.0	5.2
4 years	14.1	14.2
College:		
1 to 3 years	5.4	23.5
4 years or more	4.6	55.5
Not reported	1.4	
Total	100.0	100.0
Four years of college, or graduate work:		
Four years of college		26.1
Graduate work without degree		10.3
Master's degree		5.3
Doctor's degree or equivalent		13.8
Total		55.5

Table 3
Comparison, by major occupational group, of the assessed populations with the total population of the United States

Major occupational group	U.S. %	S (N = 1,967) %
Professional workers	6.5	38.9
Semiprofessional workers	1.0	4.7
Proprietors, managers, and officials, including farmers	19.9	20.2
Clerical, sales, and kindred workers	16.7	19.2
Craftsmen, foremen, and kindred workers	12.6	5.0
Operative and kindred workers	18.3	4.6
Domestic service workers	4.7	0.4
Service workers except domestic	7.8	4.9
Laborers	12.5	2.1
Total	100.0	100.0

Table 4
Comparison, by income,[a] of the assessed populations with the total population of the United States

Annual income (in dollars)	U.S. Male %	Female %	S (N = 1,533) Male (N = 1,463) %	Female (N = 70) %
No income (student)	[b]	[b]	34.9	18.6
Other 0–799	[b]	[b]	4.0	8.6
Total 0–799	43.0	66.8	38.9	27.2
800–1,199	17.7	18.7	5.1	17.1
1,200–1,599	16.3	8.7	10.1	22.8
1,600–1,999	8.8	3.0	7.2	14.3
2,000–2,499	6.9	1.6	8.8	7.1
2,500–2,999	2.7	0.5	6.3	5.7
3,000–4,999	3.2	0.6	11.3	2.9
5,000 and over	1.4	0.1	12.3	2.9
5,000–9,999	[b]	[b]	8.1	2.9
10,000 and over	[b]	[b]	4.2	0.0
Total (excluding first two and last two lines)	100.0	100.0	100.0	100.0

[a] For the assessed populations, income is defined as the largest amount received by an individual from his primary job before 1940.
[b] Data not available.

high-grade craftsmen explains the high income earned by the assessed candidates before they came to OSS. . . . [They] were in general
of upper socioeconomic status.[35]

It was evidently one thing to ask people like these to go
around culturally naked; they could be expected to survive,
even flourish in that condition, by virtue of inherent strengths
of personality. It was quite another matter to encounter ordinary Americans in similar states of cultural undress; their need
for clothes was apparent. That at least was what the findings of
the Research Branch were taken to show when they became
widely known in the late 1940s and early 1950s. By then the
Cold War was underway, and in that context *The American Soldier* evoked two seemingly contradictory responses, both of
which lent powerful support to social scientific pretensions.
The lesson the Army drew was that if the principles and causes
at stake in wars were "singularly unreal" to the combat soldier,
"in contrast to the issues and exigencies of his day-to-day existence," then there was little to be said for trying to make any
particular conflict especially meaningful to any particular soldier, and a great deal to be said for strengthening those daily
commitments that apparently did help to motivate men to
fight.[36] This meant pursuing one of the more promising positive lines opened up by the Research Branch's studies of behavior in combat, the discovery of "the sense of power and security
which the combat soldier derived from being among buddies
on whom he could depend."[37]

In *The American Soldier* that insight was presented as an example of what could be learned from the sociology of informal
groups, and after the war much military-supported social
scientific research was accordingly devoted to the study of small
group dynamics and leadership. The idea, as John Keegan
writes in his remarkable book *The Face of Battle*, was that the
Army could no longer proceed on the now demonstrably mistaken assumption that ordinary soldiers "in life-and-death situations" conceive of themselves as their senior officers do, as
subordinates in whatever formal command structure they happen to be assigned to. Since they evidently were always going to

35. *Assessment of Men*, pp. 498–501.
36. *TAS* II, p. 167.
37. Ibid., p. 149.

think of themselves instead as "equals within a very tiny unit," it followed that the Army should seek to identify (which is where the small-group sociologists came in) and then foster the kinds of units most likely to maximize their members' fighting capacities. That was the inference drawn by the most influential military proponent of the small-group approach to military sociology, General S.L.A. Marshall, an eminent historian of the European theater. Convinced that victory in battle depended upon organizing an army properly, he argued after the war for "a new structure of small groups or 'fire teams' centered on a 'natural fighter'." As is well known, the argument was persuasive, so much so that, as Keegan concludes, by the time of the Vietnam War Marshall had had "the unusual experience, for a historian, of seeing his message not merely accepted in his own lifetime but translated into practice."[38]

Given the outcome in Vietnam, Marshall cannot ultimately have found the experience satisfying. The widely reported phenomenon of "fragging" suggests that, without some sense that a war can be placed in a meaningful context, moral or otherwise, the fire teams waging it became almost as dangerous to the natural fighters leading them as to the enemy.

There was in fact something very odd, although not necessarily wrongheaded, about the whole small-group dynamics explanation of combat behavior and leadership. This was apparent all along to those who understood that the approach was at right angles to the main axis of prewar debates about the determinants of social adjustment, that is, nature versus nurture. As Paul Lazersfeld appreciated immediately on reading the manuscript version of *The American Soldier*, General Marshall's image of the "natural fighter" notwithstanding, the striking fact about informal groups in the Army was that their leaders were neither born nor made in the sense usually understood by sociologists. "We are used to the idea that if people grow up in different environments they will come to see the world differently," he wrote to Stouffer in 1948, "but that is usually considered a long-lasting process involving early childhood experiences." In the leadership materials gathered by the Research Branch, however, "you have a difference in social position which has been artificially created . . . and within a few

38. John Keegan, *The Face of Battle: A Study of Agincourt, Waterloo and the Somme* (New York: 1976), pp. 53, 74.

months it makes for tremendous differences in attitudes between privates, noncoms, and officers."[39]

This insight opened up possibilities for social engineering largely unimagined before the war. The SSRC Committee on Social Adjustment had of course been interested in what were called "manipulable factors" affecting behavior, but the limited hopes held out for them are suggested by their place in the committee's general analytic scheme. They appear at the end of a sequence of dichotomies that started with personal versus situational factors. Personal factors were then divided into "congenital and environmental components" in order to pose the obvious question about their respective roles. Only after a further pause to insist on the additional constraints set by "nonmanipulable situation factors" was consideration given, briefly, to the situational variables making "social control possible."[40] Critics of *The American Soldier* saw that the logic behind this taxonomy had been seriously undercut by the Army's apparently successful "research on how to turn frightened draftees into tough soldiers who will fight a war whose purposes they do not understand." They expected that the patent analogies between what the military had accomplished and what might be done in, for example, the area of labor relations would soon be exploited, with the result that social science would become even more "an instrument of mass control, and thereby a further threat to democracy."[41]

But the social scientific future did not belong wholly or even in its largest part to the social engineering fantasies of small-group sociologists. Their response to the Army's success in getting its soldiers to fight a war whose meaning they never appreciated was to see how much further it was possible to go in detaching the study and control of behavior from the social, cultural, and perhaps even biological baggage that individuals carried around with them. The much more conspicuous orthodox reaction was almost exactly the contrary: faced with evidence that Americans lacked principled commitments to causes, the true social scientist should seek to correct that deficiency, not exploit it.

39. Paul F. Lazarsfeld to Samuel Stouffer, 1 May 1948, Stouffer Papers.
40. Horst, et al., *Personal Adjustment*, pp. 12, 19.
41. Robert S. Lynd, "The Science of Inhuman Relations," *New Republic* 29 (August 1949): 22.

This was the tack that the Research Branch took during the war. Their understanding of the dual message that orientation and indoctrination programs had to convey left them little choice, especially after the Normandy invasion. In addition to helping American soldiers "make some association between their personal experiences and the larger political and economic implications of the war," Army propaganda had to lead them to see that enemy soldiers were walking exhibits of that connection. It was not enough for "our men [to] learn to hate the Germans as men who kill their buddies and shoot at them," because the kind of hatred Americans were capable of was "short-lived" and not like "the deep continuing hatred toward the Germans exhibited by elements of the French, Belgian, English and other European peoples."[42] Building on simple hatred or anger was risky, moreover, because direct contact with the enemy was likely to cause "propaganda which does not ring true to boomerang."[43] The solution was to acknowledge that observed behavior might not necessarily reveal the cultural forces shaping it, while denying the reality of that appearance. Even though German soldiers might turn out to resemble Americans when they were encountered as casualties of the war, they nonetheless had to be treated as agents of a repellent system of beliefs and dispatched accordingly. As the Research Branch proposed to tell the American infantry man:

The indivdual enemy soldier (especially the German) on superficial impression is often a nice home-loving fellow who carries photographs on him of his mother, or wife or children, or his dog. But don't be a sucker, Yank. He might not have been a bad guy if the Nazis or Tojo hadn't taught him lies all of his school years, but now with all [his] human exterior he carries disease germs just like [a] mad dog with rabies, even though you can't see the germs. He must be exterminated until . . . Berlin and Tokyo surrender unconditionally or the world won't be a safe place for us to live in.[44]

For obvious reasons, whether Germans and Japanese carried their political culture around like germs was not a question that fell within the Research Branch's scientific purview. It was

---

42. Kimball Young, "Learning and Changing Attitudes," 24 August 1944; Robin Williams, "Why Soldiers Fight (Notes based in part on observations among American combat troops in the European Theater of Operations)," n.d.; Stouffer Papers.

43. "Suggested Topics for Recommendations," n.d., Stouffer Papers.

44. Ibid.

within the domain of the Foreign Morale Analysis Division, however, and the answer forthcoming from that organization was essentially yes, Japanese beliefs are deeply ingrained, but no, killing the soldiers who carried them is not the only, or even the most efficient, way to induce the enemy to surrender. The point about well entrenched beliefs was not exactly a finding. As Alexander Leighton cheerfully acknowledged in *Human Relations in a Changing World,* it was an assumption, the eighth in a long line of assumptions that guided him, Clyde Kluckhohn, and their staff in their work: "All people everywhere have systems of belief which influence their behavior." Similarly, the idea that nothing short of the extermination of the Japanese army would bring Tokyo to surrender was not a piece of propaganda for American consumption: "There were many men in top command and policy-making positions," not to mention various individuals with many years' experience in Japan who were serving as "expert advisors" in the War Department, "who felt that Japanese morale was a solid wall of uniform strength which nothing could destroy except the actual killing of the men who displayed it."[45]

That feeling was itself an excellent example of a deeply ingrained belief, as Leighton had good reason to know by 1945. Starting in March 1944 the Foreign Morale Analysis Division had been at work attempting to do for Japan what Stouffer and the Research Branch were doing for American soldiers: assessing morale and trying to determine "whether it had flaws, what changes could be expected, and what could be done to influence it." Obviously unable to conduct surveys, the division had to rely on other materials: reports of prisoner of war interrogations, captured diaries and letters, official documents, accounts from neutral observers in Japan, and transcriptions from Japanese newspapers, periodicals, and radio broadcasts. Reports were written summarizing the results, some of which reached the Army Chief of Staff, members of Roosevelt's cabinet, and the President himself. One, dealing with Japanese home-front morale, never got out of the Office of War Information, at least not in its original form. Initially presented on January 5, 1945, it summarized mounting evidence indicating a marked deterioration in Japanese "will to fight," to the point that "a blowup of some kind" was extremely likely, or else "a

45. Leighton, *Human Relations,* pp. 46, 78.

decline into a state of chronic inefficiency." The document was promptly suppressed by the Japan Section of the Office of the Deputy Director of the OWI, not to be released until June, and then only after being substantially "toned down." As Leighton recalled four years later, it "had little effect except to anger some of those responsible for planning the OWI directives" because it conflicted with "prevailing ideas and beliefs." Among American policymakers, the "stereotype of fanatical and suicidal" Japanese resistance "hung like a spectre over planning and discussion."[46]

"The possible importance of these matters" was soon clear to Leighton. By December 1945 he was in Hiroshima directing a research team sent there by the United States Strategic Bombing Survey to study the "attitudes" of survivors.[47] The atomic bomb had been dropped four months earlier, for at least one reason that his division could have shown was dubious: in July 1945 Secretary of War Stimson still believed that there was "as yet no indication of any weakening in Japanese determination to fight rather than accept unconditional surrender."[48] Leighton was ready to concede that on this point the contrary evidence available then was only suggestive, but this did not mean that it should not have been taken into account:

The important point, however, is the fact that the findings were suggestive, strongly suggestive. As such they could have been followed up and put on a firmer foundation by overcoming the numerous difficulties in securing information and achieving communication. . . . The methods, the materials, and the personnel existed and could have been brought together in a project that would have shown results and either established or refuted the Division's contentions.[49]

None of that happened, although it would have involved efforts "infinitesimal in size and expense compared to other activities related to the atomic bomb." As a result, the decision to drop the bomb was made by men who had "very likely never heard of those findings by the Division that had bearing on the matter."[50]

46. Ibid., pp. 43, 58–60, 121.
47. Ibid., pp. 13, 14.
48. Ibid., p. 126.
49. Ibid., p. 127.
50. Ibid.

This was "the most outstanding failure" of the Foreign Morale Analysis Division, and it left Leighton convinced that the "imperious need" for social science research was in the area of American "attitudes and capacities for action toward other nations." Nor had he any doubts about the nature of the problem. It was not that Americans lacked larger frameworks of beliefs and values in which to set their experiences, but rather that they viewed the world like everyone else, through "stereotyped images" that were "starkly simple and exceedingly inaccurate."[51]

The Research Branch turned up some evidence that might have supported this argument. It concerned "the differences in hatred" that American soldiers felt toward their Japanese and German enemies. As the charts from volume II of *The American Soldier* indicate (charts 1, 2), on being asked how they expected to feel about killing the enemy, over forty percent of the men polled said that they thought they "would really like to kill a Japanese soldier"; less than ten percent had the same enthusiasm for killing Germans. Over sixty percent of the enlisted men stationed in Europe or the United States thought it would be a good idea to "wipe out the whole Japanese nation" after the war; less than thirty percent wanted to see that happen to Germany. *The American Soldier* was careful to note that these differences could not be explained by any parallel variations in enemy behavior. Indeed, they seemed to be inversely related to experience: soldiers in the Pacific were "strikingly less vindictive toward the Japanese" than men still in training in America or fighting the Germans in Europe. This was all very puzzling, and Stouffer and his staff acknowledged that they did not have any very good explanation for either "the high degree of hatred of the Japanese expressed by soldiers in training" or its decline once they engaged the enemy. It appeared to have something to do with "endemic American attitudes toward the Japanese as a race," but that line was left wholly undeveloped— which was not surprising, since it was introduced only three sentences after a remark to the effect that American soldiers were unusual in having no "ideological basis" for whatever anger they directed toward their enemies.[52]

51. Ibid., pp. 102, 126, 131.
52. *TAS* II, p. 157.

Chart 1

Percentage Giving Indicated Responses

QUESTION   COMBAT PERFORMANCE GROUPS

"How do you think you would
feel about killing a Japanese
soldier?"

Below average   Average   Above average

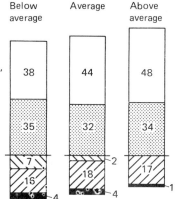

"I would really like to kill a Japanese soldier"

"I would feel that it was just part of the job,
without either liking or disliking it"

Some other idea or no answer

"I would feel that it was part of the job,
but would still feel bad about killing
a man even if he was a Japanese soldier"

"I would feel I should not kill anyone,
even a Japanese soldier"

QUESTION

"How do you think you would
feel about killing a German
soldier?"

"I would really like to kill a German soldier"

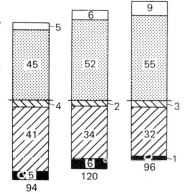

"I would feel that it was just part of the job,
without either liking or disliking it"

Some other idea or no answer

"I would feel that it was part of the job,
but would still feel bad about killing a
man even if he was a German soldier"

"I would feel I should not kill anyone,
even a German soldier"

Chart 2
Vindictiveness toward enemy peoples (veteran infantry enlisted men and
company grade officers, and infantry enlisted men in training)

QUESTION

"What would you like to see happen to the (Japanese) (Germans) after the war?"

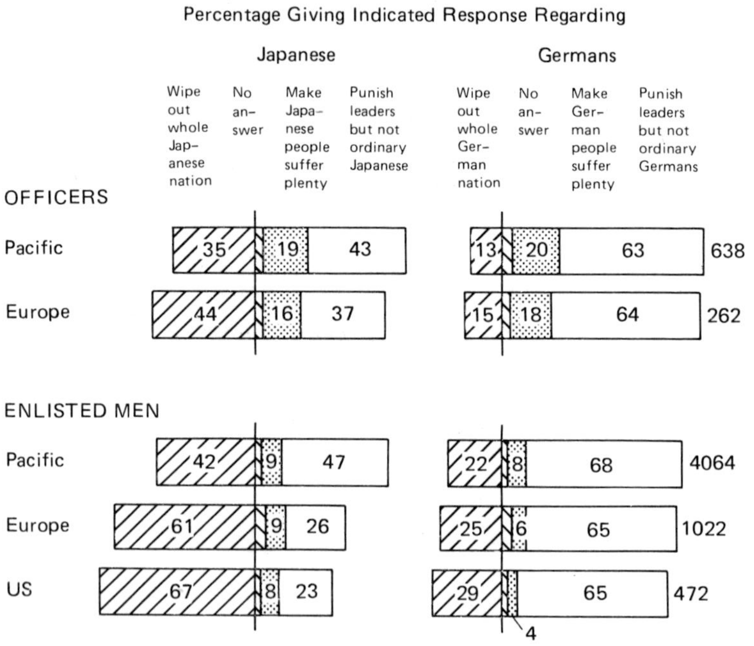

Percentage Giving Indicated Response Regarding

	Japanese				Germans			
	Wipe out whole Japanese nation	No answer	Make Japanese people suffer plenty	Punish leaders but not ordinary Japanese	Wipe out whole German nation	No answer	Make German people suffer plenty	Punish leaders but not ordinary Germans

OFFICERS

Pacific	35 / 19 / 43		13 / 20 / 63	638
Europe	44 / 16 / 37		15 / 18 / 64	262

ENLISTED MEN

Pacific	42 / 9 / 47		22 / 8 / 68	4064
Europe	61 / 9 / 26		25 / 6 / 65	1022
US	67 / 8 / 23		29 / / 65	472
				4	

Leighton was a native of Canada, which may explain why he
was not as impressed by American exceptionalism as his
counterparts in the United States. After a brief run of generally
positive reviews that took due note of the possibility that social
scientists might not have used the atomic bomb against Japan,
Human Relations in a Changing World largely disappeared from
view, except among anthropologists who gamely continue to
cite it as proof that an understanding of Japanese culture
helped shorten the war in the Pacific.[53] Its suggestions about
the influence of American cultural stereotypes on American
foreign policy were not pursued. Instead, the wartime practice
of regarding enemies of the United States as ineradicably con-
taminated by their society's habits and values, while citing the

---

53. See, for example, George M. Foster, *Applied Anthropology* (Boston: 1969), pp. 34–36.

results of social science research to show a kind of immunity on the part of Americans to their society's cultural traditions, carried over into the postwar period. The essentially propagandistic reasons for drawing the contrast in the first place were conveniently forgotten, and the assumed primacy of cultural factors in determining the behavior of other people reemerged as one of the main findings of social sciences whose domestic function became one of meeting the needs of Americans for beliefs to live by.

These themes were sounded with increasing urgency as the Cold War intensified, but the basic connections were apparent as early as 1947. Speaking then on "American Education and Soviet Propaganda," Fredrick Osborn reported that the Army had done a reasonably good job during the war of designing programs "to teach the soldier why we were fighting." "We knew that this teaching was necessary," he explained, because clear though it was that the war was being waged in defense of the "treasured beliefs of a free society," it was equally evident early on that the American soldier "took these things so much for granted that he was quite unaware . . . that they would have to be fought for." The situation then had been rectified by transmitting a "simple creed to 14 million Americans in areas all over the world," but two years after the end of hostilities something more was required to prepare Americans for a "great debate now going on all over the world, the debate being forced on the peoples of Western European civilization by the new feudalism of the Soviet Union." This realization was forced on Osborn by eight months of service on the Atomic Energy Commission of the United Nations, and it underscored the importance of promoting basic social scientific research. In the debate with the Soviet Union, the American position could "only be that of reason, reiteration of the objective facts, clear and honest presentation"; for those qualities the only recourse was to the social sciences where "we have been building up that critical quality long since accepted as fundamental in the physical and biological sciences."

Particularly in the field of psychology, but also in anthropology and sociology, trained men are gathering quantitative objective data, analyzing this data and using it to check their various hypotheses. It is beginning to be possible for them to draw a clear line between statements supported by true evidence and statements which are insufficiently supported.[54]

If this trend was sustained, as it had to be, by American universities, they would be able to meet their responsibility for supplying "information and trained advisers in all social and public affairs," "objective thinking" would take hold "among the mass of our people," and Americans could "expect to win in this great debate with Soviet Communism."[55]

Osborn's sense that the Army had managed to explain the point of the war to its soldiers placed an optimistic reading on the matter. Following hard on the publication of *The American Soldier*, the Korean War drew the attention of American social scientists to a more disturbing possibility. As Clyde Kluckhohn, by then the director of Harvard's Russian Research Center (which Osborn had played an important role in creating), put it, Americans had ceased "to be explicitly creative and forward looking in the realm of ideology and our national values." The significance of this was clear to Kluckhohn, who, as an anthropologist, knew that "one of the broadest and the surest generalizations that anthropology can make about human beings is that no society is healthy or strong unless that society has a set of common values which give meaning and purpose to group life." But what really brought the issue home to him was his experience in Japan at the end of World War II, when he found the Japanese "shocked by the fact that few of us could utter even a coherent sentence or two beyond a vague reference to the Declaration of Independence or something of that sort."[56] Kluckhohn offered those thoughts as part of his contribution to a 1950 conference at the Babson Institute of Business Administration on "Revitalizing Democracy." In case anyone missed the point he was making, the symposium proceedings were introduced by an italicized paragraph neatly expressing the crucial proposition that what World War II and Korea together showed was the inherently cultural nature of the most urgent needs of the United States:

54. Fredrick Osborn, "American Education and Soviet Propaganda," Founders' Day Dinner Address, Washington and Jefferson College, Washington, Pennsylvania, 1 November 1947.

55. Ibid.

56. Clyde K. M. Kluckhohn, "The Dynamics of American Democracy," in *Revitalizing Democracy: A Plan of Action, Babson Institute of Business Administration Bulletin* II (1950): 8, 9.

"I never saw such a useless damned war in all my life," was one GI's summary of American reaction in Korea. Army studies showed that only 5% of our World War II soldiers were able to provide any idealistic justification for the war, and one's own observations readily bear out the fact that Americans are ideologically as unprepared for victory as they are temperamentally unprepared for conquest.[57]

In these specimens of Cold War rhetoric was a case for the social sciences that important figures in the natural science establishment found they could endorse. Stouffer made that useful discovery soon after he moved to Harvard to become director of the university's new Laboratory of Social Relations and join Kluckhohn, Murray, Talcott Parsons, Gordon Allport, and others in the equally new Department of Social Relations. By 1948 he was reporting to the Rockefeller Foundation how "pleased" he was that President Conant had abandoned his skepticism about the social sciences "and especially about the very kind of social science which I [have been] attempting to support."[58] What kind of social science that was, and what made it attractive to a former chemist, is suggested by a speech of Conant's, from which Stouffer proceeded to quote at some length, that touched on the importance of "fundamental or basic research in the nature of human relations."

Human relations in our half of tomorrow's world are the key to the survival of democracy as we understand the word. A totalitarian socialistic nation may be able to get on with old-fashioned methods. The empiricism of the past may be a sufficient guide for the masters of a police state. But an open society with our ideals requires other instruments and a wider understanding of modern man. We need to put to use what has already been learned by the scientists concerned with the study of man. But even more important, we need to support the efforts of those who are trying to push forward the boundaries of knowledge in this vast area. . . . I am sure that those who are supporting the applied work in the field of human relations will understand from the analogy with the physical sciences how vastly important it is to strengthen the bonds of those in our universities who are working far away from what may be called practical results.[59]

57. Quoted in John D. Montgomery, "A Way to Believe in America," *Revitalizing Democracy*, p. 5.
58. Samuel A. Stouffer to Joseph H. Willits, 12 May 1948, Stouffer Papers.
59. Ibid.

Interestingly enough, having passed on these remarks, Stouffer promptly dissociated himself from any thought that *he* might be about to take responsibility for reexamining "our ultimate goals." Satisfied with the "ideals of Lincoln as a good working statement of the aims of our society," he was content to make his "little contribution with such technical equipment as I have toward the attainment of those ends."[60] This may be because he never quite bought the argument that American soldiers had failed to make some association between their personal experiences and the larger political and moral implications of the war. When he first took charge of the Research Branch, he thought that a statement of war aims along the lines of "the war is being fought in defense of a society in which relatively free vertical mobility is possible" would mesh with what his staff's first attitude surveys were showing was important to men in the army.[61] Toward the end of the war, when he read the notes on "Why Soldiers Fight" that later formed the basis for the crucial chapter in *The American Soldier* on "Combat Motivations among Ground Troops," he reacted with real skepticism to the suggestion that "the element of sheer identification with national symbols, as standing for a group and its ideals, over and above the concrete intermediary identifications with individual persons and organizations" was weakly developed among Americans. "Doubt this," he wrote in the margin; "think it stronger."[62]

Public Opinion, National Values, and Technological Change

There was little to choose between a small-group sociology that looked forward to more effective social engineering and a cultural anthropology that promised to ignite ideological creativity. Both grew out of the wreckage of a prewar program for political reform in which the social adjustment and survey research traditions that the Research Branch inherited were linked to the problems involved in managing the consequences of technological change. Within that program, opinion polls and attitude surveys were expected to function as instruments

---

60. Ibid.
61. Samuel A. Stouffer, "Notes on Research Discussions of Carl I. Holand and John Dollard," 25 April 1942, Stouffer Papers.
62. Williams, "Why Soldiers Fight."

for articulating the common concerns of an informed citizenry. They were also assumed to be sensitive indicators of the conflicts inevitably found in a society where individuals born with different biological endowments and raised in equally varied social circumstances had to accommodate themselves to the divisive impacts of technological change. World War II and the Cold War dissolved this particular compound vision of politics as shaped by nature, nurture, and technology. Culture and personality settled out as precipitates; adjustments to technology all but vanished from the stock of reagents analyzed by survey research; and the divisions in American society were transmuted into elements that seemed to be safely handled by social scientists.

Although these outcomes were all visible in *The American Soldier,* they still lay in the future when the Research Branch was established. American social scientists entered World War II convinced that there was a broad reservoir of enlightened public opinion in the United States. This emergent civic culture was thought to extend throughout the society. It insured that individual energies could always be enlisted in support of policies designed to serve the common interest. The only problem was to discover beforehand what the constituencies for public policies really required. Markets and elections had traditionally been counted on to provide the requisite evidence, but the Great Depression had shown that those classic signaling devices were inadequate for determining which needs represented the public good. New Deal social scientists had accordingly turned to opinion polls and attitude surveys for signs of the wants that ordinary economic and political contrivances failed to express.[63]

With the coming of the war, the logic of that prospect was extended to the problem of managing a civilian army. Despite initial opposition from military officers who assumed that insight into the needs of soldiers came automatically with the authority to command them, the case for survey research proved compelling. In part it required only recognition of the obvious fact that there was no other way for the War Department to acquire information about the wants of enlisted men. It

---

63. Karl, *Charles Merriam,* passim; Richard Kirkendall, *Social Scientists and Farm Politics in the Age of Roosevelt* (Columbia, MO: 1966), p. 186; and Hans Speier, "The Historical Development of Public Opinion," in his *Social Order and the Risks of War* (Cambridge: 1952), pp. 323–38.

could not let them vote on questions of policy or display their preferences by means of some sort of market mechanism. But surveys and polls were also attractive as guides for War Department planning and policymaking because of their more tendentious promise to elicit a kind of *public* opinion from soldiers, where that was understood to be a body of opinion pertaining to and informed by a grasp of the common good, in contrast to *private* opinions which were seen as expressing only individual self-interest. It was public opinion in this sense that Stouffer and his staff spent the war trying to extract from their questionnaires. They tacitly appealed to it in fashioning their one great accomplishment, the point system for demobilization. The plan was plausible just insofar as the point scores produced by their weighting of the key variables—length of service, combat experience, and so on—yielded an order of discharge that was more than a statistical artifact and embodied instead a genuinely public judgment about how demobilization should proceed.

This did not mean that each and every soldier was expected to agree fully with the result. As the remark about point scores having "a close correspondence" with wishes that they "did not exactly reproduce" indicates, the Research Branch did not anticipate a uniform response to its questions. Just as New Deal social scientists had believed that the enlightened public opinion they were trying to tap would display the heterogeneity of a society divided along class, ethnic, religious, and regional lines, so too Stouffer and his staff assumed that their attitude surveys would capture something of the social diversity of the Army. Nor had they any serious doubts, either at the beginning or at the end of the war, about the categories they needed to use in interpreting their results. In 1942 Stouffer was writing enthusiastically about Hadley Cantril's success in "explicitly relating" opinion data to "religion, economic status, educational level, and region."[64] When Leonard Cottrell, the head of the surveys section, took up the problem of planning research on discharged veterans in 1945, the "exploratory" effort he proposed to start focused on "differential reactions of veterans of different social and economic class." Veterans were "not a homogenous group," he wrote, and it seemed "especially important" for the branch to design its surveys so that "a rough social

64. Samuel A. Stouffer to Herbert Blumer, 8 July 1942, Stouffer Papers.

economic class breakdown" of its samples could be made.[65] As far as the kinds of data required were concerned, this would not have represented much of a departure from standard operating procedures. Army records provided information about each soldier's "age, schooling, occupational history, earnings, skills, athletic activities, associations, and preferences." The branch routinely congratulated itself on being able to produce on demand, for example, "a sample of all enlisted men, properly proportioned as to race, age, marital status, home section, rank, area of service, and length of service."[66]

After the war, when the public opinion surveyed by pollsters came to mean simply opinions disclosed to others rather than kept to oneself, the political rationale for these sampling schedules was obscured. By the time *The American Soldier* was published, demobilization seemed to have been a relatively straightforward achievement with no particular roots in the culture of American politics. That was just as well, because the logic behind the point system was at odds with what *The American Soldier* reported about attitudes toward the war in general. The discovery that enlisted men had not responded well to principled statements about American war aims implied that they had held no public opinions, in the prewar sense of opinions about the public good, only private ones. This made demobilization either, paradoxically, a social engineering success based on assumptions that its own architects' social scientific findings strongly suggested were mistaken or a practical demonstration that there was something wrong with the interpretation placed on those findings.

The issue was never joined. But judging by the account given in *The American Soldier,* when Stouffer's associates got down to analyzing their survey data after the war, they were in fact looking for something quite different from the "public" opinion they had drawn on in devising the point system. The consensus on war aims that they could not locate would have had to transcend most of the categories that during World War II had seemed so obviously useful: religion, economic status, and region; social class, occupational history, and race. These terms did not figure in the discussion of whether American soldiers

65. Leonard Cottrell to Dr. Stouffer, "Planning Research on Discharged Veterans," 11 June 1945, Stouffer Papers.
66. "Basic Record on Research Branch," 16 August 1944, Stouffer Papers.

had well-developed identifications with this, that, or the other set of national symbols, political beliefs, cultural frameworks, or moral contexts that might have served to give reality to the larger implications of the war. Stouffer and his colleagues understood that rather than asking if some particular statement of war aims could be made meaningful to a cross section of enlisted men, it was more reasonable to break the sample down into its component parts and then see which people were attracted to which statements. In *The American Soldier*, elaborate tables were accordingly presented to show how orientations toward the war and patterns of adjustment to military life varied according to what were called "background characteristics of the soldiers." But the characteristics used were invariably the same, not religion or race, not class or region, but age, marital status, and education.

Although the correlation between level of education and social class was duly noted, that was not why *The American Soldier* paid so much attention to the number of years a soldier had spent in school. Like age and marital status, education was interesting because it had an honored place in the literature on social adjustment. There was a striking coincidence between the Research Branch's preferred postwar background characteristics and the topics that the SSRC Committee on Social Adjustment had identified in 1941 as either best understood by people in the field or most worth their future attention. In the first category were success in school, happiness in marriage, and recidivism in crime; in the second, adjustment to population change, aging, technological change, and "the existing national emergency."[67] Since the last item, World War II, was the central concern of *The American Soldier* and since aging and population change turn out on examination to have been treated as one subject, the only entries from the 1941 list that failed to find a prominent position after the war in the Research Branch's analytic scheme were recidivism and technological change. Recidivism needs no comment, but the disappearance of questions about technology was of a piece with the transformation of the public opinion that everyone knew existed before and during the war into the national consensus that no one could quite find afterward.

The Research Branch might well have treated adjustment to

---

67. Horst, et al., *Personal Adjustment*, pp. 1–11.

military life as an instance of adjustment to technological change. As a former student and then a colleague of William Fielding Ogburn, Stouffer could appreciate the importance of technology in shaping modern life. Moreover, Ogburn's work on cultural lags in the wake of technological advances had a prominent place in the literature on adjustment and adaptation. There was more than a hint of these concerns with the impacts of technology in the way that the SSRC Committee on Social Adjustment initially viewed the Army, namely as an institution in need of assistance from experts familiar with "prediction problems in the vocational field."[68] The idea was that the basic problem involved, finding the right soldier for the right military job, was a function of the "highly specialized character" of modern armies, which in turn was a product of the "demands of mechanized warfare."[69] As Stouffer explained, the armed forces, like the rest of society, had "gone through an industrial revolution."[70] That being so, it was only necessary to note the obvious differences between military and civilian technologies to see that the Army's personnel problems and the citizen soldiers' adjustment problems had a common source in the mismatch between the now quite differentiated occupational structure generated by the tools of war and the skills that draftees had acquired serving the machines of an equally specialized industrial economy.[71]

Apart from a very few scattered references to "intricate weapons requiring highly specialized skills for their management," there was no trace in *The American Soldier* of this view of the Army as a technologically transformed institution.[72] The Research Branch abandoned it in favor of a picture of the military as authoritarian, stratified, tradition bound, and isolated from democratic society.[73] In addition to being more consistent with what citizen soldiers said they disliked about military life, that picture had another great advantage over accounts centered on mismatched, technologically determined

---

68. Ibid., pp. vii–viii.

69. Ibid., p. 161.

70. Stouffer, "Social Science and the Soldier," p. 109.

71. See, for example, Joseph Rosenstein, "A Study of the Adjustment of the Civilian to the Army Occupational Structure," typescript accompanying E. W. Burgess to Samuel Stouffer, 10 March 1942, Stouffer Papers.

72. For one of the few references, see *TAS* II, p. 59.

73. *TAS* I, pp. 54–56.

occupational structures. It offered a satisfactory, and satisfying, explanation of the difficulties that the branch encountered in its dealings with the Army high command: social scientists in an organization characterized by rigid hierarchies, taken-for-granted rules and regulations, rigid discipline, and so on will (so the argument went) always arouse hostility and anxiety, especially among "strategic individuals in central positions of power and influence." The obvious reason was that such organizations always develop "strong pressures against innovative, or otherwise nonconforming, behavior," while social scientific research always carries with it the presumption of "innovations which have the potentiality of 'disturbing' the existing social structure."[74]

Arguments about the transforming effects of technological change on the conditions of modern warfare were reintroduced into military sociology in the late 1950s when it became clear that a "permanently expanded military profession" was a fact of American life. This made the Army itself, rather than the soldiers conscripted into it, an important subject for study. The sociologists who made that field their own were not especially concerned with examining the particular social contexts and consequences of particular technological innovations. They were content to offer some general pronouncements linking a presumed "radical transformation" of the military establishment to a "fantastic revolution in weapons," mostly by way of setting the scene for the points they really wanted to establish. The first of these was the contrast between "the simple division of labor of the feudal armed force" and "the complex skill structure of a modern professional military organization." Second came the consequent need for the military to develop new patterns of command, control, and communication appropriate to the changed technological situation. Finally, and most importantly, from the first and second points a third followed: new and better research was required on the "organizational factors" conducive to "adaptive change."[75]

In the immediate aftermath of World War II, this strategy of invoking technological change to demonstrate the utility of or-

---

74. Robins Williams, "Some Observations on Sociological Research in Government during World War II," *American Sociological Review* 11 (1946): 574, 577.

75. See Morris Janowitz, *Sociology and the Military Environment* (New York: 1959), pp. 8, 9, 26, 38, 86.

ganizational sociology was not open to American social scientists. They assumed that the most urgent social problems of the day had "close connections" with technological developments, but to stress that theme was only to invite those self-proclaimed "experts in technology," the natural scientists, to intervene in the field; and in the ensuing competition for public attention and resources, they were likely to win out. Social scientists might know that "*scientific* competence in the field of social problems [could] only be the result of a professional level of training and experience in the specific subject-matter," but that knowledge was no match for the popular prestige of the natural scientists, "their pronouncements on almost any subject" being "widely considered as oracular."[76]

This suggests an important part of the attraction that *The American Soldier* had for sociologists and social psychologists. Besides showing that their fields had contributed to the war effort, Stouffer and his associates had found a way of discussing a war that a novel technology was often popularly believed to have won without making any significant mention of any technological innovation. Even better, their findings pointed toward a national need that no amount of technological gadgetry could meet: more creativity "in the realm of ideology and our national values." In the not-very-long run that may have been an unfortunate discovery, especially because it came at a time when to imagine that the stakes in international politics were purely or even primarily ideological was to miss a significant point. The problem with the Soviet Union in the late 1940s was not that it was a communist country populated by ideologues, but rather that it was about to become a nuclear power. In a small way *The American Soldier* contributed to that misapprehension of an important reality.

Identifying ideological creativity as a national need meant more than 'bracketing out the technological dimensions of social problems from social scientific consideration. It also required, with one salient exception, ignoring the continued division of American society along class, ethnic, racial, religious, and regional lines. For once the exception proved the rule: black soldiers were singled out for special attention, but in *The American Soldier,* as in the Army itself, the effect was only to segregate them from the rest of the military, although that was

76. Parsons, "The Science Legislation," p. 665.

certainly not Stouffer's intention, convinced as he and his staff were that "the shame of America during this war has been its Jim Crow army."[77]

This last remark was contained in a Research Branch memorandum dealing with "post-war intolerance for minorities," including Jews, Mexican- and Japanese-Americans, and conscientious objectors, as well as blacks, all of whom seemed "likely to be objects of hostility."[78] Two proposals advanced in the memorandum suggest that the branch was not so much uninterested in questions of race, religion, and ethnicity as deeply worried about what they might find if they used those categories to analyze their data. The first proposal concerned demobilization and its impact on the factor that appeared most likely to lead to "minority persecution," unemployment. In recognition of the fact that "during certain periods of our history" minority oppression had "reached ugly heights and gotten completely out of control," it was urged that demobilization should be keyed to the state of the civilian labor market, with each soldier having to have "a specific job or full-time activity to go to before the army relinquishes its hold on him." But when it came to telling the soldiers why they were not being released immediately, there was obviously to be no mention of the underlying rationale. Instead, practical considerations were to be cited: "Large numbers will have to remain in service in occupied countries. Others will have to continue supply functions. The shipping problems, etc., etc., will all have to be gone into in detail." Beyond that, it seemed reasonable to urge that the various techniques developed by the Research Branch "to mobilize the G.I.'s for war . . . be used to mobilize them for peace." Hence, the second proposal: prepare a *How We Did It Series* to take the place of the *Why We Fight Series*, "all the time stressing we did it only because Joe Palotsky, Tony Marzoni, Mani Cohen, Pat McCarthy, John Henry, Willie Chang, Tom Henderson and many others were all in there working and fighting—*together*."[79]

Demobilization did not proceed in this fashion, and the *How We Did It Series* never got produced. But it is hard to avoid concluding from these proposals that racial, religious, and

---

77. "Proposals to Curb Post-war Intolerance for Minorities," n.d., Stouffer Papers.
78. Ibid.
79. Ibid.

ethnic issues were regarded as too important to be left to social science. A final example from early in the war reinforces the image of social engineers frightened of their data, and for good reason. In late May of 1942 the "Office of Facts and Figures" in Washington received a letter from the Massachusetts Committee on Public Safety reporting a "widespread rumor that a large number of Jewish men have, through dishonest methods, escaped the draft." The report was passed on to Stouffer with a request for "a check-up." His response was a near classic piece of evasion: The Army had no data on the number of Jewish soldiers; it could not get any; it kept no records on religion; the information it had was unreliable; when asked, Jewish soldiers were hesitant to say they were Jewish; doing a study was a bad idea; releasing the results would be a worse one; this was all unofficial.[80]

Conclusion

The Research Branch was a more honorable agency for some of the things it did not try to discover. But the consequent omissions made *The American Soldier* a poorer contribution to knowledge, although a better received one than it probably would have been otherwise. Those omissions also made it less socially useful than it might have been, because they helped sustain the profound illusion that the more public policy questions can be abstracted from wider social, political, and cultural conflicts, the more likely they are to be resolved on the high ground commanded by scientific rationality. That was about the only position open to social scientists who proposed to accept a self-denying ordinance concerning issues of race, religion, and class—especially when they combined that pledge with a studied indifference to technological change. Once those topics are ruled out of bounds, there is little else of much practical significance left to discuss. It was of course hoped that general sociological ideas and processes might turn up. But the practical applications that were expected from basic research on them look almost wholly ideological—as is suggested by the trajectory running from the rabid dog theory of fascism, via James Conant's search for values fit for "an open society,"

---

80. S. A. Stouffer to R. Keith Kane, 1 June 1942, Stouffer Papers.

to Fredrick Osborn's projected great debate with Soviet communism.

This is not how matters seemed at the time to sociologists and social psychologists who still had fresh memories of what it had been like to be told what to study by the Army. With no general hypotheses of their own to test, they had had no defense against the charge that the studies they wanted to do were really, as well as apparently, useless. The parallel with the situation of ordinary soldiers was clear: their lack of principled commitments to larger causes would have left them wholly adrift in the sea of military life had they not found security from being among their buddies. Once that analogy was recognized, it seemed plausible to try to solve both problems at once. Fundamental research on human relations would produce a replacement for the prewar public opinion that no longer existed, and the new national values would have a distinct advantage over the ones lost to view. If the Manhattan Project was any guide, the raising of social scientific sights to the level of pure theory would produce objective knowledge, not mere ideology.

That turned out to be a false prospect. Objective knowledge was hard to come by, as were common beliefs that could give uniform meaning and purpose to postwar life. To search for both at the same time was to run the risk of mistaking the one for the other when either came into sight. It also increased the likelihood that certain observations would pass by unnoticed because they seemed neither objective nor constructive. The G.I.'s comment about the Korean police action—"I never saw such a useless damned war in all my life"—is a case in point. It was not a symptom of cultural malaise, personality disorder, or failed small-group interaction, although all three ailments may have been in evidence. Nor was it an appeal for any of the remedies that social scientists had to offer: ideological creativity, psychotherapy, or social engineering. Right or wrong, it was a simple statement of the facts as one man saw them.

Unfortunately, so were the rumors about Jews evading the draft, Japanese soldiers looking for opportunities to die for their emperor, and World War II serving no larger moral purpose. The different ways in which these propositions were handled shows something of why postwar social scientists were not inclined to take the attitudes of the people they were studying very seriously, at least not as reports about reality. The experience of working in the military had taught them to distinguish the opinions they were surveying, those of enlisted men, from the opinions that were authoritative, those of their superiors in the War Department. The latter were not necessarily correct opinions, but the only relevant question about them was whether they were, since they directly affected planning and policy. Enlisted men were in a different category altogether. Their opinions were real enough, but whether they were factually correct was of little consequence. It was much more important to know whether they were conducive to good soldiering and, if not, how they could be changed. This made for certain asymmetries in the relations among social science, soldiers, and the War Department. What the Research Branch passed up to higher echelons of the military was knowledge; what it passed down to the civilian soldier was propaganda. There was no thought of surveying "attitudes" inside the War Department, just as there was no thought of asking soldiers to vote on policy issues and then abiding by the results.

Alexander Leighton's account of the Foreign Morale Analysis Division mirrors these social scientific commitments to objectivity and theory; to rising above conflicts of race, religion, and class; and to the tacit proposition that ordinary people need values, their leaders knowledge. As with all mirror images, in this one only left and right are reversed, not top and bottom, and the connections between the various parts of the picture remain visible. Thus, despite a strongly worded statement about how, "in a democratic society," the results of applied social research "must not remain as knowledge restricted to transient administrators nor to powerful groups that try to set national policy," the main emphasis in *Human Relations in a Changing World* was on social scientists, policymakers, and their respective roles in "the structure of administrative and policy-making organizations."[81] As a predictable consequence, the familiar questions about how and by whom problems for research should be formulated, what weight should be given to theoretical and practical concerns, and so on are framed—as in *The American Soldier*—with due consideration of bureaucratic politics.

But the Foreign Analysis Division had lost its one great battle in the War Department, over how to end hostilities in the Pacific, while the Research Branch had won its fight about de-

81. Leighton, *Human Relations,* pp. 100, 174.

mobilization. Leighton consequently came away from the war convinced that social scientists had to do more than provide policymakers with information about the attitudes of the other ranks. The opinions of men with power had to be subjected to social scientific scrutiny, and the findings of social scientists had to be subjected to the ordinary tests of practical politics. This meant raising rather than lowering the level of controversy surrounding applied social research. Looking back, Leighton concluded that his division should not have toned down its reports on the deterioration of Japanese morale in order to avoid disputes with the upper reaches of the Office of War Information. There was, he wrote in 1949, no "escape from the existence of conflicting interests," whether inside a bureaucracy or outside in the wider world of domestic and international politics. The choice was always "between being ineffective and fighting," and the social sciences were in no way exempt from that truth, even when they sought to stand above particular policy choices and move in the more rarefied air where values flourish. "Extensive conflict between many different kinds of values" was the norm; "social science includes one kind that is in the jumble with the rest."[82]

Stouffer and his staff might well have observed, with some justice, that Leighton was less than entirely persuaded about the power and uses of information simply because his division had produced so much less of it than the Research Branch. He never had the opportunity to make the "remarkable discovery" that, according to General Osborn, so impressed the branch: "The Army gave little weight to our personal opinions; but when these opinions were supported by factual studies, the Army took them seriously."[83] The obvious rejoinder is that, demobilization aside, the matters on which the branch had opinions were of relatively little moment compared to the question of how to proceed against Japan in early 1945. Stouffer was right to describe the work of the branch as social engineering, in the sense of being concerned with means rather than ends. For all that *The American Soldier* was at some pains to distinguish conceptually between attitudes and behavior, in order to show that knowledge of the former allowed predictions of and perhaps control over the latter, both had an instrumen-

---

82. Ibid., pp. 106, 138, 205.
83. *TAS* I, p. ix.

tal relation to the sorts of problems that are usually thought to require policy decisions. As Osborn noted, the question for the branch was how to "direct human behavior";[84] "toward what" was someone else's problem—except, as the difficulty over finding a suitable statement of war aims suggests, insofar as "ends" themselves turned out to have an instrumental relation to behavior, as motives.

Whether that exception was very large or very small depended, once again, on whose actions were under discussion. To repeat, since enlisted men were not making the decisions about what World War II was for, the Research Branch could and did treat the formulation of war aims as a question of means, not ends. But when the Foreign Morale Analysis Division tried that tack with regard to whether the war in the Pacific was being waged to dislodge the Japanese emperor, the distinction between means and ends proved harder to maintain. The division saw promises to retain the emperor as devices for lowering Japan's determination to continue the war, but officials in the War Department could not treat the issue exclusively on that level, and not just because some of them regarded the Japanese as suicidal fanatics immune to American psychological warfare, although that probably helped. For policymakers, the question of the emperor's future was a genuine issue, not merely an occasion for the odd propaganda coup.[85]

Leighton never quite took the point as it applied to the Japanese emperor, but he accepted the general principle. In the changing world he described, there were choices to be made about what to do as well as how to do it, and getting those choices right depended on more than understanding people's attitudes toward them. It necessitated looking directly at the questions of race, class, and technological change—especially but not exclusively as represented by atomic energy—that Stouffer and his associates were so reluctant to touch.[86] Leighton had nothing very novel to say on these issues. Given his experience as research leader of the Strategic Bombing Survey in Hiroshima, his emphasis on technological change was predictable, and so were his observations on the subject. By his own account they fell well within the prewar tradition that en-

84. Ibid.
85. Leighton, *Human Relations*, pp. 54, 118.
86. Ibid., pp. 106, 111–16.

couraged spotters of new technologies to look immediately for cultural lags: "It did not require Hiroshima to bring realization of the slowness of man in rising to meet his needs and his opportunities."[87]

Nor, it may be added, was the mere presence of atomic weapons sufficient to focus social scientific attention where Leighton thought it should be, on the practical "problems of human relations" rather than on the theoretical issues involved in understanding them. Yet, rather like the prospect of hanging, the bomb had the power to concentrate the mind wonderfully, at least for men who were in positions where they had to think about it. When *The American Soldier* was nearly finished, Stouffer wrote to Osborn asking him for a foreword to the first volume. Osborn was then serving on the Atomic Energy Commission of the United Nations and that gave him a vantage point for reflecting on the significance of what the Research Branch had done. If its approach "could be developed and more widely used," he wrote, "it might provide further impetus for a great advance in the social relations of man." On receiving Osborn's draft, Stouffer emended the last sentence to read "great advances in the study of human relations." His old chief vigorously objected: "the advances I hope for are not advances in the study of human relations, but actual advances in the relations between human beings."[88] Osborn won this small argument, but the Cold War and World War II conspired with the exigencies of academic life to insure that the lesson American social scientists drew from *The American Soldier* was Stouffer's: instead of changing human relations, the point was to interpret attitudes toward them.

---

87. Ibid., p. 38.
88. *TAS* I, p. ix; Fredrick Osborn to Samuel A. Stouffer, 12 September 1948, Stouffer Papers.

6

Military Needs, Commercial Realities, and the Development of the Transistor, 1948-1958
Thomas J. Misa

Of the many technological projects that the military has supported since World War II, none have proved more important than the transistor. In this essay, Thomas Misa examines the means the Army Signal Corps used to advance the new field of solid-state electronics. In treating the development phase of the transistor, Misa emphasizes technology as expanding knowledge and places the subject in an institutional context. The essay well illustrates the complexities of initiating high technology enterprises, particularly the problems that can arise when innovations begin to move from military applications to commercial use.

It was in the context of the search for an effective replacement for mechanical telephone relays that scientists at Bell Telephone Laboratories invented the transistor.[1] The device unveiled in 1948 was fragile, cumbersome, and clearly ill-suited for service outside the laboratory; yet within a decade the transistor had become the core of a rapidly growing sector of the electronics industry with annual sales exceeding $100 million. This essay will argue that the transition between invention and mass marketing, the development phase,[2] was to a large extent guided and funded by the United States military and in particular by the Army Signal Corps. This active role was not without its drawbacks for the emerging industry, however, as we shall show by examining the conflict between the needs of the military and of civilian industries, including those of the Bell System itself.

More generally this essay maintains that institutions such as military agencies can act as entrepreneurs and hence shape the process of technological change. Historians typically conceive

For comments on versions of this manuscript, I wish to thank David K. Allison, Thomas P. Hughes, Robert E. Kohler, Alex Roland, Merritt Roe Smith, Ed Todd, and several anonymous reviewers. Norma McCormick and Ruth Stumm of Bell Laboratories Archive, Short Hills, New Jersey, and Dr. Kenneth Clifford of U.S. Army Communications-Electronics Command, Fort Monmouth, New Jersey were of special help with sources.

1. One exception to the general neglect of the history of modern electronics is the invention of the transistor. See Charles Weiner, "How the Transistor Emerged," *IEEE Spectrum* (January 1973): 24–33; Lillian H. Hoddeson, "The Discovery of the Point-contact Transistor," *Historical Studies in the Physical Sciences* 12 (1981): 41–76. An important study is Ernest Braun and Stuart MacDonald, *Revolution in Miniature: The History and Impact of Semiconductor Electronics* (London: Cambridge University Press, 1978); see also idem, "The Transistor and Attitude to Change," *American Journal of Physics* 45 (November 1977): 1061–65. Of several accounts of the transistor written by Bell Laboratories personnel, one merits mention here: William Shockley, "The Path to the Conception of the Junction Transistor," *IEEE Transactions on Electron Devices* ED-23 (July 1976): 587–620. See also the special issue of the trade journal *Electronics* 53 (17 April 1980); M. Gibbons and C. Johnson, "Science, Technology and the Development of the Transistor," in B. Barnes and D. Edge, eds., *Science in Context* (Cambridge, MA.: The MIT Press, 1982), pp. 177–85; S. Millman, ed., *A History of Engineering and Science in the Bell System: Physical Sciences (1925–1980)* (Murray Hill, NJ: Bell Laboratories, 1983), 4: 71–107; Hoddeson, "The Roots of Solid-state Research at Bell Labs," *Physics Today* 30 (March 1977): 23–30; and idem, "The Entry of the Quantum Theory of Solids into the Bell Telephone Laboratories, 1925–40: A Case-Study of the Industrial Application of Fundamental Science," *Minerva* 18 (1980): 422–47. On the integrated circuit, see Michael F. Wolff, "The Genesis of the Integrated Circuit," *IEEE Spectrum* (August 1976): 45–53; and Jack S. Kelly, "Invention of the Integrated Circuit," *IEEE Transactions on Electron Devices* ED-23 (July 1976): 648–54.

2. Several economists have studied the transistor but have failed to appreciate the wide gulf separating invention from marketing. For example, one study stresses the scientific basis for the invention of the transistor: Richard R. Nelson, "The Link Between Science and Invention: The Case of the Transistor," in National Bureau of Economic Research,

of entrepreneurs as individuals responsible for inventing a technology, presiding over its development, and contributing to its eventual shape and style.[3] In the case of the transistor, the Army Signal Corps had a marked effect on the content and even the style of the technology. By sponsoring applications studies, conferences, and publications in the late 1940s and early 1950s, the military services ensured a rapid dissemination of the new technology to the electronics industry. By subsidizing the construction of manufacturing facilities and overseeing the setting of standards, they influenced the size and structure of the emerging transistor industry. And finally, military requirements biased the industry toward the development of specific types of transistors. Military sponsorship helped shield the new technology from undue criticism and economic constraint and also provided the necessary momentum to push it through the development stage to commercialization.

The Search for the Transistor

The development of the transistor occurred in a period of rapid growth and qualitative change in the American electronics industry. Whereas in 1930 radios had accounted for 90 percent of the industry's total sales of $103.5 million, after World War II radio's share of sales dropped to 20 percent, owing to the rapid acceptance of television and the expanding needs of industry and the military. In 1950 total civilian sales of electronics equipment reached $1.1 billion, and by the end of the decade civilian plus military sales topped $10 billion. Wallace B. Blood, business manager of the trade journal *Electronics,* pro-

---

The Rate and Direction of Inventive Activity: Economic and Social Factors (Princeton: Princeton University Press, 1962), pp. 549–83. Another study stresses the role of market forces in the dissemination of semiconductor technology: John E. Tilton, *International Diffusion of Technology: The Case of Semiconductors* (Washington, D.C.: Brookings Institution, 1971). There are problems with both of these approaches. During development many organizational factors external to science are critical. Further, in the United States a great deal of transistor technology was disseminated before any vigorous market existed.

In order to explore the important but often neglected phase between invention and mass marketing, I will focus explicitly on the development phase of the transistor. For concepts I have drawn on John M. Staudenmaier, "Design and Ambience: Historians and Technology, 1958–77," (Ph.D. dissertation, University of Pennsylvania, 1980), pp. 138–48; and Thomas P. Hughes, "The Development Phase of Technological Change," *Technology and Culture* 17 (July 1976): 423–31.

3. For a classic discussion of entrepreneurs, see Thomas P. Hughes, "The Electrification of America: The Systems Builders," *Technology and Culture* 20 (1979): 124–61.

claimed that electronics manufacturing had undergone "a metamorphosis unique in industrial history."[4] Yet, despite this striking sales spurt, the really fundamental change in the industry was technical—the introduction of solid-state electronics, a radically new technology that broke the vacuum tube's half-century monopoly.

As the vanguard of this revolution, the Bell Telephone Laboratories were well prepared to translate innovative technical concepts into industrial realities. Scientific curiosity, technological utility, and corporate goals had successfully mixed in the Bell System before, most notably around World War I in the exploitation of Lee De Forest's original patent for the vacuum tube.[5] With the transistor, a complex of institutional and cognitive factors were once again to influence technological development.

The impetus to develop a solid-state amplifying device came from Bell's director of research, Mervin J. Kelly. Kelly had long desired to replace the mechanical relays in telephone exchanges with electronic relays. In 1939 he had placed two physicists, experimentalist Walter Brattain and solid-state theoretician William Shockley, on a project to construct a solid-state amplifier with the semiconductor copper oxide. The device failed to behave as predicted, however, and the plans were set aside to make room for war-related work. Between mid-1941 and 1945 Bell turned over nearly three-quarters of its facilities to such military projects as radar, radio and wire-based communications, aircraft training simulators, antisubmarine warfare, proximity fuzes, electronic computers for gunfire control, electronic countermeasures systems, and research on materials for the atomic bomb. Brattain and Shockley left the Laboratories for separate assignments, but Bell continued solid-state research by sponsoring a team of chemists and metallurgists who worked under the direction of MIT's Radiation Laboratory to purify the semiconductor material silicon for use in radar. Meanwhile a group at Purdue University

4. *Electronics* 53, p. 276.

5. For contrasting interpretations of the development of the vacuum tube, see Leonard S. Reich, "Industrial Research and the Pursuit of Corporate Security: The Early Years of Bell Labs," *Business History Review* 54 (Winter 1980): 504–29; and Lillian Hoddeson, "The Emergence of Basic Research in the Bell Telephone System, 1875–1915," *Technology and Culture* 22 (July 1981): 512–44. See also Arturo Russo, "Fundamental Research at Bell Laboratories: The Discovery of Electron Diffraction," *Historical Studies in the Physical Sciences* 12 (1981): 117–60.

headed by physicist Karl Lark-Horovitz also conducted research on yet another semiconductor material, germanium. The experiments of the Purdue group were so similar to those done later at Bell that some writers have speculated that had Lark-Horovitz been looking for a solid-state amplifier, instead of exploring general physical phenomena, his group would have invented the transistor.[6]

One important outcome of the war was that new and mutually satisfying relationships were forged between the communities of science, technology, and government. Vannevar Bush, an MIT electrical engineer; James B. Conant, a chemist and the president of Harvard University; and other members of the country's scientific and engineering elite had created new federally funded but civilian run organizations that had proved capable of enlisting and directing the nation's technological expertise.[7] Yet immediately after the end of hostilities, researchers found themselves without a federal patron. Designed as a purely wartime institution, Bush's Office of Scientific Research and Development (OSRD) was disbanded in 1945, and it was not until 1950, with the act authorizing the National Science Foundation, that Congress created a nonmilitary mechanism to fund basic research. In the interim, several military agencies stepped into the breech. A Joint Research and Development Board was organized in 1946 to coordinate military research and development, but it remained weak, and the Office of Naval Research soon emerged as the largest patron of the post-

6. My discussion of the invention of the transistor follows Braun and MacDonald, *Revolution in Miniature*, chapters 1–4 and Hoddeson, "Discovery." For Kelly's motivation for doing semiconductor research, see Hoddeson, "Discovery," pp. 45, 47, 52–54. Kelly joined the research division of the engineering department of Western Electric, the predecessor of the Bell Telephone Laboratories, in 1918; preceding his appointment as director of research in 1936, he was director of vacuum tube development and director of transmission instruments and electronics. He became executive vice president of the Laboratories in 1944 and president in 1951. See *Bell Telephone Magazine* 32 (Summer 1953): 70–71; and Michael F. Wolff, "Mervin J. Kelly: Manager and Motivator," *IEEE Spectrum* 20 (December 1983): 71–75. For Kelly's views on research and development, see his essay, "The Bell Telephone Laboratories—An Example of an Institute of Creative Technology," *Proceedings of the Royal Society* 203A (1950): 287–301. For a technical account of Bell's wartime projects, see M.D. Fagan, ed., *A History of Engineering and Science in the Bell System: National Service in War and Peace (1925–1975)* (Murray Hill, NJ: Bell Laboratories, 1978), 2: 3–351.

7. For background concerning science and technology in World War II, see Daniel J. Kevles, *The Physicists* (New York: Knopf, 1977), pp. 287–348; Carroll Pursell, "Science Agencies in World War II: The OSRD and Its Challengers," in N. Reingold, ed., *The Sciences in the American Context* (Washington, D.C.: Smithsonian Press, 1979), pp. 359–78; and James Phinney Baxter, *Scientists Against Time* (Cambridge, MA: The MIT Press, 1968).

war period.[8] While this more complicated system of support was being set up, university efforts, including the transistor group at Purdue,[9] were left in limbo. In contrast, private industrial efforts benefited from the relative stability of their own (mostly internal) funding.

Bell was thus not only well prepared but also well situated in the postwar period to make the most of the accumulated experience with semiconducting materials. With stable research funding, a multidisciplinary staff of over 2000 scientists and engineers, and a ready market in the massive Bell System, the Laboratories had the resources and the incentive to mount a major technological effort in the new field of semiconductor electronics.[10] The final element contributing to its program was a goal-oriented approach: the Laboratories would not repeat the near miss of Lark-Horovitz at Purdue. Having been promoted to vice-president in charge of research, Kelly signed the authorization to begin solid-state work in June 1945, two months before V-J Day. Shockley and physical chemist Stanley Morgan headed a new solid-state physics department that included Brattain; John Bardeen, a theoretician from the Naval Ordnance Laboratory; experimentalist Gerald Pearson; physi-

8. On the postwar science policy debate, see Kevles, "The National Science Foundation and the Debate over Postwar Research Policy, 1942–1945: A Political Interpretation of *Science—The Endless Frontier,*" *Isis* 68 (1977): 5–26; and idem, "Scientists, the Military, and the Control of Postwar Defense Research: The Case of the Research Board for National Security, 1944–46," *Technology and Culture* 16 (January 1975): 20–47. On the ONR see Harvey M. Sapolsky, "Academic Science and the Military: The Years Since the Second World War," in Reingold, *American Context,* pp. 379–99; and Mina Rees, "The Computing Program of the Office of Naval Research," *Annals of the History of Computing* 4 (1982): 102–20.

9. To at least one person, the Purdue and Bell efforts were competing. When Bell announced the transistor in June 1948, William Shockley, one of the Laboratories' three physicists who were to share in the 1956 Nobel Prize in physics for the transistor, reportedly cornered a Signal Corps officer to ask, "Tell me one thing, have Lark Horowitz and his people at Purdue already discovered this effect, and perchance has the military put a TOP SECRET wrap on it?" See *Microwave Journal* 8 (July 1966): 96. No all-encompassing security restriction was placed on the transistor, although this option was discussed during 1948.

10. Incorporated in 1925, Bell Telephone Laboratories was owned jointly by American Telephone and Telegraph (AT&T) and Western Electric. AT&T authorized and paid for basic research; Western Electric authorized and paid for the development of technology applicable to its products. In early 1952 the total Laboratories' staff numbered 6900, of whom 2500 were engineers and scientists; 2100 were draftsmen, technical assistants, and mechanics; and the remaining 2300 were nontechnical personnel and managers. For military work, Western Electric was usually the prime contractor, while the Laboratories carried out projects as a subcontractor. See Robert N. Anthony, *Management Controls in Industrial Research Organizations* (Boston: Harvard University Press, 1952), pp. 382–84. Throughout this essay, unless explicitly noted, the terms "Bell" and "the Laboratories" refer to Bell Telephone Laboratories.

cal chemist Robert Gibney; and Hilbert Moore, an electronics specialist. By Christmas 1947 the group had built their first device. With an ungainly laboratory apparatus, Bardeen and Brattain demonstrated convincingly that electrical amplification could occur between two closely spaced contacts on the surface of a sample of germanium (figure 1). Now began the complex and difficult process of developing a laboratory curiosity into an electronic device capable of functioning in the real world.

Several changes in Bell's organization expedited the development effort. At the time the Laboratories' physicists, chemists, and mathematicians tended to associate only with colleagues in their own discipline. As Bell's vice-president Ralph Bown observed, "Such a grouping plan tends to create dividing walls of thought, and alongside such walls often are moats in which good ideas may sink out of sight."[11] To prevent its newly invented device from slipping into such a moat, Bell formed a special project group early in 1948. A three-man committee coordinated the effort: physicist Shockley headed transistor research, physical chemist Addison H. White led research on electronic materials, and electrical engineer Jack A. Morton directed fundamental development. In this position and later as head of the entire transistor development effort, Morton had critical perspective on the difficulties Bell would encounter with developing transistors for both civilian and military use. His project reports provide the basis for a middle level of analysis between detailed technical reports and publicity-oriented articles.

Morton's group played a key role. Its specific tasks included examining the factors controlling the device's amplification bandwidth and noise level, improving the energy gain per stage of amplification, and conducting studies of the basic materials, manufacturing processes, and precise structures needed in order to produce transistors for specific applications. Organizationally separate from the physicists, chemists, and metallurgists, this group had the general responsibility for coordinating the work of these specialists. This division of labor preserved a degree of autonomy for the scientists while ensuring that Bell's broad spectrum of resources would be fully utilized in the process of development. Capitalizing on its postwar organization

11. Ralph Bown, "The Transistor as an Industrial Research Episode," *Scientific Monthly* 80 (January 1955): 45.

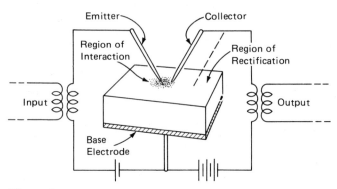

Figure 1
Point contact transistor as amplifier. Through electronic phenomena not well understood when the transistor was invented, changes in the voltage across the input leads produced changes in the voltage across the output leads. A microphone, for example, could be connected to the input and a loudspeaker to the output. *Source: Bell Laboratories Record* 50 (December 1972): 352.

and its early jump on competitors, Bell dominated transistor technology into the mid-1950s. As late as 1955 the Laboratories still collected 37 percent of the patents issued in the semiconductor field (table 1).[12]

Although the initial model device seemed to hold great promise, it also posed a formidable array of technical problems. The first was its name. During the spring of 1948 the staff debated a variety of imaginative names, including semiconductor triode, surface states triode, crystal triode, and iotatron, before "transistor" became widely accepted.[13] A more significant problem was that the early devices were a curious blend of abstract quantum mechanics and cut-and-try tinkering. Solid-state physicists could explain reasonably well the phe-

12. Tilton, *International Diffusion*, p. 57; and Francis Bello, "The Year of the Transistor," *Fortune* (March 1953): 132. For general relationships between organization and technological efforts, see Bruce Parrott, "The Organizational Environment of Soviet Applied Science," in L. L. Lubrano and S. G. Solomon, eds., *The Social Context of Soviet Science* (Boulder: Westview Press, 1980), pp. 69–100, esp. 70–71; and essays by Douglas and Allison in this volume.

13. L. A. Meacham, et al., "Terminology for Semiconductor Triodes—Committee Recommendations–Case 38139-8," Memorandum for File, MM-48-130-10, 28 May 1948, located in binder: Publications, Jack A. Morton Collection, Bell Laboratories Archive, Short Hills, New Jersey (collection hereafter cited as JAM Collection).

Table 1
Breakdown of semiconductor patents by firms (percent)

	1952	1953	1954	1955	1956
Bell Laboratories	56	51	46	37	26
Tube Firms [a]	37	40	38	42	54
New Firms [b]	7	9	16	21	20
Total Number of Patents Granted	60	92	79	73	186

a. Includes Radio Corporation of America, General Electric, Westinghouse, Sylvania, Philco-Ford, and Raytheon, all firms with previous experience manufacturing vacuum tubes.
b. Includes International Business Machines, Motorola, Hughes, International Telephone and Telegraph, and Clevite, firms that had been set up explicitly to manufacture semiconductor components or that had no previous experience manufacturing vacuum tubes.
Source: John Tilton, *International Diffusion of Technology* (Washington, D.C.: Brookings Institution, 1971), p. 57.

nomena that occurred *within* materials, but the point-contact transistor exploited poorly understood *surface* phenomena. Accordingly, the group relied more on empirical practice than physical theories to guide their work. For example, an enigmatic step called "forming" was used to attach the two closely spaced metal wires to the purified semiconductor pellet. Passing a burst of current through the germanium-metal contact attached the wires to the transistor and also, for reasons that were unclear, improved its overall performance. Finally, reliability was a problem. After manufacture, which consisted of hand assembly under a microscope, the devices were tested to determine if they would amplify electronic signals. Most did not. Four-fifths of the earliest devices were rejected, and even those that passed had serious weaknesses. Early transistors had operating characteristics that varied with the ambient temperature; they suffered from extremely high electrical noise; and they had painfully modest maximum power ratings and limited frequency ranges.[14]

Despite the transistor's many obvious shortcomings, its potential advantages were widely discussed in the trade literature.

14. B. N. Slade, "Survey of Transistor Development (Part 1)," *Radio and Television News* 48 (September 1952): 43; idem, "Survey of Transistor Development (Part 2)," *Radio and Television News* 48 (October 1952): 65, 112–14; Millman, *History of Engineering and Science*, pp. 119–22; Mervin J. Kelly, "The First Five Years of the Transistor," *Bell Telephone Magazine* 32 (Summer 1953): 77; and Hugo Gernsback, "Transistor Trends: Transistor Evolution has only Begun," *Radio-Electronics* 29 (May 1958): 33.

Even a miniature vacuum tube was much larger than the tiny transistor. Further, tubes required a bulky auxiliary power supply to heat the electrodes, which tended to burn out after a few thousand hours of use.[15] Pictures of the rooms filled with hot, glowing tubes needed for the first electronic computers offer striking testimony of the acute need for smaller substitutes. Bell Telephone Laboratories clearly saw the transistor's promise and committed resources toward its realization, but at the same time the transistor benefited from a timely fit into a preexisting program to miniaturize electronic military equipment.

The Military Promotes the New Technology

A conscious program to miniaturize military electronics began in the late 1930s when the Army Signal Corps Engineering Laboratory (SCEL) at Fort Monmouth, New Jersey, designed the first "walkie-talkie." A football-size, two-tube transmitter-receiver with a separate telephone handpiece, the walkie-talkie was meant for infantrymen on reconnaissance missions and for forward-observer fire-control personnel. The device allowed individual soldiers to remain in contact with commanders without having to drag a telephone wire or carry a heavy, full-size radio. In the immediate prewar period, the Signal Corps Engineering Laboratory developed a six-pound, completely integrated "handie-talkie," which served during the war as the workhorse of battlefield communications.

The SCEL made further refinements to Army communications during the war, but they also encountered several technical and organizational difficulties. As the number of items of electronic equipment increased and as evolving military tactics required ever more complex gear, the Signal Corps' procurement system proved incapable of coordinating the process. Moreover, reports received by the SCEL during and immediately after the war revealed many inadequacies in Army electronics. One major problem was an inability to stand up to environmental extremes such as fungus, moisture, and corro-

15. For a systematic comparison of the transistor and the vacuum tube circa 1953, see Emerick Toth and William N. Keller, "Principles of Transistor Application and System Design," in *Symposium on the Application of Transistors to Military Electronics Equipment*, held at Yale University, 2–3 September 1953, sponsored by the Committee on Electronics (Research and Development), Office of the Secretary of Defense. Copy in Library of Moore School of Electrical Engineering, University of Pennsylvania, Philadelphia, Pennsylvania.

sion in the jungle and freezing in the arctic. Other problems could be traced to the battlefield rigors of shock, vibration, temperature changes, and weather. And the sheer size and bulk of the equipment hampered operations. In response to these problems and because many Signal Corps personnel anticipated that electronics would assume even greater importance in postwar communications, surveillance, fire control, countermeasures, and intelligence, the Corps undertook a long-range research and development program to produce an integrated system of communications equipment. Miniaturization received particular emphasis.[16]

From the start the SCEL focused its miniaturization efforts on circuit-assembly techniques. One avenue it explored was the ceramic-based circuit developed for the National Bureau of Standards by the Centralab Division of the Globe-Union Corporation. Although this technique was successful in the miniature circuits of the Army's proximity fuze, it required a large investment in specialized production equipment and could be used only for simple, resistor-capacitor circuits. Complex circuits remained a problem. The labor-intensive process of hand soldering components was the only well-tested mass production method, but there were limits to reducing the size of circuits because the wiring became chaotic and prone to fail.

Working with industry, the SCEL invented, patented, and refined an automatic soldering system that bypassed the wiring problem. Individual components were plugged into a plastic board on whose backside a wiring diagram had been etched in copper. When the copper side was dipped into a molten solder bath, the components were automatically attached to the board and also properly connected. To underscore the potential of the process for mass production, the Signal Corps named it "Auto-Sembly." Perfecting this new production method oc-

---

16. In 1946 the SCEL Ad Hoc Committee on Miniaturization stated that "miniaturization should and will be a major objective in the design of future Signal Corps' equipment." William R. Stevenson, "Miniaturization and Microminiaturization of Army Communications—Electronics, 1946–1964," U.S. Army Electronics Command, Fort Monmouth, New Jersey, Historical Monograph 1 (Unpublished manuscript, 1966), pp. 1–12, quote p. 12 (volume hereafter cited as Army Miniaturization Monograph). For the Signal Corps in World War II, see its three-part official history: Dulany Terrett, *The Signal Corps: The Emergency* (Washington, D.C.: Government Printing Office, 1956); George Raynor Thompson, et al., *The Signal Corps: The Test* (Washington, D.C.: Government Printing Office, 1957); and George Raynor Thompson and Dixie R. Harris, *The Signal Corps: The Outcome* (Washington, D.C.: Government Printing Office, 1966).

cupied the SCEL into the early 1960s.[17] Auto-Sembly was an important step forward in miniaturization that would become even more useful when paired with the tiny solid-state components emerging from Bell Telephone Laboratories.

By the spring of 1948 Bell was satisfied that their preliminary work on transistors would be patentable, and a public demonstration was set for June 30. A week before the unveiling, Oliver E. Buckley, the president of the Laboratories, invited the military services for an advance look. Buckley conducted this special briefing, and Ralph Bown demonstrated the ability of the transistor to serve variously as a telephone amplifier, a radio receiver, and a circuit oscillator. There were six people in the group, two each from the Army, Navy, and Air Force. The two Army representatives—Colonel E.R. Petzing, commanding officer of the SCEL, and Harold A. Zahl, the new director of research of the SCEL—talked later that afternoon with Bell executives Bown, James McRae, and Donald Quarles about a possible Signal Corps contract in the transistor field. Quarles, who later became Assistant Secretary of Defense for Research and Development, is reported to have replied bluntly that Bell's research was not for sale. Nevertheless, arrangements were made to keep the services informed about new applications. On July 2 Zahl submitted an enthusiastic report about the meeting to his commanding officer: "The phenomena . . . will have great significance in the Signal Corps research and development program. Of particular interest is apparent promise of reduction in power requirements for electronic gear, miniaturization aspects, new current techniques, etc."[18]

Describing his "immediate course of action," Zahl planned to inform all SCEL personnel of the new device, to arrange visits to Bell for key staff, and to procure sample transistors from Bell. Within the month the SCEL formed a transistor group composed of two engineers and one physicist, in its thermionics branch, which, like Bell's, was part of the vacuum tube department. Although the transistor effort grew in a few years to full branch status, with approximately forty engineers, physicists, and chemists, the limited availability of transistors hampered early work. In an attempt to increase the number of available

17. Army Miniaturization Monograph, pp. 18–114.

18. For a reprint of Zahl's 1948 report and his later recollections of the meeting with Bell, see *Microwave Journal* 8 (July 1966): 94, 96.

devices, the SCEL created a small manufacturing facility, which produced fifty point-contact transistors in 1949.[19]

Following Zahl's enthusiastic report and the Signal Corps' initial tests, the military services persuaded Bell to sign a contract for study of applications. As parties to the June 1949 contract, all three services charged the Laboratories with investigating the usefulness of the device in switching circuits such as those of digital computers. Specifically, the project was to examine the feasibility of using miniature, standardized, plug-in transistor packages in a 370-tube data transmission set. By May 1951 Bell completed initial work and demonstrated the potential of point-contact transistors, which had an expected lifetime of 70,000 hours and yielded a fourfold reduction in volume and an eightfold reduction in power requirements over the tube version of the data set. The work produced two significant, concrete results. First, the eight reports for this contract provided the first published research on transistor applications to digital computers. Second, a later model of the data set became the first military equipment produced by Bell's manufacturing affiliate, Western Electric, using large numbers of transistors.[20]

Concurrent with this feasibility study, Bell also conducted a project for the Navy Bureau of Ordnance that may have been the first application of transistors. W.H. MacWilliams, Jr., an engineer trained at Johns Hopkins who had worked on fire control for the Navy during the war, successfully transistorized a component of a Bell simulated warfare computer in early 1949. The simulator used forty transistors, nearly all that were available at this early date.[21] MacWilliams's project prefigured the close interplay later to emerge between Bell's work on military systems and transistor development.

Two political events contributed to the military patronage of

19. Army Miniaturization Monograph, p. 118.

20. Bell Telephone Laboratories, "An Appraisal of Military Transistor Development, 1948–1957," 7 August 1957, n.p., copy in folder: Solid State Devices—Semiconductors, Subject Files Box 90, Bell Laboratories Archive (document hereafter cited as BTL Report, 1957). See also J. P. Molnar, "Military Applications of Transistors," speech given at press conference, 17 June 1958, copy in folder: Transistor—Tenth Anniversary, Lloyd Espenschied Collection, Bell Laboratories Archive; and Fagan, *History of Engineering and Science,* pp. 549–50. The activities within Bell between the initial military briefing and the signing of the first transistor contract deserve further study.

21. W. H. MacWilliams, "A Transistor Gating Matrix for a Simulated Warfare Computer," *Bell Laboratories Record* 35 (1957): 94–99.

the transistor. The National Security Act of 1947 centralized the military services under a new Secretary of Defense. The first secretary, James Forrestal, was confronted by a tradition of interservice rivalry, competition, and duplication. By creating a Research and Development Board under the secretary to supplant the ineffective Joint Research and Development Board, the act also attempted to improve the coordination of military research and development. As an arm of the Secretary of Defense, however, the board reflected the department's bureaucratic weakness. The board consisted of a civilian chairman and two representatives from each of the three services, and it operated through a complex web of committees and subcommittees. Further, it had no control over money and could only coordinate the projects that the services had already initiated.[22]

As elsewhere, the board's activities in the transistor field were loosely structured. An Ad Hoc Group on Transistors was organized in the summer of 1951. Chaired by a Bell vice-president, James McRae, the group was supposed to be a high-level body that would set broad policies for coordination of the transistor programs of the three services. It was joined shortly by a permanent Subpanel on Semiconductor Devices, but the group retained decision-making power. Although the military's transistor program was officially a joint service undertaking, the board's weakness permitted the Army Signal Corps to draw on its greater experience to become the military's center of transistor expertise. The Corps' prominence became manifest in 1951, when the Electronic Production Resources Agency, with the concurrence of the three services, assigned to it the responsibility of developing the new technology for military purposes.[23]

A second political event, the Korean War, strengthened Bell's ties to the military. This was demonstrated particularly in

22. Demetrios Caraley, *The Politics of Military Unification* (New York: Columbia University Press, 1966); Edwin A. Speakman, "Research and Development for National Defense," *Proceedings of the Institute of Radio Engineers* 40 (July 1952): 772–75; and Booz, Allen and Hamilton, Inc., *Review of Navy R&D Management, 1946–1973* (Washington, D.C.: Department of the Navy, 1976), pp. 16–25. For the administrative difficulties of the Research and Development Board, see Don K. Price, *Government and Science* (New York: New York University Press, 1954), pp. 144–52; and James L. Penick, Jr., et al., eds., *The Politics of American Science* (Chicago: Rand McNally, 1965), pp. 191–93.

23. I. R. Obenchain and W. J. Galloway, "Transistors and the Military," *Proceedings of the Institute of Radio Engineers* 40 (November 1952): 1287–88; and Army Miniaturization Monograph, p. 119. See also Jack A. Morton, copy of talk given to Generals Partridge and Doolittle, 7 January 1953, copy in binder: Semiconductor Devices—General, JAM Collection.

the work on air defense systems. In the closing months of World War II, the Army Ordnance Corps had asked Bell to study the feasibility of using ground-launched guided missiles against attacking bombers. By the outbreak of hostilities in Korea in 1950, Bell had nearly completed the design for the Nike air defense system, and the Army instituted a crash production program.[24] This expanded Nike program revealed a 250 percent increase in national military research and development expenditures between 1950 and 1953 (from $600 million to $1.6 billion).[25]

During the 1950s the military was also involved in a number of publicity efforts aimed at disseminating the new technology. As an explicit task of its second military transistor contract, Bell Laboratories held a symposium on transistor characteristics and applications at its headquarters in Murray Hill, New Jersey, in September 1951. Staff members presented twenty-five lectures and demonstrations to over three hundred representatives of the military services, universities, and electronics firms. Far from being an academic gathering, 139 industrial and 121 military personnel soundly outnumbered the 41 university representatives. In November the symposium proceedings appeared in a widely circulated 792-page volume. Each participant received a copy, the military services distributed 5500 copies at government expense, and Bell transistor licensees received an unknown number.[26]

A conference for Bell licensees in April 1952 disseminated more detailed information about the new technology itself.

24. Charles C. Duncan, "Communication and Defense," *Bell Telephone Magazine* 37 (Spring 1958): 16.
25. Arthur D. Little, *Basic Research in the Navy* (Cambridge, MA: Arthur D. Little, 1959) 1: 13. See also Samuel P Huntington, "NSC-68 and Rearmament, 1950–1952," in *The Common Defense: Strategic Programs in National Politics* (New York: Columbia University Press, 1961), pp. 47–63; and Paul Y. Hammond, "NSC-68: Prologue to Rearmament," in Warner R. Schilling, Paul Y. Hammond, and Glenn H. Snyder, *Strategy, Politics, and Defense Budgets* (New York: Columbia University Press, 1962), pp. 267–378.
26. The Bell Telephone Laboratories report on the first conference listed the participants by name and institutional affiliation. No branch of the military services received preferential treatment either in attending the symposium or in receiving the published proceedings. See "Final Report on Task 3—Transistor Symposium," 1 February 1953, copy in an unmarked box in the Robert M. Ryder Collection, Bell Laboratories Archive. The general theme of the diffusion of technology has been most fully explored by Nathan Rosenberg; see his classic article, "Technological Change in the Machine Tool Industry, 1840–1910," *Journal of Economic History* 23 (December 1963): 414–43; and his collection, *Perspectives on Technology* (London: Cambridge University Press, 1976), pp. 141–210.

This conference, held at the urging of the military services and with funding from Western Electric, resulted in two fat volumes that became the canon of transistor technology (they were known within Bell as "the Bible"). As late as 1957 an internal Bell report described these volumes as "the first and still only comprehensive detailed treatment of the complete material, technique and structure technology." The same report noted that the information provided "enabled all licensees to get into the military contracting business quickly and soundly."[27]

Several factors account for the rapid assimilation of the new technology by the military bureaucracy. Because the Army Signal Corps had institutionalized the goal of miniaturizing electronic communications gear, it was primed for the announcement of the point-contact transistor. The transistor also complemented the Corps' new component-oriented mass production process. Finally, the rearmament effort following 1950 released new research and development funds for projects of military interest. These organizational, technological, and political factors combined to make the military a vigorous patron and promoter of the new technology. By sponsoring applications studies, organizing bureaus for production development, and disseminating the new technology to industry, the military assumed responsibility for presiding over the process of technological development and hence began its activities as an institutional entrepreneur in this new field. Much remained to be done with transistor development in the early 1950s. Before any large-scale production runs could be accomplished, a number of technical problems and manufacturing bottlenecks had to be overcome.

Technological Advances

By the end of World War II, scientists at Bell Laboratories had produced several metallurgical innovations that were to aid the invention of the transistor and exert an important influence in its subsequent development. First, J.H. Scaff and H.C. Theuerer had discovered that nearly pure silicon ingots could be prepared by melting silicon in a vacuum. These purified

27. BTL Report, 1957. See also the proceedings of military-sponsored conferences cited in notes 15 and 53.

ingots possessed a curious property: some would rectify current—that is, they would act as a one-way valve for the passage of electricity—only when they were in a negatively charged electrical field; others would do so only when they were in a positive field. Scaff and Theuerer named the former "n-type" and the later "p-type." Tipped off by a slight odor of phosphorus when the ingots were removed from the oven, the two metallurgists determined that extremely small amounts of impurities, below the level of spectroscopic detection, were responsible for the peculiar behavior. They found that elements on either side of the fourth column of the periodic table (the column that contains the semiconductors silicon and germanium) most actively produce the rectification effect. Elements from the fifth column, including phosphorus and arsenic, donate their excess electrons to the semiconductor's crystal lattice and make it n-type, whereas elements from the third column, including boron and indium, induce a deficit of electrons and make the crystal p-type. To prepare semiconductor materials for transistors, then, one simply had to dope a pure sample with a tiny amount of the desired impurity— approximately one atom in one hundred million. Using techniques developed at Bell by Gordon Teal and J.B. Little, one could grow a large single crystal (typically 8 cm. long and 2.5 cm. in diameter) from the doped sample. This crystal could then be diced into the small pellets needed for point-contact transistors.[28]

A refinement of the crystal-growing apparatus allowed the Laboratories to realize a radically new type of transistor. William Shockley had proposed the idea of a "junction transistor" early in 1948 and had elaborated its theory in a book, *Electrons and Holes in Semiconductors,* in 1950. A transistor consisting of three sandwiched layers of p- and n-type germanium was an elegant conception, but with the crystal-growing techniques then available, it simply could not be made. Not until 1951 did Teal and Morgan Sparks manage to modify their crystal-growing apparatus to accept pellets of impurities. This innovation made it possible to build Shockley's germanium sandwich. While a mechanical apparatus continuously pulled a solid bar out of a crucible of molten n-type germanium, it was doped

28. *Electronics* 53, pp. 223, 226; Slade, "Survey of Transistor Development (Part 1)," pp. 44–45; and Millman, *History of Engineering and Science,* pp. 417–22.

with a small amount of p-type impurity and then quickly re-doped with an excess of n-type impurity. The resulting n-p-n wafer in the bar was cut out and diced, and tiny leads were attached to its three regions, producing a "grown junction" transistor (figure 2). The key was the center layer, the "base," which controlled the passage of current across the device from the "emitter" to the "collector." Junction transistors had the advantage of relying not on the poorly understood surface phe-nomena that the point-contact transistor exploited, but on the less complex, better understood interactions of the two internal p-n junctions. They were also electronically less noisy and me-chanically less fragile.

A year later work at General Electric and at the Radio Corpo-ration of America yielded a second method of constructing junction transistors. Two p-type pellets were placed on oppo-site sides of a thin n-type wafer. When heat was applied, the pellets melted slightly into the wafer, producing an "alloy junc-tion" transistor.[29]

Bell Laboratories announced two further advances in 1954. Purifying the semiconductor material had continued to be a problem, since the process required an extremely low level of impurities controlled to within a few percent. Purification was now greatly aided by the introduction of "zone refining," which Bell's W. G. Pfann had adapted from aluminum technology. Perfected after a three-year effort, the new procedure utilized the fact that impurities are more soluble in the liquid than in the solid phase. A heating apparatus slowly swept a narrow band of molten material across a horizontal bar of solid semicon-ductor material, carrying impurities to one end of the bar, which was then cut off. Zone refining could be repeated several times to reduce unwanted impurities to less than one part per billion.

The second major invention was the diffusion technique for manufacturing transistors. Junction transistors had been re-stricted to low-frequency uses because of difficulties in control-ling their dimensions and, in particular, in reducing the thickness of the base layer of the triple-decker semiconductor

29. *Electronics* 53, p. 239; Robert M. Ryder, "Ten Years of Transistors," *Radio-Electronics* 29 (May 1958): 34; B.N. Slade, "Survey of Transistor Development (Part 3)," *Radio and Television News* 48 (November 1952): 69; Millman, *History of Engineering and Science*, pp. 575ff; and Shockley, "The Path to the Conception of the Junction Transis-tor."

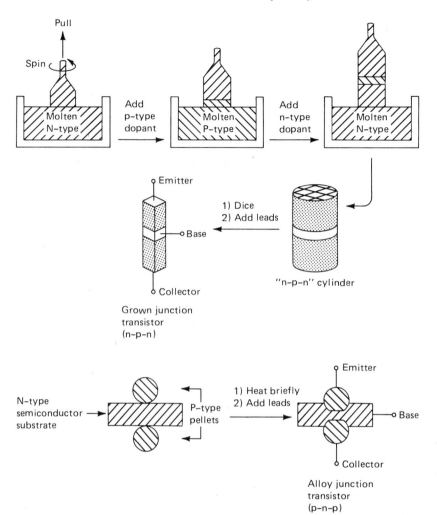

Figure 2

Fabricating junction transistors. Junction transistors were constructed by forming a three-layer sandwich of n-type and p-type semiconducting material. For grown junction transistors the layers were made while growing a large single crystal by adding a p-type dopant and then quickly redoping with an n-type dopant in excess. Alloy junction transistors were made by melting two p-type pellets onto an n-type substrate. In either method the center layer became the base, and the top and bottom layers became the emitter and collector. *Source:* R.L. Pritchard, *Electrical Characteristics of Transistors* (New York: McGraw-Hill, 1967), pp. 24–25.

sandwich. The diffusion technique solved this problem by exposing a solid semiconductor substrate to an atmosphere of vaporized doping agents, which by diffusing into the substrate produced a very thin surface layer. By carefully controlling the temperature and the duration of the exposure, the thickness of the transistor's layers could be dramatically reduced and, more important, precisely controlled. Diffusion technology extended the maximum frequency germanium transistors could amplify by approximately two orders of magnitude, from 10 to 1000 million cycles per second. The new manufacturing technique thus created a family of well-understood transistors capable of amplifying high frequencies.[30]

By the mid-1950s, then, transistors were no longer fragile laboratory curiosities. Innovations in metallurgical techniques and solid-state theories had allowed Bell Laboratories to build the first junction transistor. This new device had many specific forms, and all shared the advantages over point-contact transistors of being better described by contemporary physical theories, electronically less noisy, and mechanically more robust. Further, the new diffusion technology produced transistors that could amplify high frequencies and be mass produced. These advances, in turn, made practical the emerging efforts to build a large production capacity for the new technology.

Industrial Mobilization

The problem of procuring adequate numbers of transistors persisted well into the 1950s. Although the transistor appeared promising to many people, as indicated by the interest in the Bell symposia, its potential could not be realized until researchers had a sufficient supply to allow experimentation, the building of prototypes, and the manufacture of electronic equipment. Because of the clear military importance of the transistor and because the services were generally concerned with ensuring a war-ready industrial base,[31] the military under-

30. Ryder, "Ten Years of Transistors," pp. 35–37; Millman, *History of Engineering and Science,* pp. 423–28, 597–600; Jack A. Morton, "Bell System Transistor Program," 1958, copy in binder: Papers and Talks, JAM Collection; and Jack A. Morton and William J. Pietenpol, "The Technological Impact of Transistors," *Proceedings of the Institute of Radio Engineers* 46 (June 1958): 955–59.

31. See Seymour Melman, *The Permanent War Economy* (New York: Simon and Schuster, 1974); and Robert D. Cuff, "An Organizational Perspective on the Military-Industrial Complex," *Business History Review* 52 (Summer 1978): 250–67. For

took to help build a national production capacity for this technology. Not only did they increase their research and development contracts and their procurement, but they also started to underwrite the construction of private manufacturing facilities.

Again, the Army led the way. Two Signal Corps officers wrote revealingly in 1952 of the military's interests in the new technology:

In more normal times the military services would embark on only a modest program of "transistorization" leaving the broad general problem of the maximum utilization of these devices to the ingenuity of our industry and research institutions. Now, however, in this period of international tension the services consider the possible benefits of transistors to military equipments [sic] as sufficient to warrant substantial programs in this field and to include concurrently not only research and development, but the planning and preparation of facilities for producing large quantities of these devices.[32]

Military support of transistor research at Bell rose from a small level in 1950 to 20 percent of total funding in 1952 and to 50 percent in 1953, a level sustained through 1955. Bell's second military contract, signed in May 1951, also provided for an expanded role for military priorities. Whereas the first contract had been limited to application and circuit studies, the second specified that services, facilities, and materials were to be devoted to studies of military interest, while work continued on applications and circuits.[33]

Bell now began to coordinate transistor development with military requirements. Indeed, Bell's military systems laboratory at Whippany, New Jersey, would generate most of the military projects for which transistors were required. Under this contract, Bell and the military services jointly chose to de-

technological change in this political context see Kent C. Redmond and Thomas M. Smith, *Project Whirlwind* (Bedford, MA: Digital Press, 1980); and Clayton R. Koppes, *JPL and the American Space Program* (New Haven: Yale University Press, 1982).

32. Obenchain and Galloway, "Transistors and the Military," p. 1288. Lt. Colonel Obenchain was Assistant to the Commanding Officer for Research, SCEL; First Lieutenant Galloway was a member of the Office of the Director of Research of the SCEL.

33. From 1948 to 1957 Bell's transistor development program cost $22.3 million, 38 percent of which was funded by the military. During this period, Bell spent an additional $12.5 million of its own funds for physical, chemical, and metallurgical research in areas related to semiconductors. BTL Report, 1957.

velop twelve electronic prototypes for military systems.[34] Coordinating the development of devices to match the requirements of specific military systems proved a task that would tax the resources and organization of the Laboratories throughout the 1950s.

Significantly, at this same time Bell was experiencing difficulties in introducing transistors into the telephone system. In the fall of 1952 the Laboratories conducted a trial installation of transistorized direct-dial switching equipment in Englewood, New Jersey; and the first all-transistor telephone system was tested a year later in rural Georgia. Nevertheless, Mervin Kelly urged caution. "The transistor," he conjectured, "will come into large-scale use in the Bell System only gradually. Other fields of application—military electronics systems, home entertainment, special services—may well have the larger initial uses."[35] As a carefully integrated complex of sophisticated electronic equipment, the telephone system could add transistors only as old equipment was retired and as the newcomer demonstrated its reliability and economy. Rural telephone systems, which previously were without vacuum tubes or other amplifiers, were now the first systems to be transistorized, and it was not until the early 1960s that transistors were in large-scale use throughout the telephone system.[36]

High cost also severely constrained civilian applications of transistors. Commenting on the "discouragingly slow" introduction of the new technology into the telephone system, Jack Morton, Bell's director of transistor development, wrote: "Even though we realize the larger complexity of Bell systems as a contributing factor, we are impressed with the fact that economic difficulties, particularly the cost of components, looms as a large factor in this situation."[37] When Raytheon introduced one of the first transistor radios in 1955, the firm gave it a price of $80 and aimed promotion at the luxury market. Hearing-aid users formed one group willing to pay for the

34. Ibid. The twelve prototype devices are described in this report.

35. Kelly, "The First Five Years of the Transistor."

36. For example, the total value of the Bell System's transistorized equipment by 1963 had reached only $150 million—approximately 5 percent of the annual sales of Western Electric, the manufacturing branch of AT&T. See Jack A. Morton, "Application of Transistor Technology to the Bell Communications System," copy in binder: Material Prepared and Used by J. A. Morton in connection with British Patent Case, JAM Collection.

37. Morton, "Bell System Transistor Program," p. 20.

transistor's small size and low power requirements. The first hearing aid, with two vacuum tubes and one transistor, sold for a smart $229.50.[38] Military users were also ready to meet the steep costs of the new device in order to obtain its notable advantages. Unlike the hearing-aid users, however, the military services could afford to help pay the developmental costs needed for increased production.

The Army employed three related strategies to build up a large production capacity for transistors. One was to finance new plants directly. In 1953, for example, the Signal Corps underwrote the construction of a huge Western Electric transistor plant at Laureldale, Pennsylvania. Altogether the Army spent nearly $13 million in underwriting the construction of pilot plants and production facilities. In addition to Western Electric, General Electric, Raytheon, Radio Corporation of America, and Sylvania benefited from such military support.[39]

A second Army program stressed engineering development. Whereas work at the fundamental level of development translated concepts and inventions into usable prototypes, engineering development carried these prototypes to the point where they could be manufactured in production quantities efficiently and economically. The Army intensively funded this industrial mobilization effort. Before 1956 the Signal Corps' contracts for research and fundamental development in semiconductors— with Bell Laboratories alone before 1955, and thereafter also with Radio Corporation of America and Pacific Semiconductor—averaged $500,000 annually. After 1956 these contracts averaged approximately $1 million annually. In comparison, for the more expensive process of engineering development, the Army let contracts from 1952 to 1964 totaling $50 million, for an annual average of over $4 million.[40]

A third Army initiative influenced the cohesion of the emerging transistor industry. In mid-1953 the Signal Corps sponsored a conference aimed at standardizing the operating characteristics of transistors. The details were hammered out in meetings with representatives of the Navy and Air Force, leaders in the industry, and the Radio Electronics Television Manu-

---

38. See "Transistors: Growing Up Fast," *Business Week* (5 February 1955): 86; *New York Times,* 14 January 1953, 18 March 1954; *Wall Street Journal,* 17 April 1953; Bello, "The Year of the Transistor," p. 132.

39. Bello, ibid., p. 129.

40. Army Miniaturization Monograph, pp. 116, 128–30.

facturers Association.[41] Historians have evoked both practical and ideological factors to explain the military's frequent efforts to standardize procedures and technology.[42] In this case, the complexity of the electronic systems for which transistors were developed suggests a compelling objective reason for the Signal Corps' initiative. Nevertheless, the standardization of components for transistor circuitry was limited throughout the 1950s because the industry was unable to agree on standard shapes and sizes, and firms in turn were reluctant to invest in new production tooling in the absence of such consensus.[43]

The overall effects of the military's wide-ranging support were pronounced but complex. In its role as an institutional entrepreneur presiding over technological change, the Army undoubtedly increased the pace of transistor development. Historians Ernest Braun and Stuart MacDonald have even argued that military support produced a sizable overcapacity. In 1955, for example, 3.6 million transistors were manufactured in the United States, yet the industry's capacity was over four times larger: 15 million units.[44] Placing these production figures in their national political context helps to resolve this anomaly. Seymour Melman's thesis that in the years following World War II the American defense industry remained in a state of permanent mobilization for war suggests an explanation for these figures.[45] Even though the excess capacity wasted production capital, it was there for rapid mobilization in the event of a large-scale war. Although this strategy may have served the military's program, the cost- and resource-conscious electronics industry moved in the late 1950s to coordinate supply more closely with demand.

41. *Electronics* 53, p. 240.

42. See Lewis Mumford, *Technics and Civilization* (New York: Harcourt, Brace and World, 1934; reprinted 1963), pp. 84–96; and Merritt Roe Smith, "Military Entrepreneurship," in O. Mayr and R. C. Post, eds., *Yankee Enterprise: The Rise of the American System of Manufactures* (Washington D.C.: Smithsonian Press, 1981), pp. 63–101, revised and reprinted as the first essay in this volume.

43. Army Miniaturization Monograph, p. 150.

44. Braun and MacDonald, *Revolution in Miniature*, p. 81.

45. Melman, *The Permanent War Economy*. For difficulties in mobilizing the war economy during World War II, see Alan S. Milward, *War, Economy and Society* (Berkeley: University of California Press, 1977); and William M. Tuttle, "The Birth of an Industry: The Synthetic Rubber 'Mess' in World War II," *Technology and Culture* 22 (January 1981): 35–67.

✗Military Requirements and Technological Development

By 1954 the transistor development effort at Bell had shifted to an emphasis on applications. In March Jack Morton observed that "over the last year and one-half, Bell Laboratories systems applications have grown at an almost explosive rate."[46] Figure 3 shows the striking increase in the circuit and systems development staff from 1951 to 1954. More important, the earlier dependence on the military transistor market appeared to be ending. "When last year's forecast for 1954 and beyond was made," Morton continued, "Military items accounted for the bulk of the orders even through 1956. . . . However, this year's forecast can be seen to depend almost entirely on Bell applications."[47] The updated forecast indicated the requirements of the Bell System for transistors during 1955 would eclipse those of Bell's military projects by a factor of ten. Two large telephone projects alone, Rural Carrier and Line Concentrator, accounted for 500,000 transistors in 1955 and over one million in 1956. By contrast, Bell's total military transistor sales were scheduled to be 60,000 in 1955, increasing to only 175,000 in 1956. It appeared the extensive use of transistors predicted by Mervin Kelly in 1940 was finally a fact. But events at Bell during 1955 were to revise the optimistic forecast of the previous year and reemphasize the role of the military in transistor development. This continued presence extended the military's entrepreneurial influence beyond presiding over the development process. Military enterprise now began to shape the style of the technology.

Bell Laboratories had several reasons to be wary of dependence on military patronage. One was the unreliability of military contracting. For example, in November 1952 the Laboratories had begun designing a solid-state photosensor for the Naval Research Laboratory (NRL). By December 1954 Bell had delivered nine prototypes to the Navy and had completed the contract except for final engineering development. Although Bell scientists felt that the prototypes fully met the stringent electrical and mechanical specifications, the NRL abruptly dis-

---

46. Bell Telephone Laboratories, "Semiconductor Devices: Research and Development Report—March 5, 1954," n.p., copy in binder: Semiconductor Devices, Box 67, Bell Laboratories Archive (document hereafter as BTL R&D Report, 1954).
47. Ibid.

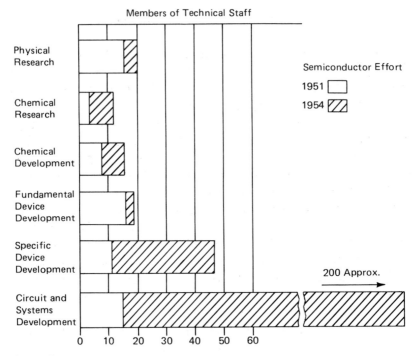

Figure 3
Bell Telephone Laboratories semiconductor effort 1951 and 1954. *Source:*
Bell Telephone Laboratories, "Semiconductor Devices: Research and Development Report—March 5, 1954," Bell Laboratories Archive.

continued the project for unclear reasons. Aside from the waste of time and resources, morale suffered when two years of effort were discarded without explanation. A second related reason for caution was the instability of the military transistor market. As table 2 shows, Western Electric's production for military applications fluctuated wildly throughout the 1950s. Cost reductions were difficult to achieve under such conditions because the uneven production runs required frequent changes in tooling, leading to additional expenses and unproductive down time.[48]

After 1955 Bell's problems with the Army centered on difficulties with the development of the Nike II antiballistic missile system. The Army's air defense program had shifted from its post–World War II concern with enemy bombers to

48. BTL Report, 1957, see "Task 6." For further difficulties with the military transistor market, see Jack A. Morton, "Military Device Development," 1958, copy in binder: Papers and Talks, JAM Collection.

Table 2
Transistors and diodes for military applications manufactured at Western
Electric plants at Allentown and Laureldale, Pennsylvania

Year	Total Military Diodes	Total Military Transistors
1952	196,800	35,200
1953	96,500	75,700
1954	220,200	146,200
1955	143,000	44,800
1956	179,000	166,300
1957 (first 6 months)	258,800	36,500

Source: Bell Telephone Laboratories, "An Appraisal of Military Transistor
Development—1948–1957," 7 August 1957, Subject Files, Box 90, Bell Laboratories Archive, Short Hills, New Jersey.

building a system that would guard against attacks by long-range missiles. In February 1955 the Army Ordnance Corps asked Bell to begin planning a major new missile system. The Whippany military systems laboratory immediately began work on the Nike II, a program that resulted in the design of the Nike-Zeus and Nike-X missiles. The project lasted for twenty years and was "the largest and most extensive program in depth and breadth of technology carried out by the Bell System for the military services."[49]

Unlike the original Nike study of a decade earlier, the Nike II program began when the transistor's small size and low power requirements were readily available. The transistor development staff was already overextended and could take on new projects only at the expense of current ones. Although the number of staff members working on applications had increased dramatically from its level in 1951, Morton still complained of a shortage of development engineers and of a "serious curtailment of fundamental development."[50] To alleviate the problem he recommended that forty-four more engineers and scientists be added to the development team. He

49. Fagan, *History of Engineering and Science,* p. 394. On the original Nike project see 370ff. and on the Nike II and its subsequent projects see 394ff.

50. Morton identified three long-range problems: "(1) For any given specific device there may be a number of alternate technologies which might be used. Lack of fundamental analytical technology development forces the project engineer to choose a technology purely on an expediency basis. (2) Lack of device manpower and lack of fundamental analytical technology work prevents the rapid exploration of new and promising structures. . . . (3) Exploration of the device applications of new materials, even silicon, is lagging and our competitive bargaining position may be challenged seriously." BTL R&D Report, 1954.

also warned that "any rescheduling of military devices will require rescheduling of large blocks of major systems in both Bell and Military areas. This should only be done," he added, "if it can be proved that the new devices are much more urgently needed for actual major military systems work."[51] In fact, one of the two major telephone projects was cancelled during this period. Although this project was already in trouble because of its high cost, the demanding work on Nike II almost certainly hastened its demise.

In addition to complicating the application of transistors to the Bell telephone network, the effort to coordinate device development with military systems work significantly influenced the style of the emerging technology. The high performance requirements of military electronics systems exacted the utmost from their components, and the transistor was no exception. An examination of specific characteristics of the transistors the military chose to promote reveals the tension between military performance and commercial economy.

Military applications frequently required electronic components to withstand high temperatures.[52] The ambient temperature within jet aircraft and guided missiles, for example, often exceeded 75° C, the maximum operating temperature for germanium transistors. Equally important, military applications required the equipment to be of the small size so easily achieved by using solid-state devices. Consequently, silicon transistors, with a maximum operating temperature above 150° C, sold briskly to the military services, despite their much higher cost as compared to germanium transistors. The military's preference for silicon transistors allowed their chief manufacturer, Texas Instruments, to carve out a niche in the semiconductor market.

A secondary reason for the military interest in silicon devices was their resistance to radiation. With the expansion of the Navy's nuclear-powered fleet and the Air Force's plans to develop a nuclear plane, the procurement of transistors able to withstand radiation became a stated goal.[53]

51. Morton, "Bell System Transistor Program." Quote from BTL R&D Report, 1954.

52. Two Signal Corps officers had noted in 1952: "The large variation with temperature of the present type of transistors' [operating characteristics] is . . . of particular difficulty since so many military equipments require operation over extended temperature ranges." Obenchain and Galloway, "Transistors and the Military," p. 1288. See also George R. Spencer, "Transistors: Past, Present, Future," *Radio-Electronics* 29 (May 1958): 40.

53. For Navy needs see W.I. Bull, "Transistor Reliability and Military Requirements,"

But what the military wanted most was a transistor capable of amplifying high-frequency signals. High-frequency radios, high-speed data transmission equipment, and high-speed computers all required high-frequency amplifying devices. The use of transistors would allow dramatic reductions in size, weight, and power requirements.[54] Because junction and point-contact transistors simply could not fulfill this need, the Signal Corps strongly supported the development of new types of devices. For example, in 1954 researchers at Bell invented the intrinsic barrier transistor in a project supported by its Signal Corps contract. This exotic four-layer device was so named because the fourth layer formed an internal barrier that supported the base layer of the semiconductor sandwich, whose thickness controlled the high-frequency response. This allowed the base to be made much thinner without breaking down at high voltages. Even though this device was difficult and expensive to manufacture, it was used in several military applications.

The diffusion process described above was the real breakthrough in high-frequency devices. The Signal Corps' support of diffusion research at Bell was striking. Indeed, in the late 1950s, the Signal Corps' support of fundamental transistor development exceeded Bell's own in-house support in only a few cases, and all of these were devices produced by the diffusion process.[55] Moreover, the Signal Corps coordinated its support to push the new technology from fundamental development into manufacture. In late 1955, when Bell released the diffusion process to industry,[56] the Signal Corps was ready with a hefty dose of engineering development funds for those who

---

and for Air Force needs see Palmer E. Koenig, "Transistor Reliability and Air Force Requirements," in *Proceedings of the Transistor Reliability Symposium,* sponsored by the Working Group on Semiconductor Devices of the Advisory Group on Electron Tubes, Office of the Assistant Secretary of Defense Research and Engineering, 17–18 September 1956 (New York: New York University Press, 1958).

54. BTL Report, 1957, see "Task 4"; and Ryder, "Ten Years of Transistors," pp. 36–37.

55. Morton, "Military Device Development," pp. 6–7.

56. Political considerations, in the form of an ongoing antitrust suit, persuaded Bell to release its technology to industry quickly in order to avoid appearing to monopolize the transistor field. The Justice Department had initiated a suit against AT&T in 1949 with aims of splitting Western Electric away from AT&T. The proceedings were halted in 1956, however, when AT&T agreed to several concessions. One was being enjoined from selling semiconductor devices on the commercial market (the military and space markets, as well as the Bell System, remained open); another was that Western Electric was forced to license all its existing transistor patents with minimal royalties. See Tilton, *International Diffusion,* pp. 50, 76.

would undertake to manufacture transistors. In fiscal year 1956 the Corps placed the largest engineering development contracts up to that time, totaling over $15 million. Nearly all the semiconductor firms in the country participated. A Corps historian noted that the program's purpose was "to make available to military users new devices capable of operating in the very high frequency (VHF) range which was of particular interest to the Signal Corps communications program."[57]

The Signal Corps was not the only service interested in diffusion technology. This became evident in a Bell program to identify "preferred devices" whose development was to be expedited. The purpose of this program was to combat the rapid proliferation of transistor types that marked the mid-1950s. Whereas in 1953 there were 60 different types of transistors, by 1956 there were 275, and by 1958 there were more than 900.[58] This variety greatly complicated the development of large-scale military systems. Each manufacturer would supply several transistor types, and integrating these into the same system often proved difficult. To streamline the Laboratories' work, W.C. Tinus and R.R. Hough, managers of the Nike program with extensive experience in military systems engineering, appointed a committee to address the possibilities for standardization. The committee's report, issued in July 1957, called for the creation of a Military Semiconductor Program to expedite the development of transistors needed specifically for military systems. The program was quickly enacted and several computer and missile projects, including Nike II, received this special attention. Describing the military's priorities, Morton observed that the preferred transistors were all diffused. These included germanium transistors for high-frequency service and silicon devices to meet high-temperature requirements.[59]

57. Army Miniaturization Monograph, pp. 125, 130.

58. Gernsback, "Transistor Trends," p. 33; and Spencer, "Past, Present, Future," p. 39.

59. Morton, "Military Device Development," pp. 14–17. The committee was chaired by E.H. Bedell; its other members were heads of transistor development projects. The family of "preferred devices" was designed to: "(1) Meet the performance and reliability requirements of military systems, (2) Gain wide acceptance and use such that the manufacturing level would be high with the resulting economies in manufacture, and (3) Be made available on a stockroom basis to provide early support of systems development." Uses for the preferred devices included, "Stretch," a military project on high-speed computers at the University of Illinois; "Lightening," a military-directed computer research project done by Remington-Rand Univac designed to produce the ultimate in high-speed computers; several military projects of Bell Telephone Laboratories, including the Ballistic Missile Early Warning System (BMEWS) and the Nike-Hercules and Nike-Zeus missiles; the inertial guidance system developed by ARMA for

Bell's Military Semiconductor Program received favorable attention from other firms, and the Laboratories soon moved to implement a similar program to streamline transistor development for Bell System applications. To this end, Walter A. MacNair, a former director of military systems engineering and vice-president of the Bell-affiliated Sandia Corporation,[60] organized in February 1958 a "Preferred Codes" program. Like its counterpart for military applications, this program required extensive cooperation between those engaged in development work and those concerned with applications. MacNair appointed a committee of division chiefs chaired by J.J. Ebers, the head of Device Development who had also served on the original committee that designed the Laboratories' Military Semiconductor Program, to carry out the program. Comparing the preferred transistors of the military with those of the Bell System, Morton noted that the Bell list included several types of alloy germanium transistors. "Unlike the Military applications," Morton explained, "there are a large number of Bell System applications which do not require the high-performance diffused devices." He concluded with an important point concerning the relative need for diffused devices. Even though the preferred list for Bell System applications included diffused germanium devices, Morton forecasted that Bell would need only small numbers of them. In contrast, he expected diffused transistors to be used in large numbers for military applications, "primarily because of the very high switching speeds required in Nike-Zeus."[61]

The production figures for diffused transistors demonstrate the accuracy of Morton's report. By the end of 1958 Western Electric had manufactured 171,000 diffused transistors for military applications, but *none* for consumption by the Bell System (see table 3). The close coordination of device development with military systems work had effectively translated military performance needs into manufactured devices, but at the same

---

the Atlas and Titan missiles; and the Polaris submarine-launched missile system. For information on Hough and Tinus, see Fagan, *History of Engineering and Science*, pp. 24, 400–401, 676, 690, 697.

60. The Sandia Corporation was a former branch of the Los Alamos Scientific Laboratory that the University of California had divested in 1949. Sandia was owned by the Atomic Energy Commission and managed by Bell executives. See Fagan, ibid., pp. 650ff.

61. Morton, "Bell System Transistor Program," pp. 8–11.

Table 3
Total transistors manufactured by Western Electric through 1958 by type and application

	Application	
Transistor Type	Military	Bell System
Point Contact	291,000	259,000
Grown Junction	115,000	265,000
Alloy Junction	116,000	98,000
Diffused Germanium	145,000	–
Diffused Silicon	26,000	–
Total	693,000	622,000

Source: Jack A. Morton, "Bell System Transistor Program," p.1, Jack A. Morton Collection, Bell Laboratories Archive, Short Hills, New Jersey.

time it had complicated and perhaps even compromised Bell's overall production efficiency.

Military Enterprise and Technological Style

The activities of the military services, specifically the Army Signal Corps, during the late 1940s and throughout the 1950s had several notable effects on the emerging transistor technology. A close-working combination of industrial firms and military agencies presided over the development phase of this technology. If industry's strengths were technological, the military's were organizational. Military-sponsored conferences and publications rapidly disseminated the new technology to industry. Moreover, by subsidizing engineering development and the construction of manufacturing facilities and by leading the movement to standardize operating characteristics, the military catalyzed the establishment of an industrial base. Finally, military requirements for transistors that could withstand high temperatures and amplify high frequency signals promoted the development of certain types of high-performance devices, including silicon transistors and transistors constructed by diffusion technology. The Army Signal Corps not only underwrote the fundamental development of this technology, but also, as soon as it was released to industry, expedited the translation of prototypes into manufacturing processes and devices by strongly supporting engineering development.

In addition to presiding over the development process, the military services contributed notably to the technology's eventual shape and style. As the electronics industry moved into the 1960s, the cost of transistors came down and, consequently, commercial sales rose and eventually surpassed military sales. Although it might be tempting to conclude that military patronage had merely allowed the technology to mature until costs could be reduced, this simplistic "pump priming" interpretation needs to be examined closely.[62] As the case of the Signal Corps' intensive promotion of the high-performance diffused transistor illustrates, military patronage could be tightly tied to specific variants of the new technology that filled requirements virtually unique to the military. Further, the subsequent development of solid-state technology suggests that military patronage has had several enduring aspects. A complex of characteristics suggesting a technological style, including the structure of the industry and the technology appearing at its cutting edge, were linked to the military in the 1950s and have continued to be associated with military enterprise.

The structure of an industry significantly contributes to its technological style. One specific characteristic of use to historians is standardization. Examining when the specifications for a technology are standardized on a national basis, and by whom, emphasizes the technology's cultural context. For the American transistor industry, the military services orchestrated standardization relatively early, in mid-1953. In contrast, the British semiconductor industry remained without national standards until the 1960s.[63] The attempts to designate preferred devices needed for specific projects at Bell Laboratories in the late 1950s was another milestone in standardization. Significantly, it was Bell managers who had had extensive experience with complex military projects who initiated these programs, first for the Laboratories' military systems and then for those of the Bell System itself.

---

62. The pump-priming model is maintained by Norman J. Asher and Leland D. Strom, "The Role of the Department of Defense in the Development of Integrated Circuits," Institute for Defense Analyses Report P-1271, May 1977, available from NTIS, Springfield, VA, as report No. ADA048610; and Air Force Systems Command, *Integrated Circuits Come of Age* (Washington, D.C.: Air Force, 1966). A more sophisticated view informs Daniel P. Jones, "From Military to Civilian Technology: The Introduction of Tear Gas for Civil Riot Control," *Technology and Culture* 19 (April 1978): 151–68; and John H. Perkins, "Reshaping Technology in Wartime: The Effect of Military Goals on Entomological Research and Insect-Control Practices," ibid., pp. 169–86.

63. See Jerome Kraus, "The British Electron-tube and Semiconductor Industry, 1935–62," *Technology and Culture* 9 (October 1968): 544–61.

Examining the cutting edge of an emerging technology and the interests pushing its development provides another clue to the cultural context shaping technological style. Military patronage of the most technically sophisticated, high-performance semiconductor technology has been a recurring pattern in the United States. When coupled to the development process, military needs promoted the high-performance diffused transistor in the 1950s, the integrated circuit in the early 1960s, and the very-high-speed integrated circuit in the early 1980s.[64] The driving force behind the semiconductor industry in Japan, in contrast, has been the powerful Ministry of International Trade and Industry (MITI). In electronics MITI has emphasized not ultra-high-technology product lines but rather the large-scale, coordinated expansion of the semiconductor market.[65]

If the American military's entrepreneurship did in fact increase the overall rate of development of the transistor, this increase was not achieved without cost.[66] Bell Telephone Laboratories in particular felt acutely the tension between military performance and commercial economy. Even though the transistor had been invented for use in the telephone system, this use was realized only in the early 1960s. In part, Bell Laboratories' military transistor work throughout the 1950s compromised this effort by draining manpower from development work on Bell System applications. Scientists and engineers with experience in the new field were a valuable resource and could not be replaced at will.[67] Further, since the specific characteristics of the transistors developed for military applications were

64. For a military view of these developments, see Asher and Strom, "The Role of the Department of Defense" and Air Force Systems Command, *Integrated Circuits Come of Age*. On military support of very-high-speed integrated circuits, see Robert DeGrasse, "The Military and Semiconductors," in J. Tirman, ed., *The Militarization of High Technology* (Cambridge, MA: Ballinger Publishing Co., 1984), pp. 77–104; and *Science* 219 (11 February 1983): 750. For speculation on the next generation of military support of ultra-high technology, see "Nanocomputers from Organic Molecules?" *Science* 220 (27 May 1983): 940–42.

65. Chalmers Johnson, *MITI and the Japanese Miracle: The Growth of Industrial Policy, 1925-1975* (Stanford: Stanford University Press, 1982).

66. I use the term "cost" here in the economist's sense of "opportunity cost," referring to an opportunity foregone or postponed.

67. Concern remains on the part of science policy analysts about the adverse effects of the military's draining of technical personnel from commercial applications; see Michael Schrage, "Defense Budget Pushes Agenda in R&D," *Washington Post*, 12 August 1984, p. Fl. For an informed discussion of the current problems of U.S. electronics, see Charles H. Ferguson, "The Microelectronics Industry in Distress," *Technology Review* 86 (August–September 1983): 24–37.

frequently different from those developed for the Bell System, the spillover from military to commercial uses was incomplete at best.

The style of high-technology innovation illustrated by the transistor has been an enduring force in the postwar era. In areas as diverse as computer graphics, artificial intelligence, and numerically controlled machine tools, military enterprise has significantly altered the technological landscape.[68] In setting priorities for research and development, serving as a large consumer of new products, and influencing the structure of an entire industry, the military has become a de facto architect of high technology policy. This mission-agency style of innovation has entailed selecting specific variants of emerging technologies. By definition this has biased technological change. If pervasive, the distinction between military performance and commercial economy has not been absolute. For the transistor, neither design criterion, performance, nor economy, completely dominated the other. Rather, these two technological characteristics formed a complex matrix of possibilities and outcomes that shaped the development of this important technology.

68. See Tirman, *The Militarization of High Technology;* and David Noble's essay in this volume.

7

U.S. Navy Research and Development since World War II
David K. Allison

Research and development occupy an important place in modern military organizations. Indeed, more than half of this country's engineers and scientists are currently engaged in defense-related activities. A significant portion of that work is sponsored by the United States Navy, and in this essay David Allison describes how the Navy's research program has evolved since World War II. Allison's intimate knowledge of the subject provides us with a close look at how the naval bureaucracy works, how power is distributed within it, and how those in command influence the pace and direction of technological change. Although his association of technology and management favors an interpretation of technological change as expanding knowledge, his analysis also suggests that technological innovation is inextricably connected with the politics of research and development.

Introduction

During World War II developing new weapons became a major element in American defense strategy. The government enlarged its few existing military laboratories and established many new ones. It created a National Defense Research Committee, later expanded into an Office of Scientific Research and Development, to mobilize scientists and engineers in universities and private industry.[1] It negotiated military contracts with Bell Telephone Laboratories, RCA, General Motors, and thousands of other corporations. The products that resulted— the proximity fuze, microwave radar, electronic warfare, and the atomic bomb, to name a few—convinced military leaders that substantial investment in military research and development should not end with the war. It had become crucial to maintaining military strength.

Listen, for example, to the counsel of Army Chief of Staff Eisenhower to other Army leaders in 1946:

> The lessons of the last war are clear. The military effort required for victory threw upon the Army an unprecedented range of responsibilities, many of which were effectively discharged only through the invaluable assistance supplied by our cumulative resources in the natural and social sciences and the talents and experience furnished by management and labor. The armed forces could not have won the war alone. Scientists and business men contributed techniques and weapons which enabled us to outwit and overwhelm the enemy. Their understanding of the Army's needs made possible the highest degree of cooperation. This pattern of integration must be translated into a peacetime counterpart which will not merely familiarize the Army with the progress made in science and industry, but draw into our planning for national security all the civilian resources which can contribute to the defense of the country.[2]

Fleet Admiral King gave similar advice in his final report on the war to the Secretary of the Navy. "Only by continuing vigorous research and development," he concluded, "can this country hope to be protected from any potential enemies and maintain the position which it now enjoys in possessing the greatest ef-

---

1. See James Phinney Baxter III, *Scientists Against Time* (Boston: Little, Brown & Co., 1946), and Irwin Stewart, *Organizing Scientific Research for War* (Boston: Little, Brown and Co., 1948).

2. A copy of the memorandum appears in Seymour Melman, *Pentagon Capitalism: The Political Economy of War* (New York: McGraw-Hill, 1970), pp. 231–34.

fective naval fighting force in history."[3] Vannevar Bush, head of the Office of Scientific Research and Development and chief spokesman to the government for the civilian scientific community, told the President in his influential report *Science: The Endless Frontier:*

> We cannot again rely on our allies to hold off the enemy while we struggle to catch up. There must be more—and more adequate military research in peacetime. It is essential that the civilian scientists continue in peacetime some portion of those contributions to national security which they have made so effectively during the war.[4]

Because this consensus formed as the Soviet Union changed from ally to adversary, military research and development became a high national priority. The National Security Act of 1947, which merged the Army and Navy into a single National Military Establishment, also created a Research and Development Board "to advise the Secretary of Defense as to the status of scientific research relative to the national security, and to assist him in assuring adequate provision for research and development on scientific problems relating to the national security."[5]

As the years have passed, research and development have remained important in the American defense establishment. The unqualified enthusiasm of 1946, however, has evaporated. In 1961 Dwight Eisenhower, at the end of his second term as President, singled out ties between science and technology, private industry, and the defense community as cause for special concern. He cautioned the nation, "In the councils of government we must guard against the acquisition of unwarranted influence, whether sought or unsought, by the military-industrial complex. The potential for the disastrous rise of misplaced power exists and will persist."[6] Many other public

3. Ernest J. King, *U.S. Navy at War 1941–1945; Official Reports to the Secretary of the Navy* (Washington, D.C.: Department of the Navy, 1946), p. 281.
4. Vannevar Bush, *Science: The Endless Frontier; Report to the President on a Program for Postwar Scientific Research* (Washington, D.C.: GPO, 1945), pp. 1–2.
5. U.S. Congress, *National Security Act of 1947,* Pub. L. 253, 80th Cong., sec. 214, "Research and Development Board"; reprinted in Alice C. Coe, et al., *The Department of Defense: Documents on Establishment and Organization 1944–1978* (Washington, D.C.: Department of Defense, 1978), p. 47.
6. Dwight D. Eisenhower, "Farewell Radio and Television Address to the American People, January 17, 1961." U.S. Presidents, *Public Papers of the President of the United States: Dwight D. Eisenhower, 1960–61* (Washington, D.C.: GPO, 1961), pp. 1035–40.

figures have since expressed similar worries, especially during the Vietnam War and more recently in the wake of the military expansion advocated by the Reagan administration.

Managing research and development has been as problematic as determining its role in defense policy. Research and development are risky and expensive. Military leaders expect a higher return on investment during peacetime than during war because time is not as critical. Although managers have tried to make careful choices among technological options, many programs have still failed. Finally, politics has become an important factor in deciding whether to conduct programs in universities, private industry, nonprofit corporations, or the federal defense laboratories.[7]

This essay examines one portion of American military research and development since World War II, the work of the U. S. Navy. The first part describes evolving management practices as the Navy tried to focus attention on areas of greatest promise and to make the R&D process more efficient. The second part reviews two specific Navy R&D programs. No major programs are ever alike, and the examples are not "average" or "typical"—especially because both were successes. Nonetheless they do illustrate some of the administrative trends and illuminate the R&D process the Navy has worked to foster and control over the last thirty-five years.

Administration of Navy Research and Development

The United States Navy of 1946 was the largest, most sophisticated fighting fleet in the world. Wartime developments, however, raised important technical questions that had to be answered in a peacetime era. What effect would the new atomic bombs have on the configuration and use of naval forces? Could aircraft carriers become platforms for the delivery of atomic weapons, given the large, heavy aircraft this mission required? Were nuclear-powered submarines possible? What forms of rockets and missiles should the Navy develop? What changes would the rapid progress in electronics bring in command, communications, control, and sensor technologies? Looking further ahead, what forms of basic scientific research

7. John R. Fox, *Arming America: How the US Buys Weapons* (Cambridge, MA: Harvard University Press, 1974).

into atomic structure, properties of materials, space science, and other fields would pay dividends comparable to those obtained from atomic physics and electronics during World War II?

Answering these questions required both technical expertise and effective administration. The search for better management of research and development soon began stressing the Navy's basic organizational structure, which had changed little since the early 1900s. Figure 1 depicts the organization in 1945.[8] As the heavy outlining indicates, it was broken into three principal parts: the Navy Department, which included the bureaus, boards, and offices located in headquarters; the operating forces, primarily the fleets; and the shore establishment, the Navy field facilities that provided the logistical, material, and other support needed by the operating forces.

Functionally, the Navy was a bilinear organization. The Secretary of the Navy had final responsibility for the entire agency. Below him, however, power was split. The Chief of Naval Operations, a Naval officer, exercised military control, while the under and assistant secretaries, civilians, exercised administrative control or business management. The division was not arbitrary. It separated producer logistics, the job of the business side of the Navy, from consumer logistics, the job of the fleet. The consumers were to inform the producers about what they needed with written operational requirements for equipment and supplies.

Informing was not controlling. The most characteristic feature of the bilinear system was the independence of the producers, the bureaus of the Navy, from their customers. The bureaus had direct access to the Secretary of the Navy, had their own appropriations, and had independent responsibility for research, development, evaluation, procurement, and distribution of materials and facilities. A Material Division within the Office of the Assistant Secretary of the Navy was responsible for establishing general Navy procurement policy, but it had only the power to coordinate. The bureaus met the material needs of the Navy as they saw fit.

---

8. The information used in this section comes principally from Booz-Allen and Hamilton, *Review of Navy R&D Management 1946–1973* (Washington, D.C.: Department of the Navy, 1976) and *Review of Management of the Department of Navy* (NAVEXOS P-2426B) (Washington, D.C.: Department of the Navy, 1962), cited hereafter as the Dillon Report.

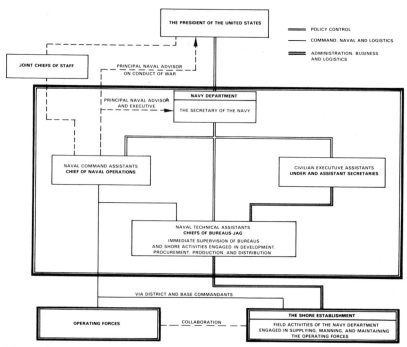

Figure 1
Overall operating organization of the naval establishment, 1946. *Source:*
Booz-Allen and Hamilton, Inc., *Review of Navy R&D Management, 1946–1973* (Washington, D.C.: Department of the Navy, 1976), p. 4.

The bureau system originated in 1842. From then until World War II, the number and type of bureaus varied, but, as one analyst has stated:

Their independence of action, in terms of each other and the Operating Forces, remained intact. . . . Each [bureau] was under the administration of a flag officer with line authority over his organization. By virtue of the clear lines of authority and accountability, the bureau chiefs enjoyed virtual autonomy in technical and business management matters.[9]

At the end of the war, the Navy had seven bureaus: Ordnance, Ships, Aeronautics, Supplies and Accounts, Naval Personnel, Medicine and Surgery, and Yards and Docks. The first three, often called the "material bureaus," bore principal responsibility not only for ships, aircraft, and equipment, but also for

9. Booz-Allen and Hamilton, *Review of Navy R&D Management*, p. 5.

research and development in the physical sciences and engineering.

Joining them in 1946 was a new organization, the Office of Naval Research (ONR). Admiral Harold Bowen, Director of the Naval Research Laboratory, and several other ranking naval officers had lobbied for such an office even before World War II. They had argued that the material bureaus focused too much on the present to be responsible for the Navy of the future.[10] The success of the Office of Scientific Research and Development (OSRD) had demonstrated the value of linking scientists engaged in basic research to the military. When OSRD expired at the end of the war, both the Navy and Congress wanted to maintain the relationships it had fostered and also to promote basic research in the Navy's own laboratories. The Office of Naval Research was created to accomplish these goals.[11] Organizationally, it was under the Assistant Secretary of the Navy for Air, who had responsibility for overseeing all Navy research and development. Thus it was free from control by the material bureaus. Figure 2 depicts the general structure of the Navy research and development establishment in 1946.

Research and Development in the Immediate Postwar Years
The bilinear structure of the Navy and the control of the material bureaus over procurement of naval material determined the character of Navy research and development during the immediate postwar period. But change had already begun, particularly in the creation of the Office of Naval Research and the establishment of a formal system of administrative control of research and development. By current standards, that system was simple. It required only three formal types of planning documents: *planning objectives*—broad statements of a scientific or operational problem whose solution might require new scientific knowledge or the development of new equipment; *operational requirements*—outlines of the estimated operational

---

10. David K. Allison, *New Eye for the Navy: the Origin of Radar at the Naval Research Laboratory* (Washington, D.C.: Naval Research Laboratory Report 8466, 1981), pp. 161–74.

11. The Office of Naval Research succeeded another office, the Office of Research and Inventions, which had been created in May 1945 for similar purposes. See *Administrative History: Office of Research and Inventions*, July 31–December 1945 (Unpublished history in the series "World War II Naval Administrative Histories," in the Navy Department Library, Washington Navy Yard, Washington, D.C.). On the Office of Naval Research, see James L. Penick, et al., eds., *The Politics of American Science, 1939 to the Present* (Cambridge, MA: The MIT Press, 1972), pp. 180–88.

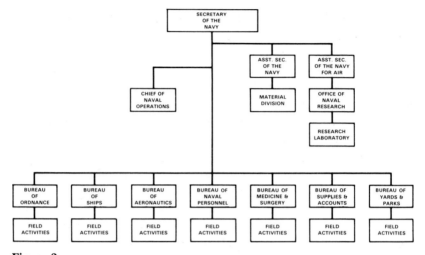

Figure 2
Organization for Navy R&D in 1946. Each of the bureaus had field activities appropriate to its mission. The principal research, development and test laboratories for naval material were under the Bureaus of Ordnance, Ships, and Aeronautics.

performance of systems or equipment designed to solve an operational problem; and *research requirements*—statements of need for scientific knowledge in physical, psychological, sociological, and earth sciences.[12]

Although officials throughout the Navy Department could prepare the first two types of requirements, the Office of the Chief of Naval Operations alone could issue them. In accord with the bilinear structure of the Navy, the material bureaus were responsible for satisfying the requirements. Only in the area of research did a single part of the Navy, the Office of Naval Research, both establish needs and execute the program required to meet them. A Navy policy statement of the era summarized the duties of the bureaus and ONR as follows:

Research and development programs are planned by the Chiefs of the Bureaus and Offices to contribute to the fulfillment of planning objectives and operational requirements. An essential feature of this system is the freedom of the Bureaus and Offices to initiate any and all research and development projects which they consider necessary in light of their assigned responsibilities and the guidance furnished them, in order to insure consideration of all potentially valuable approaches. Decisions as to which of these projects so initiated will be

12. Booz-Allen and Hamilton, *Review of Navy R&D Management*, pp. 185–97.

prosecuted, and what level of effort, is made by the Chiefs of the Bureaus and Offices subject to any specific direction by the Chief of Naval Operations and budgetary considerations.[13]

Real power is often exercised not through assigned responsibilities, but, as the policy subtly suggests, through funding. Oversight of Navy R&D in the budgeting process, however, was slight. Congress made appropriations directly to the bureaus, and research and development programs often did not even appear as separate line items. The Office of the Chief of Naval Operations conducted annual budget reviews, but only rarely did they touch on particular research and development efforts.[14] Bureau chiefs usually had freedom to distribute R&D funds as they wished.

The Navy used four principal means to execute its research and development program. It performed work in its own laboratories, it contracted with scientists and engineers in universities, it contracted with nonacademic research institutions, and it contracted with private corporations.

World War II bequeathed the Navy many research and development laboratories. As peace returned, these institutions turned their efforts increasingly toward the earlier phases of research and development: the generation of new ideas or working models of new equipment. They also conducted acceptance tests for industry, improved existing equipment, and solved problems related to use of the equipment in the fleet. The bureaus controlled most of the laboratories. Figure 3 lists the principal Navy research and development facilities in 1946.

The Office of Naval Research was largely responsible for Navy funding of research in universities and nonprofit institutions.[15] Indeed, during the immediate postwar period, ONR sponsored far more basic research than any other government agency. It continued to act as the government's major patron of basic science until the National Science Foundation was established in 1950.

Private industry conducted some research and designed some new equipment, but its principal role was manufacture.

13. OPNAV Instruction 0390.1 of 25 May 1951, Subj: Coordination of Research and Development, as quoted in ibid., pp. 192–93.

14. Ibid., pp. 243, 255–68.

15. Penick, *Politics of American Science*, pp. 185–89. See also Daniel S. Greenberg, *The Politics of Pure Science* (New York: New American Library, 1967), especially pp. 133ff.

OFFICE OF RESEARCH AND INVENTIONS/OFFICE OF NAVAL RESEARCH
Naval Research Laboratory, Washington DC
Special Devices Division, Long Island NY

BUREAU OF SHIPS
David Taylor Model Basin, Carderock MD
Engineering Experiment Station, Annapolis MD
Naval Boiler and Turbine Laboratory, Navy Yard, Philadelphia PA
U.S. Navy Underwater Sound Laboratory, New London, CT
Naval Electronics Laboratory, San Diego CA
U.S. Navy Material Laboratory, New York NY
Industrial Testing Laboratory, Navy Yard, Philadelphia PA
U.S. Navy Metals Laboratory, Munhall PA
Navy Radiological Defense Laboratory, San Francisco CA
Naval Mine Countermeasures Station, Panama City FL

BUREAU OF AERONAUTICS
Naval Air Material Center, Philadelphia PA
    Naval Air Experimental Station, Philadelphia PA
    Naval Aircraft Modification Unit, Johnsville PA
    Naval Auxiliary Air Station, Weston Field PA
    Naval Aircraft Factory
Naval Air Station, Patuxent River MD
Naval Air Test Center, Point Mugu CA

BUREAU OF ORDNANCE
Naval Proving Ground, Dahlgren VA
Naval Powder Factory, Indian Head MD
U.S. Naval Ordnance Test Station, Inyokern CA
Naval Ordnance Laboratory, White Oak MD
Explosives Investigation Laboratory, Indian Head MD
Naval Torpedo Stations: Newport RI, Keyport WA, and Alexandria VA
Ordnance Aerophysics Laboratory, Daingerfield TX
Naval Aviation Ordnance Test Station, Chincoteague VA
Naval Mine Warfare Test Station, Solomons MD

Figure 3
Principal Navy research and development activities, 1946 (excludes medical laboratories). *Source:* Booz-Allen and Hamilton, Inc., *Review of Navy R&D Management, 1946–1973* (Washington, D.C.: Department of the Navy, 1976), p. 116.

Except for the ships built in government shipyards, industry produced virtually all the Navy's equipment. Among the most important jobs of the material bureaus were writing, negotiating, and overseeing industrial contracts.

The evolution of research and development systems continued for the next thirty-five years as the Navy sought to maximize gains and minimize risks in its programs. Three of the most important structural trends were consolidation and centralization of authority over research and development, formalization and systematization of the R&D process, and evolution of the methods for doing the work.

Centralization

The centralization of authority was gradual but eventually brought about the abolition of both the bilinear structure and the material bureaus. The growth of central offices in the Department of Defense was the principal cause of the trend, for as their power waxed, that of the individual services waned. Explaining what happened in the Navy, therefore, requires glancing at what occurred on higher levels.

When the National Military Establishment was formed in 1947, the Secretary of Defense had a coordinating instead of a controlling role. The law stated:

> The Department of the Army, the Department of the Navy, and the Department of the Air Force shall be administered as individual executive departments by their respective secretaries and all powers and duties relating to such departments not specifically conferred upon the Secretary of Defense by this Act shall be retained by each of their respective Secretaries.[16]

By 1949, however, this organizational approach had failed, and Congress reorganized the National Military Establishment as a fully centralized Department of Defense. The service secretaries lost their cabinet status and independent access to Congress and the President.[17] Henceforth, the Secretary of Defense spoke for the military.

Many derivative changes followed, and the centralizing tendencies soon reached research and development. In 1953 the Research and Development Board was abolished and its functions assigned directly to the secretary.[18] In 1955 Assistant Secretaries of Defense were established for research and development and for applications engineering. Then in 1957 the two positions were merged into an office of Assistant Secretary of Defense for Research and Engineering. Finally, in the

---

16. U.S. Congress, National Security Act of 1947, sec. 202; reprinted in Coe, et al., *Department of Defense*, p. 40.

17. U.S. Congress, National Security Act of 1947 as Amended by Pub. L. 216, 81st Cong.; reprinted in ibid., pp. 84–106.

18. U.S. Congress, House, Reorganization Plan No. 6 of 1953, 83rd Cong., 1st Sess., H. Doc. 136; reprinted in ibid., pp. 157–58.

Defense Reorganization Act of 1958, Congress upgraded the position to Director of Defense Research and Engineering. This official, who ranked between the other assistant secretaries and the Deputy Secretary of Defense, had full authority to approve, modify, or disapprove all R&D programs of the military departments.[19]

Centralizing authority over defense research and development under one official in the Department of Defense led to analogous changes in the Navy. In 1958 the secretary established an Assistant Secretary for Research and Development and gave him full control over all Navy R&D. His power included managing the R&D appropriation: he was the only civilian executive assistant to the secretary with such broad control. In the following year the position of Deputy Chief of Naval Operations (Development) was created within the Office of the Chief of Naval Operations to centralize its management of operational requirements for research and development. As these new officials asserted their power, they eroded the independent authority of the bureaus.[20]

Other developments also threatened it. When the Navy began building the Polaris submarine and strategic missile system, it put a special project office in charge instead of a bureau. The success of Polaris led to project offices for other major programs, casting further doubt on the effectiveness of the bureau system. In 1959 the Bureaus of Ordnance and Aeronautics were combined to form a Bureau of Weapons, primarily to end fights over which of the two should be in charge of Navy missile development. This new bureau, however, was too unwieldy for traditional bureau management, and its formation failed to stop troublesome disputes with the other bureaus over who was responsible for what.[21]

In 1948 the Navy made an early attempt to remedy shortcomings of bureau management by creating the Office of Naval Material, a successor to the Material Division established under the assistant secretary after World War II. The Chief of Naval

19. U.S. Congress, Defense Reorganization Act of 1958. Pub. L. 559, 85th Cong., sec. 203, reprinted in ibid., p. 204. This position was later changed to Under Secretary of Defense Research and Engineering.

20. Booz-Allen and Hamilton, *Review of Navy R&D Management*, pp. 53–64.

21. See "Farmington Conference Agreement" in file "Reorganization Plans, 1966" in "Booz-Allen and Hamilton Source Materials for Navy R&D Study," box 5 in Operational Archives Branch, Naval Historical Center, Washington Navy Yard, Washington, D.C. See also the Dillon Report, vol. II, study 4, pp. 30–32.

Material (CNM) who headed the new office was responsible for establishing policies for procurement, contracting, and production of material throughout the Navy and for determining the policies and methods to be followed by the bureaus in implementing them. A decade later, however, reformers who favored centralization judged the CNM's powers too limited and recommended strengthening the organization for material management above the bureau level. In 1963 the Navy created a Naval Material Support Establishment to assist the Chief of Naval Material in planning and coordinating material procurement. Even this, however, left the structure and power of the bureaus essentially unchanged, and problems continued.[22]

The final step came in 1966. The Navy abolished the technical bureaus and redistributed their functions into six "systems commands": the Air Systems Command, Ship Systems Command, Ordnance Systems Command, Electronic Systems Command, Supply Systems Command, and Facilities Engineering Command.[23] The CNM was now given complete control over them all. Since the CNM in turn reported to the Chief of Naval Operations, this action also ended the bilinear structure. In a report to Congress justifying the change, Secretary of Defense McNamara wrote:

It is the belief of the Secretary of the Navy, which I share, that the Department of the Navy should be organized in such a fashion that the Navy's senior military officer, the Chief of Navy Operations, will have the same breadth of authority and responsibility for material, personnel, and medical support functions as he now has for the operating forces of the Navy. Additionally, the Secretary of the Navy believes that the organization performing the Navy's material support functions should be so structured as to subject them to more effective command by the Chief of Naval Material under the Chief of Naval Operations.[24]

22. See Julius A. Furer, *Administration of the Navy Department in World War II* (Washington, D.C.: Department of the Navy, 1959), p. 845; *Report of the Committee on Organization of the Department of the Navy* (Franke Board) (Washington, D.C.: Department of the Navy, 1959), pp. 46–47.

23. The Bureau of Medicine and Surgery and Bureau of Naval Personnel remained. Since 1966 the Naval Sea Systems Command has replaced the Naval Ship System Command, and the Naval Ordnance Systems Command has been abolished and its functions redistributed. The Bureau of Personnel is now the Naval Military Personnel Command and the Bureau of Medicine and Surgery is the Naval Medical Command.

24. U.S. Congress, House, *Reorganization of the Department of the Navy. Communication from the President of the United States transmitting a plan for the Reorganization of The Department of the Navy*, 89th Cong., 2nd Sess., H. Doc. No. 409.

This change consolidated virtually all activities related to research, development, and acquisition of naval material under the Chief of Naval Operations. Only the Office of Naval Research remained separate. In 1980, however, even its independence was compromised. The Chief of Naval Research assumed the title and responsibilities of Chief of Naval Development, and, in that capacity, reported to the Chief of Naval Material. This put him directly in the chain of command of the Chief of Naval Operations.[25]

Systematization

As authority over research and development changed, so did methods of managing it. In the early years following World War II, the Congress, top military leaders, and even bureau administrators themselves concentrated only on allocating resources for research and development and then evaluating results. They paid little attention to what lay in between. As they consolidated and centralized their power, however, and as research and development became more expensive, their interest grew, and they began to scrutinize the whole process.

Secretary of Defense McNamara took the most important action when he introduced a new method for managing all defense resources known as the Planning, Programming, and Budgeting System (PPBS). Previously there had been little connection between the planning that the Joint Chiefs of Staff and other military leaders did and the budget decision-making of the civilian secretary and his assistants. Furthermore, as Charles Hitch, Assistant Secretary of Defense (Comptroller) under McNamara and the man who devised the system, stated:

Planning was based on missions, weapons systems, and military units or forces—the "outputs" of the Defense Department; budgeting, on the other hand, was based on "inputs," or intermediate products such as personnel, operations and maintenance, procurement, construction, etc., and there was little or no machinery for translating one into the other.[26]

25. The change was made on September 10, 1980. The chain of command then became: the Chief of Naval Development was also the Deputy Chief of Naval Material (Technology) and reported to the Chief of Naval Operations through the Chief of Naval Material.

26. Charles J. Hitch, *Decision-Making for Defense* (Berkeley: University of California Press, 1965), p. 26.

To link planning and budgeting in such a way that the latter conditioned the former, "programming" was added to the process. In this phase managers revised requirements for manpower, equipment, and installations that had come from the planning process and linked them with their costs. When stated in a prescribed, formal fashion, these conditioned requirements became "program elements." Elements combined to become complete programs for each service, and together they made up the "Defense Program," which became the fundamental document for decision-making on the defense budget.

Concurrent with this change, Hitch began requiring that all defense planning, programming, and budgeting be projected five years into the future instead of merely a single year, as had been the practice. The annual program thus became an updated five-year program.

The program had nine major categories, including, as the sixth, research and development. To show the stages of program elements within the research and development process, Dr. Harold Brown, Director of Defense Research and Engineering under Secretary McNamara, ordered the R&D category divided into five parts. Since then a sixth has been added, making the divisions research, exploratory development, advanced development, engineering development, management and support, and operational systems development.[27]

27. The Navy's *RDT&E Management Guide* defines the categories as follows:

6.1 Research includes scientific study and experimentation directed toward increasing knowledge and understanding in those fields of physical, engineering, environmental, biological, medical, and behavioral–social sciences related to long-term national security needs. It provides fundamental knowledge for the solution of identified military problems and also part of the base for subsequent exploratory and advanced developments in defense-related technologies.

6.2 Exploratory Development includes all effort directed toward the solution of specific military problems, short of major development projects. This type of effort may vary from fairly fundamental applied research to quite sophisticated "breadboard" hardware, study programming, and planning efforts.

6.3 Advanced Development includes all projects that have moved into the development of hardware for experimental or operational test. It is characterized by line-item projects, and program control is exercised on a project basis.

6.4 Engineering Development includes those development programs that are being engineered for service use but that have not yet been approved for procurement or operation.

6.5 Management and Support includes research and development effort directed toward support of installations or operations required for general research and development use. Included would be test ranges, military construction, maintenance support of laboratories, operations and maintenance of test aircraft and ships, and studies and analyses in support of the R&D program.

Although the military had used various methods for dividing research and development into phases before, planning and budgeting categories had never been formalized to this extent. The new scheme soon became very influential. The Navy for example, assigned different categories to different officials, and they managed their work in significantly different ways. The research program, for instance, became primarily the responsibility of the Chief of Naval Research; the Chief of Naval Development handled exploratory development[28] and the systems commands largely managed the remaining arms. In short, what started as a general planning scheme became the principal mechanism for dividing, managing, and controlling the R&D process.

The PPBS rationalized preparation of the defense budget as a whole. Planning and developing major new weapons systems, such as ships, aircraft or missiles, however, required special care. To focus additional management attention on these activities, Secretary McNamara introduced in 1965 a formalized weapons system acquisition process.[29] It had three principal phases: concept formulation, contract definition, and acquisition. During concept formulation, managers analyzed the basic technical, economic, and military factors related to the proposal and gave conditional approval for development. In contract definition, they set final specifications. During the acquisition phase, equipment was manufactured, tested, and, after meeting all specifications, added to the government's inventory.[30]

To provide necessary information for decision-making, McNamara introduced a new form of document: the development concept paper (DCP). This carefully standardized report summarized technical, economic, and strategic factors related

---

6.6 *Operational Systems Development* includes research and development effort directed toward development, engineering and testing of systems, support programs, vehicles, and weapons that have been approved for production and service employment. From *RDT&E Management Guide*, 7th ed. (NAVSO P-2457) (Washington, D.C.: Department of the Navy, 1979), pp. 2–7.

28. The position of Chief of Naval Development was established in 1964, partly as a result of action on the Dillon Report and partly on the recommendation of the Assistant Secretary of the Navy for Research and Development. The function of the office was to coordinate the Navy's exploratory development program. The individual who held it was also "double-hatted" as Deputy Chief of Naval Material (Development) and in this capacity had responsibility for coordinating development, test, and evaluation programs for the Chief of Naval Material.

29. Booz-Allen and Hamilton, *Review of Navy R&D Management*, p. 236.

30. Dean Roberts, "OMB Circular A-109 Impact on New Development" *Defense Systems Management Review* 2 (Autumn 1979): 27–35.

to the system and thus became useful to the Director of Defense Research and Engineering and other officials in the early stages of development.[31]

Like the PPBS system, the acquisition management system developed under Secretary McNamara has remained a basic tool of defense management, although later administrations have made many significant changes. Melvin Laird, McNamara's successor as Secretary of Defense, established a Defense System Acquisition Review Council (DSARC) of high-level defense officials to oversee major acquisitions. The council made formal reviews of all large acquisitions at three major "milestones": program initiation, full-scale development, and full-scale production. To a greater degree than in the McNamara years, lower-level officials controlled the programs between these points. Because of the role the new Acquisition Review Council assumed, major system acquisition is now frequently called the "DSARC process."

In 1976 the Office of Management and Budget, having studied procurement throughout the government for several years, issued a general administrative regulation, OMB Circular A-109, that established policies on major acquisitions by all federal agencies. Because previous experience in the Department of Defense was the basis for many of the provisions of the regulations, it had less effect on DOD than on other agencies. The main changes necessary were in the earliest stage of defense procurement.

OMB had learned in its study that government officials too often conceived requirements in the wrong way: they defined needs in terms of equipment instead of equipment in terms of needs. Consequently the regulation ordered:

Determination of mission need should be based on an analysis of an agency's mission reconciled with general capabilities, priorities and resources. When analysis of an agency's mission shows that a need for a new major system exists, such a need should not be defined in equipment terms, but should be defined in terms of the mission, purpose, capability, agency components involved, schedule, and cost

---

31. This document has remained central in the acquisition process over the years, although its form and names have changed slightly. DCP now stands for Decision Coordinating Paper. First drafted during the initial stages, the DCP is revised and updated while a system goes through the formal acquisition stages.

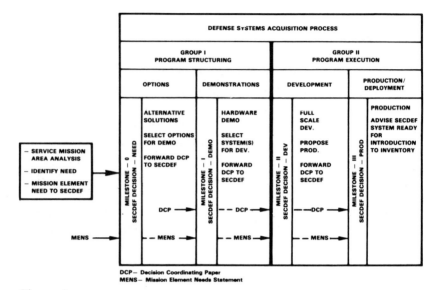

Figure 4
Acquisition management as modified by OMB Circular A-109. *Source: Department of the Navy RDT&E Management Guide,* 7th ed. (NAVSO P-2457) (Washington, D.C.: Department of the Navy, 1979), p. 2-16.

objectives, and operating constraints. . . . Mission needs are independent of any particular system or technological solution.[32]

To meet this new requirement, the Defense Department established another step in its acquisition cycle. Before investigating even the options for system development, the services would have to prepare a formal Mission Element Needs Statement (MENS). This would briefly state the need in mission terms, assess the threat, explain why existing capability did not meet the need, describe limits or boundary conditions for solutions, and propose a program for seeking alternative solutions. Figure 4 shows the redefined DSARC process.[33]

In spite of the many changes made over the years, problems have continued to plague defense acquisition: cost overruns, schedule slippages, equipment failures. When President Reagan took office in 1981 proposing increased defense expenditures, he promised further reforms to strengthen the credi-

---

32. Office of Management and Budget, "Circular A-109 to the Heads of Executive Departments and Establishments, Subject: Major System Acquisitions," April 5, 1976, p. 7.
33. Department of the Navy, *RDT&E Management Guide,* pp. 2–17.

bility of the Defense Department and its ability to manage its programs efficiently. Early in the administration, Deputy Secretary of Defense Frank Carlucci told his executive assistant, "If [Secretary] Weinberger and I do nothing else in these next four years except to straighten out the weapons acquisition process, we will have had a successful tour."[34] By July 1981 Carlucci had completed a careful study of the acquisition process and compiled a list of thirty-two new areas for action. The study reaffirmed the use of PPBS and DSARC as the principal acquisition management tools; however, it recommended better integration of the two systems so that the DSARC could not approve programs that had not been funded. To combat delays in the evolutionary cycle of new weapons from conception to deployment, the Carlucci Study advised reducing administrative oversight and formal reporting. This included lowering the dollar threshold for programs that had to be reviewed at the DOD level and cutting back the number of milestones requiring formal DSARC review from four to two (i.e., "SECDEF Decisions" in figure 5).[35] Although the results of the "Carlucci initiatives" are not yet clear, they represent the most comprehensive changes in defense acquisition since the 1960s.[36]

Formalization of research, development, and systems acquisition at the Department of Defense level also led to similar formalization in the Navy. Thus, for example, the Decision Coordinating Paper (DCP) for the DOD level led to the Navy Decision Coordinating Paper (NDCP), the Defense System Acquisition Review Council (DSARC) to the Department of the Navy System Acquisition Review Council (DNSARC), and formal program reviews at the DOD level to preliminary program reviews within the Navy. Furthermore, although systematic high-level reviews applied only to major programs, they engendered analogous reviews for less costly programs by lower-level officials. In general, the history of administrative control in the Navy since the 1960s reflects what occurred at the DOD level.[37]

34. Quoted in Vincent Puritano, "Getting Ourselves Together on Systems Acquisition," *Defense* 81 (October 1981), p. 9.

35. G. Dana Brabson, "Department of Defense Acquisition Improvement Program," *Concepts* 4 (Autumn 1981): 54–75.

36. For an interesting interpretive account of the evolution of PPBS, DSARC, and systems management in DOD in general, read Edwin A. Deagle, Jr., "Organization and Process in Military R&D," in Franklin A. Long and Judith Reppy, eds., *The Genesis of New Weapons* (New York: Pergamon Press, 1980).

37. The best general source on the current Navy R&D management system is the Navy's *RDT&E Management Guide.*

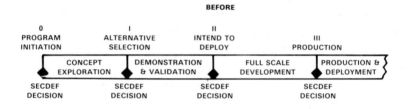

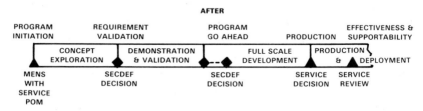

Figure 5
Acquisition management before and after Carlucci Reforms. *Source:* Dana
Brabson, "Department of Defense Acquisition Improvement Program,"
Concepts 4 (Autumn 1981), p. 60.

The systematic management that has evolved over the last
four decades is costly and time-consuming. The Carlucci initiatives, although more comprehensive than previous attempts at
reform, are only the most recent in a series of efforts to reduce
the administrative burdens of program oversight. No matter
how successful, they are not the beginning of a retreat to the
laissez-faire management styles of the immediate postwar period. The Department of Defense remains firmly committed to
systematic, high-level management of the entire process of research, development, and acquisition, and that commitment
will continue to be expressed in formal management controls.
DOD may streamline the techniques it uses, but it will not
significantly reduce them.

Performers
Centralization and formalization of R&D management strongly
affected the individuals and institutions that executed the
Navy's R&D program. For university scientists, Navy research
funding became increasingly difficult to get. After the creation
of the National Science Foundation in 1950, Navy support for
basic research in a wide variety of areas dissolved, and grants

began to focus more directly on subjects of immediate interest to the Navy. In addition, administrative controls tightened over the years as pressure increased to demonstrate the direct relevance of Navy research programs to Navy system development. The best-known example of this pressure is the Mansfield amendment to the Military Procurement Act of 1970. It stated that "none of the funds authorized . . . may be used to carry out any research project or study unless such project or study has a direct and apparent relationship to a specific military function or operation."[38] Although establishing the detailed linkage required by the amendment did not become regular practice in the Navy, the recent combination of the positions of Chief of Naval Research and Chief of Naval Development indicates that the pressure for relevance continues.

Private industry has continued to produce virtually all Navy equipment and also to conduct a substantial portion of Navy research and development. The formalization and systematization of the R&D process made defense business more risky and costly and brought greater concentration of defense work in large firms. Contractors had to spend more time and money on preparing documentation and guiding it through the review and approval process. "Free market" competition for major weapons systems became virtually impossible because only a small number of large firms could afford to engage in the process. The formal management reviews controlled the limited competition that remained.[39]

In its dealings with industry, the Navy tried various types of contracts, from fixed price to cost-plus-fixed fee to cost-plus-incentive fee. In the late 1960s, the Nixon Administration favored a technique known as total package procurement, which was supposed to reduce costs by having corporations bid on the entire program of development and production of weapons systems. Like many other ideas, however, this one was discarded after leading to several well-publicized cases of cost overruns.[40] As the recent bitter fights between the Navy and General Dynamics Corporation over production of the Trident

38. Booz-Allen and Hamilton, *Review of Navy R&D Management*, p. 223.

39. See, for example, Jacques S. Gansler, *The Defense Industry* (Cambridge, MA: The MIT Press, 1980), and General Accounting Office, "Impediments to Reducing the Costs of Weapons Systems" (PSAD-80-6) (Washington, D.C.: GPO, 1976).

40. Booz-Allen and Hamilton, *Review of Navy R&D Management*, p. 368, and Deagle, "Organization and Process," pp. 166–67.

submarines attest, thirty-five years of experience in the postwar era has led to no certain formulas for smooth DOD-industry interaction.

While industry has manufactured most new Navy equipment, the Navy's in-house laboratories have continued to play a major part in the R&D process. As the administrative and financial work related to systems development and production became more complex and time-consuming over the years, the individuals in the bureaus and system commands relied to an increasing extent on the laboratories. The laboratories not only generated new ideas, but also helped the Navy be a "smart buyer" who could fully understand the strengths and limitations of the products industry was selling.

Figure 3 showed the laboratories as they existed in 1946; figure 6 shows them in 1984. As is evident, some of the laboratories, most notably those classified as "test and evaluation" centers, have remained under the systems commands. Additionally, the Office of Naval Research has maintained control over its laboratories, including the Naval Research Laboratory. But note that there is now a group of laboratories at the top echelon of the Naval Material Command independent of the systems commands.

The formal separation of these from the material bureaus occurred on April 1, 1966, a month before the bureaus themselves were replaced by the systems commands. During previous years, improving defense laboratories was the subject of many studies, especially during Secretary McNamara's tenure. Often the studies pointed to problems in oversight by the bureaus.[41] One of them prepared by the Deputy Director of Defense Research and Engineering, Dr. Chalmers Sherwin, entitled "Plan for the Operation and Management of the Principal DOD In-House Laboratories," finally caused the separation.[42]

Sherwin's objectives were to raise the prestige of the laboratories, increase their quality, simplify management, and improve career patterns for scientists and engineers. He recommended

41. See Robert J. Mindak, "Management Studies and their Impact on Navy R&D" (Washington, D.C.: Office of Naval Research Report ACR-205, November 1974).

42. The report was dated November 16, 1964. Secretary McNamara sent it to the secretaries of the Army, Navy, and Air Force on November 20. A copy of the report is in Series 4, "Shore Establishment Realignments and Policy affecting Navy Laboratories, 1960–1982," RG 3, Archives of Navy Laboratories, David W. Taylor Naval Ship Research and Development Center, Bethesda, MD.

OFFICE OF NAVAL RESEARCH
Naval Ocean Research and Development Activity, Bay St. Louis MS
Naval Research Laboratory, Washington DC

NAVAL MATERIAL COMMAND
David W. Taylor Naval Ship Research and Development Center, Bethesda MD
Naval Air Development Center, Warminster PA
Naval Coastal Systems Center, Panama City FL
Naval Ocean Systems Center, San Diego CA
Naval Surface Weapons Center, Dahlgren VA
Naval Training Equipment Center, Orlando FL
Naval Underwater Systems Center, Newport RI
Naval Weapons Center, China Lake CA
Navy Personnel Research and Development Center, San Diego CA

NAVAL SEA SYSTEMS COMMAND
Naval Explosive Ordnance Disposal Technology Center, Indian Head MD
Naval Ordnance Missile Test Facility, White Sands NM

NAVAL AIR SYSTEMS COMMAND
Naval Air Propulsion Center, Trenton NJ
Naval Air Test Center, Patuxent River MD
Naval Environmental Prediction Research Facility, Monterey CA
Naval Weapons Evaluation Facility, Albuquerque NM
Pacific Missile Test Center, Point Mugu CA

NAVAL ELECTRONICS SYSTEMS COMMAND
Naval Space Systems Activity, Los Angeles CA

NAVAL FACILITIES ENGINEERING COMMAND
Naval Civil Engineering Laboratory, Port Hueneme CA

NAVAL SUPPLY SYSTEMS COMMAND
Naval Clothing and Textile Research Facility, Natick MA

Figure 6
Principal Navy research and development activities, 1984 (excludes medical laboratories).

consolidating laboratories into large functional centers with broad missions, putting a civilian scientist or engineer in charge, and placing the laboratories much higher in the administrative chain than was normally the case, for example, under a "Director of Laboratories" reporting to the Assistant Service Secretary for Research and Development.

Secretary McNamara and other top DOD officials strongly supported the conclusions of the Sherwin plan. Thus while top Navy officials basically opposed it, they knew they had to make changes. Consequently, in December 1965 the Navy created the position of Director of Navy Laboratories, and in April 1966 transferred fifteen laboratories from the material bureaus to him. These would henceforth report directly to the Chief of Naval Material through the new official. Since then, the fifteen have been reorganized and consolidated into nine.[43] The laboratories were granted the power to organize and conduct their programs as they saw fit but were not given significant independent appropriations. Not only did the requirements for their work continue to come from the system commands, so did the bulk of their funding. This meant that while the system commanders did not keep management control over the laboratories, they did retain a substantial amount of program control.

Navy Research and Development in 1984
The changes discussed above are all apparent in the organization of Navy research and development in 1984 (figure 7). The Planning, Programming, and Budgeting System (PPBS) and acquisition review by the Defense System Acquisition Review Council (DSARC) are the basic management systems used. Underlying all the alterations was the desire to reduce financial and technical risk and to focus programs on areas that appear to promise the greatest return on investment. The reforms have slowed the process of moving from initial ideas for defense systems to developing, producing, and implementing them in the fleet. They have favored evolutionary change. Throughout the whole period, however, the Navy's commit-

---

43. See Laboratory Management Division, Naval Material Command, "Realignment of the CNM RDT&E Activities, 1966 to 1970" (Unpublished study dated 24 May 1970) in Series 4, "Shore Establishment Realignments and Policy affecting Navy Laboratories, 1960–1982," RG 3, Archives of Navy Laboratories, David W. Taylor Naval Ship Research and Development Center, Bethesda, MD.

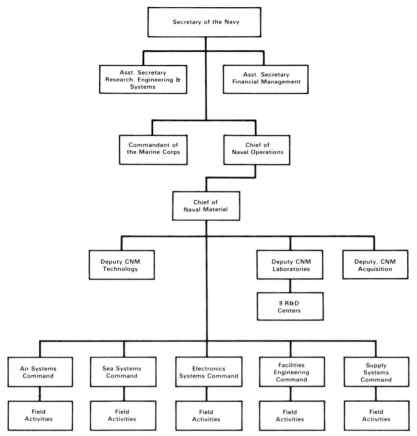

Figure 7
Basic structure of Navy research and development, 1984.

ment to research and development has remained firm. Despite the many hurdles through which any new system must now jump on its path from idea to implementation, the Navy has continued to provide a remarkable degree of freedom, flexibility, and support for the exploration and testing of new ideas to improve technology. Those who bemoan the frustration of briefing their programs to all the officials now regulating them have forgotten the much greater difficulties of their predecessors, who, in the years before World War II, had to defend the need for the Navy to sponsor any major research and development programs at all.[44]

44. See Allison, *New Eye for the Navy*, and Albert B. Christman, *Sailors, Scientists, and Rockets* (Washington, D.C.: Department of the Navy, 1971).

Examples of Navy Research and Development

Naval research and development since World War II has profoundly changed the speed, nature, and lethality of naval warfare. The advent of nuclear power made possible a "true" submarine, one designed to cruise under water, not just dive occasionally. Guided missiles have largely replaced guns and bombs. Combining the nuclear submarine and ballistic missile in the Polaris and later the Poseidon and Trident weapons systems added to the Navy's traditional objectives of controlling the seas and projecting power ashore—the strategic mission of nuclear weapons delivery. Satellites for communications, surveillance, and navigation have extended Naval Operations to outer space.

Besides advancing into new areas of technology, the Navy has made substantial progress in traditional fields. Ships are now quieter and harder to detect. Lighter and stronger materials, along with better structural design, have improved their construction. New fire-fighting techniques and damage-control procedures have made them safer in combat. Advances in guidance and navigation have enhanced weapons capabilities and new forms of explosives—conventional as well as nuclear—have enhanced their power. Sensors such as monopulse radars and towed acoustic arrays have brought substantial improvements in capabilities for locating targets above and below the sea surface. But as sensors have improved, so have means to counter them, such as chaff, paints that absorb electronic signals, and radar signal jammers. To meet the ever-increasing need for high-speed information processing, computers now monitor tactical situations and help commanders make rapid decisions. Reliance on electronic devices has become so pronounced that the Navy has built ships and aircraft equipped with nothing but sensors and computer processors for gathering tactical information or assisting in weapons delivery. In sum, although the fleet of the 1980s outwardly resembles that of 1946, the Navy has undergone more profound changes during the last four decades than in any other comparable period in its history.

The two following examples illustrate some of the administrative and organizational methods the Navy has used to manage research and development. The projects both reflect general trends and show some of the reasons for them. They also demonstrate several other points: formalizing and systematizing the R&D process has not eliminated the significance of individual initiative; nor has the importance that high-level officials now attribute to research and development meant that good ideas always find easy acceptance and approval.

Sidewinder
One of the highest Navy priorities in the immediate postwar years was developing guided and unguided missiles. Missiles for air intercept were particularly important because jet power had made aircraft faster and more maneuverable, thus harder to locate and hit. The most successful solution to these problems was the Sidewinder missile.

Sidewinder was a product of the Naval Ordnance Test Station (NOTS), China Lake, California. Situated in the Mojave desert not far from Death Valley, this facility was established in 1943 for cooperative work between the Navy and the California Institute of Technology on aircraft rockets. After the war, the Bureau of Ordnance transformed NOTS into a permanent research and development laboratory.

The bureau first asked NOTS to develop a guided missile in 1948. The bureau prescribed the approach. To save time and money, NOTS had to integrate readily available "off-the-shelf" components. As is often the case with such efforts, NOTS, under these restrictions, could do no better than others already in the field.[45] Consequently, when the Research and Development Board of the National Military Establishment pressured the Bureau of Ordnance to reduce its guided-missile programs, the bureau terminated the NOTS effort and directed the laboratory to limit itself to traditional areas of expertise, such as unguided rocket development and aircraft fire control.

This restriction, however, could not stop NOTS engineers from thinking about missile guidance. Dr. William B. McLean, head of a group investigating aircraft fire control, was especially concerned about the subject. He knew that the greatest problem with making aircraft rockets more accurate was not improving their aim before firing but correcting their course after launch in response to unexpected maneuvers by targets.

---

45. Naval Weapons Center, "Sidewinder 1 and 1A Air-to-Air Guided Missiles (AIM-9A and -9B)" (Unpublished manuscript in Archives of Navy Laboratories, David W. Taylor Naval Ship Research and Development Center, Bethesda, MD).

To make significant progress in fire control, that is, he had to study guidance systems.

In 1949 McLean prepared a report proposing the development of an infrared homing rocket. Unlike other Navy guided missiles, which depended on guidance equipment in the launch vehicle, McLean's plan was to place all guidance apparatus in the missile itself. The desirability of such a system was clear. But its feasibility was not. Not only would the rocket have to be small, lightweight, rugged, and reliable enough to stand the effects of launching, but it would also have to be easy and cheap to build, so that the cost of each missile would not be exorbitant.

The Bureau of Ordnance initially refused to fund the project. It had already struck NOTS from the list of institutions approved for guided-missile work, and the planned project failed to meet any established operational requirement. The Navy's stated need was for all-weather missiles. Because infrared light does not penetrate clouds, this one would function only in clear skies.

Despite this reproof, McLean did not stop. To support his effort, he convinced the technical director of NOTS to establish a research and development project using exploratory and foundation research funds available for expenditure at his discretion. The bureau knew what was happening but treated the project with benign neglect.[46]

To move from McLean's initial plans to a practical missile system, NOTS had to solve many technical problems. The guidance system presented the most difficulties, and making it cheap, simple, and reliable required many innovative ideas. Most missile systems of this era used two gyroscopes—one to indicate the position of the missile and the other the position of the target. McLean and his team realized, however, that as long as all guidance equipment was in the missile, it needed only one gyroscope—the one to track the target. Knowledge of the position of the missile in relation to anything else was unnecessary.

A second major innovation came in determining how to make guidance corrections by using torque balancing. The tracking device sensed target motion. The guidance system

46. Theodore F. Gautschi, *An Investigation of the Management of Research and Development Projects* (China Lake, CA: Naval Ordnance Test Station AdPub 112, 1962), p. 108. Harbridge House, Inc., "Sidewinder" (Unpublished manuscript history) in "Booz-Allen and Hamilton Source Materials for Navy R&D Study," box 8 in Operational Archives Branch, Naval Historical Center, Washington Navy Yard, Washington, D.C., p. 37.

translated this into a torque on the missile steering mechanism, which caused a change in course. This torque was balanced by torque created on the steering fins by the wind in the airstream. Once the course correction ended, the steering torque would end, and the counterbalancing torque would return the missile to equilibrium. The advantage of the torque balance was that it required no monitoring of the changing shape of the airframe because of dynamic stress, no information about missile speed, and no measure of variations in altitude or angle of attack.

These two innovations are only several of many in the Sidewinder program, but they exemplify what was characteristic of its design as a whole: NOTS engineers worked out the complexities in their minds so the missile would be simple.[47]

As efforts to crystallize the details of the design continued, the project maintained its low visibility. Although this meant that the investigators had limited support, they had almost complete freedom in organizing and managing the program. Moreover, as many involved in the project would later say, the official view that "it couldn't be done" reinforced their determination and commitment. There was a limit, however, to what they could accomplish in obscurity. Procurement of major components, system integration, test flights—the preliminaries to developing a viable weapons system—were expensive, and spending large sums of money meant obtaining official approval.

By May 1951 NOTS engineers were ready to test their ideas and asked the Bureau of Ordnance for funds to develop components. After a review in September, however, the Guided Missile Committee of the Research and Development Board again said no.[48] Finally, in October, the tide turned. Admiral William S. Parsons, Assistant Chief of the Bureau of Ordnance, visited NOTS with a study group, reviewed the missile program, and was persuaded. Parsons had been among the top Navy representatives on the Manhattan Project, and his technical judgments commanded respect throughout the Navy. After his decision, action was rapid. On October 16 a letter from the Office of the Chief of Naval Operations requested the Bureau

---

47. Naval Weapons Center, "Sidewinder 1 and 1A Air-to-Air Guided missiles (AIM-9A and -9B)," and William F. Wright and Bud Gott, interviewers, "Group Interview on Sidewinder Missile Project, (S-112), 14 March 1980" (Unpublished interview transcript available from Naval Weapons Center, China Lake, CA. Cited by permission).
48. Gautschi, *An Investigation of Management*, p. 111.

of Ordnance to support the project. Clearing up the matter of requirements, the letter stated that Sidewinder "is considered to support and be a development which may fulfill, as an interim weapon, Operational Requirement AD-14701, Aircraft against other Aircraft."[49] From that time on, adequate program funding was not a problem. Sidewinder had finally been authorized under the prevailing management system.

By April 1952 NOTS had a workable design and was ready to begin procurement of the experimental models. In early 1953 Philco Corporation became the contractor for the control and guidance components and prime contractor for the whole system. It delivered test models later in the same year. Although initial firings were unsuccessful, by September a Sidewinder came within six feet of a target drone. Then on January 9, 1954, the missile hit its first target, a remote-controlled bomber. Subsequent tests were also positive. Sidewinder was a demonstrated success.[50]

The developers froze design for a production missile in March 1954 and began perfecting the new weapon for service use. In July 1956 Sidewinder was released to the fleet, and the U.S. Air Force and other NATO nations soon adopted the weapon (figure 8).[51]

The success of Sidewinder meant that the Bureau of Ordnance had to procure many copies. In 1956 the bureau used an improved version of the missile, Sidewinder 1A, as a test case for a procurement method called "multiple-source bidding." Long used for simple products, this technique was unusual for complete weapons systems. It worked in this way: to expand the number of suppliers capable of producing a product, the government funded the acquisition of special tooling or new equipment a corporation needed to build it. The government also paid for pilot production. Then, the new source competed against the original supplier. In this case, after preliminary competition the General Electric Company was chosen from among eight interested firms. By 1957 full-scale competition between General Electric and Philco was underway. The number of orders each received depended on their unit price, and as a result of competition, this dropped dramatically. In fiscal

49. Harbridge House, Inc., "Sidewinder," p. 45.
50. Gautschi, *An Investigation of Management*, pp. 110–15.
51. Ibid., pp. 115–16; Harbridge House, Inc., "Sidewinder," passim.

A   B

Figure 8
A. The "Heat Homing Rocket," 1950. B. Sidewinder as it went to the fleet,
1956. F9F-8 with two missiles. *Source:* Naval Weapons Center, China Lake,
California.

year 1962 each missile cost one-seventh of what it had during
the original pilot production and one-third the price of original
large-quantity production.[52] Sidewinder had become a weapon
the Navy could afford.

The early history of the Sidewinder missile program reveals
many interesting features of Navy R&D administration. The
most fascinating, obviously, is how what became one of the
most versatile weapons in America's arsenal began life as a

52. Philip J. Klass, "Competition Slashes Sidewinder 1-A Price," *Aviation Week and
Space Technology* (April 9, 1962): 89–95.

bastard child, while many of the legitimate programs of the same era ended as expensive failures. The degree to which the project was "unauthorized," however, should not be overemphasized. The Navy R&D management system has always allowed lower-level managers some flexibility to pursue projects on their own initiative. And although this freedom is more restricted than it was thirty years ago, much independence remains. The Sidewinder history also demonstrates the role the in-house Navy Laboratories frequently play in the R&D process. After generating the ideas basic to Sidewinder, the Naval Ordnance Test Station transferred it to industry for production. Subsequently NOTS and its successor, the Naval Weapons Center, have participated in numerous improvement programs and helped oversee procurement of many new generations of Sidewinder. The continuing vitality of the program is evident in the fact that Navy fighters used a recent version, the Sidewinder AIM-9L, to shoot down the Libyan fighters in the Gulf of Sidra incident in 1981. Finally, the commanding role of the Bureau of Ordnance over the project was clear. Not until the bureau had accepted Sidewinder could it become a weapon destined for full-scale production and acceptance into the fleet.

R&D administrators often say that innovations can result either from "technology push" or "requirements pull." In the former case, the ideas for new systems come from scientists and engineers who perceive new opportunities during their research. In the latter, they come from planners who identify needs that new technology might satisfy and then work to develop it. Sidewinder resulted from "technology push." The Naval Tactical Data System, to which we now turn, was the result of "requirements pull."

The Navy Tactical Data System
Recording, correlating, and processing information from radars, sonars, visual sightings, and other forms of data gathering became a major problem on board warships during World War II. The creation of Combat Information Centers, specially configured rooms for evaluating tactical information, provided an interim solution. There, sailors used large, vertical plotting boards to record data, evaluate threats, calculate intercept courses and speeds, and assign gun batteries. Their tools for data exchange and notation were grease pencils, manual plot-

ters, and voice communications.[53] As jet aircraft and missiles appeared in weapons inventories in the postwar years, however, manual information processing became obsolete.

The need for improved data handling was a symptom of the Navy's broader need for improved capabilities for general surveillance and air defense. Recognizing this, Admiral F. R. Furth, Chief of Naval Research, initiated Project Lamp Light in May 1954 to make a broad investigation of the whole subject. Jointly sponsored by the Navy and the Air Force, the study was conducted by the Massachusetts Institute of Technology and involved representatives from all three armed services, many defense contractors, and representatives of allied governments. The final report, entitled "The Defense of North America," gave particular attention to mechanized data-handling systems.[54]

As a temporary solution to shipboard data handling, the report recommended improving the Electronic Data System (EDS), an automated analog data processor developed by the Naval Research Laboratory, and installing it on selected Navy ships. EDS was introduced into the fleet in 1956.[55] The study group further recommended, however, that the Navy turn from analog processors to the emerging technology of solid-state digital computers for its data-handling system of the future.

Responding to this recommendation, the Chief of Naval Research established a Committee on Technical Data Processing Systems in April 1955 to develop a statement of technical requirements for a Navy tactical data system (NTDS). Issued in August 1955, the statement called for a system based on a digital computer that would also include cathode-ray tube situation displays, radio data links, and peripheral equipment. The system was to satisfy the full range of shipboard needs for data processing related to antiair warfare, surface warfare, amphibious warfare, electronic countermeasures, and antisubmarine

53. Fred W. Kittler, "Command Control—the Navy's Link to Global Strategy," *Naval Engineering Journal* (February 1963): 32–4.

54. The report appeared in 1955. See Louis A. Gebhard, *Evolution of Naval Radio-Electronics and Contributions of the Naval Research Laboratory* (Washington, D.C.: Naval Research Laboratory Report 8300, 1979), pp. 381–84.

55. Harbridge House, Inc., "Naval Tactical Data System," in "Booz-Allen and Hamilton Source Materials for Navy R&D Study," box 8 in Operational Archives Branch, Naval Historical Center, Washington Navy Yard, Washington, D.C., p. 13.

warfare. The degree to which these requirements taxed the state of the art is significant. At the time no one had manufactured any general purpose solid-state computer, much less one ready for the rigors of duty at sea during battle.[56]

The team of individuals who had prepared the technical requirements for the Chief of Naval Research had included representatives from the Office of the Chief of Naval Operations. These men now worked to have the CNO accept them and issue the development characteristics. When they appeared in April 1956, they detailed the requirements and stated that the system must be able, among other things:

to provide current displays to assist command appraisal and evaluation of the air situation, surface situation, subsurface situation, or the deployment of units . . .

to develop information required for the assignment of areas of responsibility to specified ship, aircraft, and weapon capabilities, and to insure that all command and surveillance stations will be kept current in such assignments . . .

to receive information from a variety of sources including radar, sonar, ECM, data links, visual, orders, etc. and to process it with identifying track numbers and supplementary category information.[57]

To develop and procure NTDS, the Bureau of Ships established a special project office in May 1956, under the Assistant Chief for Electronics. Although the organizational structure of the office changed slightly over time, a special project office of some form continued to oversee the effort until the initial system was completed and approved for service use.[58]

The project management office was too small and limited to monitor the technical aspects of development. Originally it planned to use a single prime contractor to provide technical assistance and system integration. In the Navy's estimation, however, there was only one company capable of handling the job, the Bell Telephone Laboratories, and the Navy could not persuade Bell to accept it. Consequently, a tripartite division of tasks was made: 1) the project office would retain general technical control; 2) separate prime contractors would manufacture the major components of the system, with the computer con-

56. Ibid., p. 23.
57. Ibid., pp. 17–18.
58. Ibid., p. 28.

tractor serving as system engineer; and 3) a Navy laboratory would act as system integrator and technical advisor to the bureau.[59] The Navy chose Remington Rand, the UNIVAC Division of Sperry Rand, as the principal computer contractor and systems engineer; Hughes Aircraft to develop digital displays; Collins Radio to make communication data links; and Ford Instrument Company to build the analog-digital converter. The Naval Electronics Laboratory provided systems integration.

The development of the Naval Tactical Data System in many ways exemplifies systematic R&D management as it was supposed to occur. The Office of Naval Research participated in planning, the Office of the Chief of Naval Operations established development characteristics, the Bureau of Ships ran the project, a Navy laboratory gave technical advice, and several contractors developed and built the equipment. In other ways, however, the project ran in a relatively unstructured fashion. A core of key individuals administered it from the Lamp Light study through final development. The Bureau of Ships project office was always small. It used personal contact rather than formal documentation to monitor progress and did not demand consistent reporting. Because of the high priority of the effort from the outset, funding was never tight and financial management was not stringent. No total cost estimates were prepared before development was undertaken, and alternative approaches were not explored. In sum, while systematic management had begun to be part of Navy R&D in the late 1950s, when NTDS came into being, informality and dependence on the undocumented judgment of key personnel still could characterize weapons system procurement.[60]

NTDS progressed rapidly from inception to implementation. The Project Office began work in May 1956 and awarded major contracts during the same spring and summer. Delivery of test equipment to the Naval Electronics Laboratory for assembly and check-out began in December 1958, and the first tests of the completed system began in April 1959. Operational evaluation was conducted between 1961 and 1962, and NTDS received approval for service use by CNO in April 1963. It was

59. Ibid., pp. 38–39.
60. Ibid., p. 22.

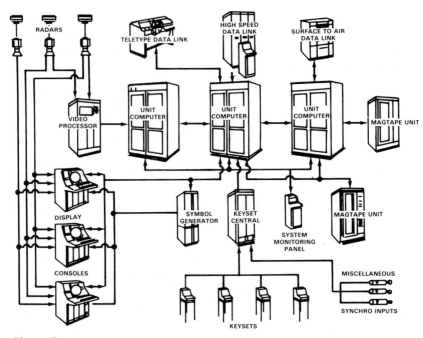

Figure 9
Typical NTDS configuration, 1963. *Source:* Fred W. Kittler, "Command,
Control—The Navy's Link to Global Strategy," *Naval Engineer's Journal*
(February 1963), p. 35.

installed on an experimental basis in the carrier *Enterprise* in
1961 and in the *Long Beach* in February 1963 (figure 9).[61]

Several factors were instrumental in fostering this quick success. Although the requirement for the system stipulated that it
would be capable of capturing, processing, and displaying data
relevant to a wide variety of ship functions, the only substantial
capability developed in the initial system was processing data
for antiair warfare. Other capabilities were left until later. This
was clever management: the broadness of the plan made it
easier to sell, and the limitations added later made it easier to
accomplish.

In addition, the project office carefully cultivated relations
not only with those officials involved in development, but also
with those involved in acceptance and use of the equipment. It
chose the Naval Electronics Laboratory to provide technical
guidance on development partly because it was located very
close to the Fleet Anti-Air Warfare Training Center and the

61. Ibid., pp. 9–10.

Fleet Programming Center, the institutions where sailors would learn to use the new equipment. This was a great help in solving detailed problems and fostering system acceptance.[62]

Over the years NTDS assumed many of the functions its original planners had hoped it would, but could only dream of. These included methods for handling information related to antisurface and antisubmarine warfare and means of accommodating input from sonars, navigation equipment, and even satellites. NTDS not only became the primary tool for maintaining a current picture of a ship's tactical environment, but also assumed roles in communicating with higher command authorities and in weapons delivery.

This success, however, had some negative consequences. In 1979 the Executive Panel of the Office of the Chief of Naval Operations met to begin a comprehensive review of the NTDS system and plan its future. The panel found that the system was beset by many problems. It concluded:

There appears to be no one in charge of top-level requirements, architecture, or configuration control. There seems to be little correlation between C3 [command, communications, and control] architectural development and considerations of NTDS capabilities and development. There is a need to examine and change the hierarchy of requirements and functions, not just to add new software. . . .

There is no formalized method for screening new tasks or functions to be placed on NTDS. Central processor (computer) capacity has been overallocated without adequate recognition being given to the resulting degradation of performance of existing functions. NTDS computer capacity (actually, the only digital processor around in earlier years) has been viewed as a "free lunch," requiring only a modest reprogramming expenditure, by those building systems to interface with it. In many cases now, there isn't any more "lunch."

There has been inadequate investment and focus in planning and evolution of the system. The current NTDS, an extrapolation of a 20-year-old system design, was originally developed to support AAW [antiair warfare] and now must support AAW, ASW [antisubmarine warfare], ASUW [antisurface warfare], EW [electronic warfare], and Amphibious missions. It was originally designed to support 15 functions and to interface with two data links and a missile system. It has since received, "bandaid-by-bandaid," modifications, adapters, peripherals, etc., to support 27 functions and to interface with [many types of equipment] and 3 data links. This has resulted in an overloaded system with slow-processing times. The consideration of the "sunk cost" of the system has stifled a much needed, fundamental

62. Ibid., p. 83.

reexamination of concepts, functions, and architecture in the light of current threats, sensors, and weapons systems capabilities, and additional new requirements. It has therefore responded to new requirements such as those imposed by Vietnam and new capabilities offered by Automatic Detection and Tracking (ADT) radars without a thorough top-down review.[63]

The first step towards solving these problems was evident: a study of NTDS to determine how the Navy should modify it to meet the needs of the late twentieth century. For our purpose the most interesting thing about the reexamination is the degree to which it reflected R&D planning style. The Office of the Chief of Naval Operations, showing its increasingly important role in Navy R&D planning, was to make the study. It was to be a thorough, top-level review. It was to be based on systematic functional analysis. And it was to focus on careful delineation of general mission needs, not particular forms of equipment or merely better performance of current tasks as presently conceived. The final report of the study group reflected this approach:

The necessity to ensure that anything that purports to be general guidance is not somehow predicated on the current structure of the system is clear. In fact, it was the emergence of new combat subsystems and capabilities that led the [CNO Executive Panel] to recommend re-examination of the basis on which functions are allocated to NTDS and to re-define the management philosophy for the system.[64]

Could one ask for a better implementation of the spirit of OMB Circular A-109 and its requirement that planners focus on missions and not equipment?

The details of the final report are unimportant here. What is interesting, however, is how the group redefined NTDS. Previously the Navy considered the system as a piece of computer equipment to which other pieces of equipment were linked. Now a mission concept predicated on the purpose of the equipment replaced this traditional, hardware-based understanding. The result was a rethinking of NTDS and related devices within the context of two distinct functional categories: the Combat

63. As quoted in Office of the Chief of Naval Operations, Surface Combat Systems Division, *Final Report of the CNO NTDS Functional Allocation Study Group*, (Washington, D.C.: Department of the Navy, 1981), pp. 6–7.
64. Ibid., p. 15.

Action Direction System (CADS) and the Combat Action Support System (CASS). The report defined them as follows:

The *Combat Action Direction System* (CADS) consists of all men and data (or information) handling capabilities employed in the execution of shipboard combat action direction functions;

The *Combat Action Support System* (CASS) consists of that portion of the CADS for which there exists an operational requirement for automation to satisfy the *information requirements of combat action decision-makers* in the CADS. . . .

[CADS] is comprised of all persons who (or machines which) make or authorize the decisions that must be made in [taking a ship into battle], together with all persons or machines that support the decision-makers by producing the specific information needed in decision-making from the resources available aboard the ship. . . . The CASS supports these decision-makers by implementing information processing functions necessary to produce answers, [and] manipulate and interpret data to produce answers.[65]

This redefinition made determining, evaluating, and prioritizing shipboard computer needs easier and more effective. Figure 10 is a diagram contrasting the traditional and new ways of conceiving NTDS.

The degree to which the new concepts will bring basic changes to NTDS remains to be seen. Regardless of the results, however, the plans provide an instance of high-level management directing the use of systematic functional analysis to cope with a fundamental technical problem of the Navy.

Conclusion

The United States emerged from World War II as the predominant world military power. It has retained this position over the last four decades only through successful military research and development. Because of the importance, cost, and risk of military research and development, Congress and top defense officials continually demanded better management, especially during the aftermath of major program failures or cost overruns. This usually meant more management and greater centralization of authority and responsibility. For most of the postwar period, the additional expense and delays the new safeguards themselves introduced were not considered impor-

---

65. Ibid., pp. 17–18. Emphasis is in the original.

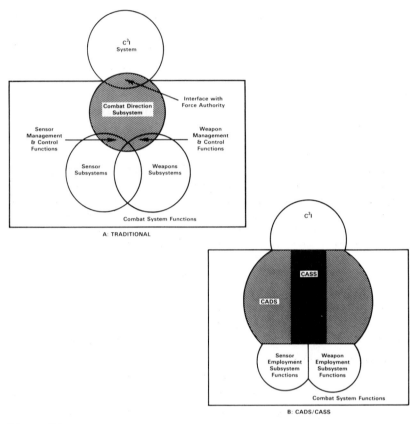

Figure 10
Conceptualization of NTDS. *Source:* Office of the Chief of Naval Operations, Surface Combat Systems Division, "Final Report of the CNO NTDS Functional Allocation Study Group" (Washington, D.C.: Department of the Navy, 1981).

tant. More recently, however, reformers seem to be accepting the fact that increased management eventually brings only more paperwork, not more control, and that a substantial portion of risk is an integral part of military R&D that no managerial technique can eliminate. At least this conclusion appears to have motivated many of the Carlucci reforms.

Although the evolution of management has been important, it should not be overemphasized. As the Sidewinder and NTDS projects demonstrate, programs are still defined primarily by the particular individuals involved in them and not by oversight by higher authority. In the new environment, initiative, advocacy, and entrepreneurship have become more disciplined and more bureaucratic, but no less necessary for program success.

8

Command Performance: A Perspective on the Social and Economic Consequences of Military Enterprise
David F. Noble

David Noble departs from the monographic format of the essays in this volume to address some of the broader social implications of military enterprise. Unwilling to accept the military's role in technological development with uncritical enthusiasm, Noble intends his remarks as a polemical commentary on military enterprise as a social force and as a corrective to popular misapprehensions that often associate technological change with human progress. He is particularly concerned with the ideological underpinnings and socio-economic ramifications of military "command technologies." By looking briefly at three instances of military enterprise, he maintains that such command technologies have inherent qualities (performance, command, modern methods) that bias them against the interests of small producers and working people. Not everyone will agree with Noble's assessment, but no one can afford to ignore his penetrating criticism.

It is well known that the military has played and still plays an important role in stimulating technological development, especially in wartime. Everyone has heard at least some of the war stories about radar, nuclear energy, electronics, computers, high-performance aircraft, and control and communication systems. All of these technologies, and the list is endless, were created to serve purposes peculiar to the military: proximity fuses for bombs, submarine and aircraft detection, combat communications, rocketry and missile warfare, gunfire control, ballistics calculations, command and control defense networks, high-speed, versatile-flight aircraft, and nuclear weapons. Yet, despite these original objectives, it is a common perception that such military-born technologies as these "spill over," as the story goes, into civilian use, where they are then adapted for a host of other more benign and even beneficial purposes. This is the conventional account of the role of the military in technological development. Essentially, this military role is characterized in two ways. First, it is what economists would call an "externality" as far as the civilian economy is concerned; military-born technologies enter the economy from "outside," so to speak, in a random, unsystematic way. (Of course, this is the way economists view all technological development; for them, military technologies merely combine two externalities, the military, or government, and technology.) Second, the military influence on the development of the technologies is only temporary and is restricted to the actual military uses of the technologies, for weaponry and the like. That is, the military influence does not permanently mark the technologies nor does that military influence in any way spill over along with the technologies when they undergo transfer from the military to the civilian arenas.

I would like to suggest that this conventional view of the role of the military in technological development is problematic on both counts. First, the military role has not been the "externality" that it appears to be when viewed through the lens of the neoclassical economist. Rather, it has been central to industrial development in the United States since the dawn of the industrial revolution. Lewis Mumford among others has been arguing this for some time, but it has been terribly difficult for us to shake the economistic habits as well as the peculiarly American blindness to the presence of the military.[1] Seymour Melman, in his writings about Pentagon capitalism and the permanent war economy, has perhaps contributed more than anyone else to our understanding of the centrality of the military since World War II. But this is not just a postwar phenomenon, as we will see. Second, the influence of the military on the technologies is not temporary, something removed when the technologies enter the civilian economy. The influence spills over in the specific shape of the technologies themselves and in the ways they are put together and used, with far-reaching economic and social consequences that have barely been examined. Professor Melman has described some of the consequences of the military's role in our economy: the vast military monopoly of material, technical, and human resources that might be used to meet human needs were they not diverted toward nonproductive, wasteful, and extremely dangerous military objectives; the corrupting influence of cost-maximizing military procurement and contracting practices, which has given rise to generations of managers who are incapable of truly independent, innovative, efficient, or economical production and to legions of technical personnel who are incompetent to produce for a competitive market or otherwise meet such nonmilitary specifications as cheapness, simplicity, accessibility, and the like.[2] But there are still other ways, more subtle perhaps, in which the military shapes technological and industrial development, with consequences that are no less profound.

In 1965 the Air Force produced a half-hour film, *Modern Manufacturing: A Command Peformance,* to promote the use of numerical control manufacturing methods in the military-oriented aerospace industry. The new technology had been developed over the previous decade and a half under Air Force aegis. The film offers a good reflection of the scope of the military's penetration into private industry, ranging from aircraft manufacture to machine tools, electronic controls and communications, and computers (Republic Aviation, Lockheed, Hughes, Giddings and Lewis, Pratt and Whitney, Sund-

1. Lewis Mumford, *Technics and Civilization* (1934; Harbinger Book, 1963), pp. 81–106. Also see Mumford's *The Myth of the Machine: The Pentagon of Power* (New York: Harcourt Brace Jovanovich, Inc., 1970), pp. 145–53, 164–69, 236–67; and *The Myth of the Machine: Technics and Human Development* (New York: Harcourt Brace Jovanovich, Inc., 1967), pp. 188–202, 212–23, 278–81.

2. See Seymour Melman's *Pentagon Capitalism: the Political Economy of War* (New York: McGraw Hill Book Co., 1970) and *The Permanent War Economy* (New York: Simon and Schuster, 1974)

strand, Kearney and Trecker, Raytheon, and IBM are all in the film). It also illustrates what I will refer to as the dominant characteristics of the military approach to industrial development, characteristics which, together, constitute a system of thinking that informs, embraces, and transcends the particular technological developments themselves. For simplicity, I have reduced these to three basic preoccupations: performance, command, and "modern" methods.[3]

By performance I mean the emphasis placed upon meeting military objectives and what follows necessarily from it. These objectives include, for example, combat readiness, tactical superiority, and strategic responsiveness and control; and these objectives demand such things as—in the case of the Air Force—fast, versatile, and powerful aircraft, keeping up with the arms race for ever more sophisticated weaponry, and worldwide communications and control. And these secondary objectives require another set of manufacturing specifications: the capability to manufacture highly complex yet reliable parts for high-speed aircraft and missiles (airfoils, variable thickness skins, integrally stiffened wing sections, etc.); short lead time and turnaround time to accommodate rapid design changes; and, perhaps most important, the interchangeability, reproducibility, and compatibility vital to an integrated system, what Merritt Roe Smith has aptly called "the uniformity principle."[4] "Tapes can be sent anywhere in the world," the narrator of the film boasts, "to produce interchangeable parts." These military performance objectives are justified in the name of national survival and so too are the product specifications and manufacturing criteria that follow from them. Note that cost has not yet been mentioned. Certainly, it is an important consideration, but only a secondary one, insofar as it is "consistent with reliability and reproducibility," to quote the film again. There is no direct concern whatsoever for meeting market demands or social or human needs.

In a talk at MIT in 1979, British design engineer Michael Cooley referred to management as "a bad habit inherited from

3. U.S. Air Force, *Modern Manufacturing: A Command Performance* (1965). Also see "The Manufacturing Technology Management Initiative," a videotape produced by General Alton D. Slay, Commander of the Air Force Systems Command, and presented to the Assembly of Engineering Planning Meeting for Manufacturing Studies Board and a Specific Study of the Air Force Manufacturing Technology Program (Washington, D.C.: National Research Council, July 27, 1979).

4. Merritt Roe Smith, "Army Ordnance and the 'American system' of Manufacturing, 1815–1861," p. 48 in this volume.

the church and the military."[5] I do not know about the church's influence upon management, but the military's is certainly clear. The military term for management is command, a rather straightforward notion that means the superior gives the orders and the subordinate executes them, with no ifs, ands, or buts. It is the ideal of managers everywhere. In the private sector it follows, with more or less legitimacy, from the pursuit of profit, control, and the will to power. In the military it follows from the need to meet performance objectives. Uniformity in manufacture, for example, presupposes command, direct and uncompromised. The film noted above well illustrates this point. First, it was made for top management, not the workforce who would actually be doing the work. The film's theme, the proposed and promised aim of modern manufacturing, was "to shorten the chain of command" and "greatly reduce the opportunities for a breakdown in communication." This means less human intervention between order and execution and those who remain will perform "reliably," according to "fixed instructions" that are "not subject to human error or emotion." Thus there will be no reliance upon the autonomy, skill, initiative, or creativity of people who stand between the commander and the machinery. The dream, the supreme management fantasy, is well depicted at the beginning of the film when the top manager voices his commands into the microphone: "orders to the plant." "Humans do what they do best," create, so long as by "humans" we mean top management; on the other hand, "machines do what they do best," the automatic following of orders, down to the last detail.

This brings us to the last characteristic military preoccupation: modern methods. Modern manufacturing means a fetish for machinery which won't talk back, a preoccupation with capital-intensive production. Modern means numerically controlled production machinery, computers, assembly robots, plotters and drafting machines, inspection, testing equipment, transfer machines, machines for automatic welding, forming, and pipe bending. These are the "elements of our plan of the future." People, except for top management and designers, disappear from view almost entirely. Indeed, "modern" specifically means machines, to be contrasted with "conven-

5. Michael Cooley, "Making Technology Serve the Worker" (Technology and Culture Seminar, MIT, December 6, 1979). Also see Cooley's *Architect or Bee? The Human/ Technology Relationship* (Boston: South End Press, 1980).

tional," meaning people. But conventional also means "backward" or "primitive," and this gives force to the modernizing drive. This is illustrated, in a racist and pictorially (though unintentionally) humorous way, by scenes of half-clad natives making iron by hand and even operating a lathe in a thatched hut. The message is clear: conventional means reliance upon people and people mean error, emotion, primitiveness. The ideal of the military, and of the managers who have inherited the military habit, is the automatic factory (factory on, factory off). In the meantime the military system of manufacture means a highly regimented system of people, temporary place holders for the robots of the future, who are paced and disciplined by the machinery which has come under direct management control.

Performance, command, modern methods—these then are the dominant characteristics of the military approach to industry, justified in the name of national security and enforced throughout industry by the system of military procurement contracts. The Department of Defense "expects defense contractors to maintain a modern base in their facilities," the film concludes. Within the military framework, all of this makes perfect sense; it is logical, supremely rational. But it becomes irrational in other contexts, as the military approach spills over, permeates, and diffuses throughout the economy, carried in the form of performance requirements, habits of command, and machine designs—irrational because in other contexts the objectives are different. The focus is on meeting social and human needs through the production of cheap goods and services; meeting the demands of a competitive market; fostering the kinds of things all Americans profess to value: self-reliance, democracy, life, liberty, and the pursuit of happiness. The rationality of the military is not always compatible with these objectives; indeed, it is often destructive of them. It increases costs while using up valuable resources; it devalues human judgment, skill, autonomy, self-reliance, initiative, and creativity; it leads to the actual depletion and atrophy of the store of inherited human skills; in its fetish for capital-intensive production, it contributes to the dislocation and displacement of untold numbers of workers and possibly to massive structural unemployment; it fosters, in its emphasis on command, what Mumford has called an "authoritarian" rather than a "democratic" technics,[6] and thus, in the name of order, creates social instability and mounting industrial tensions; in its insistence on uniformity and system integration, it fosters ever increasing complexity and, its correlates, greater inflexibility and unreliability; and, finally, it places human beings, the subject of society, of history, of production, in a subordinate role to military objectives, to the commands that flow from those objectives, and to the machinery that automatically executes those commands. Nothing could be more irrational or more frightening.

It might appear to some, by this time, that I am simply against progress, against technology, against mechanization and automation per se. This is not true. Indeed, quite the contrary. I applaud new ways of doing things as much as the next person. I find many of the particular machines illustrated in the film marvelous. But I try to discipline myself, temper my fascination and enthusiasm, by looking hard at the proposed uses of the technologies and the likely consequences, the human and social costs, the potential for greater or less happiness. This is what I understand to be rationality. Progress for what? What kind of progress? Progress for whom? With these questions in mind I would like to examine, more closely and concretely, three examples of the military influence on industrial and technological development. So as not to show favoritism or bias, I have selected one illustration from each service, the Army, the Navy, and the Air Force. My apologies to the Marines. Each example describes a major technological change in which the military played a central role: interchangeable parts manufacture, containerization and numerical control. Without question, these were major technical accomplishments, with far-reaching if ambiguous consequences for the civilian economy. In military terms, as well as in technical terms, these were magnificent departures from tradition, bold steps forward for progress. They also reflect well the dominant military characteristics that I have enumerated: performance, command, and modern methods. And they also illustrate that there have been costs unaccounted for, serious questions, begged by the ideology of progress, which have yet to be confronted fully by any of us.

6. Lewis Mumford, "Authoritarian and Democratic Technics," *Technology and Culture* 5 (1964): 1–8.

Army: Interchangeable Parts

In the first half of the nineteenth century, the Ordnance Department that ran the country's armories and arsenals evolved an ideology of uniformity. The performance objective underlying this ideology was to ensure the ready field repair of firearms through the supply of interchangeable spare parts. The impulse behind all this stemmed from the difficulties encountered in the War of 1812.[7]

The logistical problems faced in that war prompted the Ordnance Department to search for better ways of supplying troops in the field. Under the direction of officers like Decius Wadsworth, George Bomford, George Talcott and others, the Army began earnestly to promote the pioneering methods of inventors like John Hall and Simeon North, to try to establish uniformity in the manufacture of firearms at both Springfield and Harpers Ferry armories and at the plants of private contractors. This performance criterion became something of an obsession. As one self-described "soldier-technologist," Major Alfred Mordecai, explained in 1861, "my ability consists in a knowledge and love of order and system, and in the habit of patient labor in perfecting and arranging details."[8] And the performance criteria of uniformity necessitated the establishment of *command* over all productive operations, heretofore relatively autonomous. This was effected by the establishment of an ongoing bureaucracy, in Smith's words, for the "specific regulation of the total production process from the initial distribution of stock to the final accounting of costs." The whole system came to be viewed as a "complex machine." And, at the heart of the system were the *modern methods* of manufacture, the physical embodiment of "fixed orders": the hardened steel gauges, the patterns, the special machines and fixtures, which replaced human craft skill in producing, testing, and evaluating parts, and thereby eliminated human error and ensured uniformity.

---

7. This section draws primarily upon Merritt Roe Smith's essay "Army Ordnance and the 'American system' of Manufacturing" in this volume. Also see Otto Mayr and Robert C. Post, eds., *Yankee Enterprise: The Rise of the American System of Manufactures* (Washington, D.C.: Smithsonian Institution Press, 1981); Nathan Rosenberg, *Perspectives on Technology* (Cambridge: Cambridge University Press, 1976); and Merritt Roe Smith, *Harpers Ferry Armory and the New Technology* (Ithaca, NY: Cornell University Press, 1977).
8. Smith, "Army Ordnance," p. 74.

The effective use of these new methods required tighter management control and supervision, the elimination not only of traditional skills but also of traditional work patterns and routines grounded in the autonomy of the craftsman. Uniformity of parts was followed soon by uniformity in housing, in working hours, in shop discipline, presaging the scientific management of the next century (another product of the arsenals). The uniformity system was imposed on the contractors by the military contract system; Bomford "apprised private manufacturers that the issuance of future arms contracts would depend on their performance, especially the degree to which they updated their operations."[9]

As is well known, the uniformity system developed in the armories became the basis of the so-called American system of manufactures, characterized by special machinery, precise gauges, and interchangeability of parts. Men left the arms business to set up the machine tool industry and went on from there to carry the principle of uniformity into the manufacture of railroad equipment, sewing machines, pocket watches, typewriters, agricultural implements, bicycles, and so on.[10] The rest, as they say, is history, the history of progress.

But there was another side to this story, which we have not heard much about. First, not everyone was enthusiastic about the uniformity system, and that does not only mean the workers who had to buckle under to tighter discipline. "As impressive as they were," Smith reminds us, "these accomplishments tended to obscure fundamental ambiguities and tensions associated with the introduction of the uniformity system. . . . conflicting opinions existed over the need for mechanization as well as the importance of uniformity." Many people were skeptical, such as those Harpers Ferry workers and managers who for forty years successfully resisted the introduction of the full system. Smith reports that "similar feelings also existed among some contractors, although they were less willing to oppose the uniformity policy for fear of losing their contracts." It is interesting to note that, without the new system, Harpers Ferry continued for some thirty years to match the output of

9. Ibid., p. 61.
10. Edwin A. Battison, *Muskets to Mass Production* (Windsor, VT: American Precision Museum, 1976), pp. 30–31; David A. Hounshell, *From the American System to Mass Production, 1800–1932* (Baltimore: The Johns Hopkins University Press, 1984); Rosenberg, *Perspectives on Technology*, pp. 9–31.

Springfield, where the system was first installed. The benefits of the system, clear to the military, were not so clear to many manufacturers, given the high cost, uncertainties, and inescapable industrial conflict it engendered. Moreover, the system proved a disaster for those who lost their jobs, were subjected to much tighter discipline, and suffered the deskilling and degradation that always accompanies modern methods. Not surprisingly, class conflict, "seethed beneath the glowing veneer of industrial achievement." "To those who planned and orchestrated the uniformity system," Smith concludes, "such changes . . . seemed fully compatible with their ideas of rational design . . . [but] to those who worked under the system, the new regimen represented a frontal assault on valued rights and privileges."[11]

Navy: Containerization

In July 1952, at the initiative of the military—the Chief of Naval Research and representatives from the Army and Marines—the National Research Council Committee on Amphibious Operations held a joint government-industry conference on marine cargo handling and transport. The *performance* objective of the military was the improvement of cargo transport efficiency and the elimination of potential dock tie-ups, so as to improve U.S. undersea warfare capability and guarantee logistical support for military operations. The military had for some time been doing research in the area—the industry had done nothing—and the military, for its purposes, sought to promote joint government-industry research to improve maritime efficiency. A year later the NRC conducted the S.S. Warrior study to determine precisely what the inefficiencies were in cargo handling. A few years after that the NRC conducted the San Francisco Port Study, under the direction of Admiral E. G. Fullingwider, to document the inefficiencies of longshoring. Shortly thereafter, in 1960, the famous "Mechanization and Modernization Plan" was agreed upon by the Pacific Maritime Association and the ILWU, West Coast Longshoremen's union. Here modernization essentially meant the substitution of standardized containers for loose cargo, the replacement by container ships (to accommodate the containers) of conventional freighters, and the substitution of large docks and cranes for

11. Smith, "Army Ordnance," pp. 65, 79–80.

finger piers and longshoremen. In short, the M & M Plan meant the elimination of traditional longshoring. Pushed by the military, the container revolution was "greatly accelerated" by the demands of the Vietnam War. More recently, the computerization of crane-loading operations has inaugurated the second container revolution.[12]

The conventional system of longshoring was marked by the gang system, essential cooperative teamwork, and a decentralization of initiative, innovation, and skill. Since each ship was different in its construction and each load of cargo was unique to the job, there was an endless variety of tasks and inviolable autonomy for the gang as it tried to solve unique problems. The work itself was governed by specific work rules that specified manning, load limits, and safety standards. The hiring hall was an egalitarian institution which guaranteed, through its principle of low man out, a fair distribution of work among union members. Both the work rules and the hiring hall were won by the union through decades of bitter struggle, highlighted by a 1934 strike. Beyond the work itself, but essential to the longshore culture, were ongoing conversation, comraderie between partners and informal workgroups, and a lively and vital dock community. The container revolution changed all this. Cargo packaged in containers (and loaded and unloaded away from the dock) was now standardized, as were the ships designed to carry them, and therefore the dockwork now dominated by large cranes became routinized and subject to closer control and discipline. On the large container docks the workforce was atomized. The new methods broke up the gangs and informal work groups, put an end to close contact, conversation, and comradeship, and decimated the dock community. Work rules were surrendered by the union and the principle of institutionalized justice in the hiring hall was violated by the introduction of "steady men," crane operators, whose presence was justified in the name of expensive and sophisticated equipment and the skills they supposedly demanded (this remains a controversial issue). The chain of *command* was thus shortened, with the displacement of the majority

12. Stan Weir, "Effects of Containerization on Longshoremen" (U.S. Department of Labor, 1977); Herb Mills, "The San Francisco Waterfront: The Social Consequences of Industrial Mechanization," in Andrew Zimbalist, ed., *Case Studies on the Labor Process* (New York: Monthly Review Press, 1979), pp. 127–55; Lincoln Fairley, "ILWU-PMA Mechanization and Modernization Agreement" (U.S. Department of Labor, Labor-Management Services Administration).

of longshoremen and the close and relentless supervision of equipment operators. "Throughout the shift," crane operators "are simply told by radio or computer print-out where to pick up or place their next container. There is no occasion for initiative or innovation on their part nor is there any on-going operational need for their employers to in any way consult with them."[13]

Again, performance, command, modern methods. Backed by the military, employers argued "you can't hold back progress; you just can't fight the machine," and the union buckled under, holding on to the privileges for the few at the expense of the many. But for the longshoremen as a whole, their union, and their communities, as well as for their hard won dignity and workplace principles of autonomy and equality, this progress was a disaster. Lincoln Fairley, Research Director of the ILWU from 1946 to 1967, was originally a strong supporter of the Mechanization and Modernization Plan, but he has changed his mind. "As experience under the Plan developed," he writes, "I began to share with many of the longshoremen doubts about how it was working out from the standpoint of the men. Not only did the gains to the employers far outstrip the gains to the union and its members but, at least in this case, the union was weakened and the employers regained much of the ground they had lost in the 1930s. The presumed social and economic benefits deriving from a modernized and more efficient industry are no adequate offset." Even Harry Bridges, the ILWU president who made the historic deal, has had second thoughts. As Fairley reports, "it is known that Mr. Bridges believes the Plan to have been a mistake."[14]

Air Force: Numerical Control

Numerical control (NC) was largely underwritten by the U.S. Air Force, including research and development, software development, actual purchase of machinery for contractors, and training of programmers and operators. Total government subsidy of NC development and implementation was over $60 million.[15] As already indicated, the *performance* objectives were high-speed aircraft and missiles, requiring complex machining

---

13. Mills, "San Francisco Waterfront," p. 144.
14. Fairley, "ILWU-PMA Mechanization and Modernization Agreement." Also see Fairley, *Facing Mechanization: The West Coast Longshore Plan* (Los Angeles: Institute of Industrial Relations, Monograph Series 23, UCLA, 1979), pp. xi–xii.
15. For a detailed treatment of this subject, see the author's *Forces of Production* (New

capability and uniformity. The NC revolution, as it came to be called, was fueled by the Korean War and the Cold War of the 1950s and 1960s. The *command* imperative entailed direct control of production operations not just with a single machine or within a single plant, but worldwide, via data links. The vision of the architects of the NC revolution involved much more than the automatic machining of complex parts; it meant the elimination of human intervention—a shortening of the chain of command—and the reduction of remaining people to unskilled, routine, and closely regulated tasks. It is no surprise, then, that Air Force development of NC involved no worker or union participation.[16]

The way to achieve all this was through *modern methods,* that is, numerical control, which is the translation of part specifications into mathematical information that can be fed directly by management into a machine without reliance upon the skills or initiative of the machinist. The whole fantasy was the fully automatic, computer-controlled factory, still being pursued by the Air Force Integrated Computer-Aided Manufacturing (ICAM) Program. A December 1980 Air Force request for proposals on computer manufacturing systems reads: "Sources are sought which have the experience, expertise and production base for establishing a Flexible Manufacturing System for parts. This FMS should be capable of providing a technically advanced production facility for the manufacture of aerospace batch-manufactured products. The system shall be capable of automatically handling and transporting parts, fixtures, and tools, automatically inspecting part dimensional quality and incoming tool quality; integrated system control with machinabililty data analysis; computer aided process planning and scheduling and other capabilities that would provide a totally computer-integrated machining facililty." The Air Force advertisement announcing this request for proposals also points out that "extensive subcontracting to aerospace and other manufacturers, machine tool vendors, universities and other technology companies is expected."[17]All will get caught up in the military quest for the automatic factory.

York: Alfred A. Knopf, 1984). Also see "Social Choice in Machine Design: The Case of Automatically Controlled Machine Tools," in *Case Studies on the Labor Process,* pp. 18–50. For an extended version of the same essay, see *Politics and Society* 8 (1978): 313–47.

16. See NC files, MIT Archives, Cambridge, MA.

17. *Commerce Business Daily* (December 11, 1980).

Essentially, numerical control is the technical realization of management control envisioned by the directors of the Ordnance Department back in the nineteenth century. Gauges, patterns, jigs and fixtures, process planning, time studies—over the years all were designed to get the workforce to perform, manually or with machinery, in a specified way, machine-like. NC is a giant step in the same direction; here management has the capacity to bypass the worker and communicate directly to the machine via tapes or direct computer link. The machine itself can thereafter pace and discipline the worker. Essentially, this transforms skilled batch work into continuous process, assembly-line work. From the military point of view, it is the command performance, supremely rational, the dream come true. But economically and socially, it raises as many problems as it solves.

During the 1940s machine tool manufacturers and control engineers were experimenting with many forms of new equipment for metalworking, trying to put to use wartime developments in electronics and servo-control systems. They came up with improved tracer-controlled machines, plugboard-type controls, and record-playback controls—an ingenious development whereby a machinist made a first part manually while the motions of the machine were recorded on magnetic tape; thereafter the tape was simply played back to recreate the machine motions and duplicate the part. These technologies were ideally suited for small batch automatic production, where a change in set-up was required for each short run of parts. For set-up (programming) and operation, all of these systems relied upon the traditional store of machinist skills and were therefore readily accessible to most metalworking enterprises. But because of the development of NC, they never got very far in either full development or actual use. NC development was dictated by the performance and command objectives of the Air Force; these other technologies, which could not be used effectively to make highly complex parts and, most important, which relied upon the skills and resources of workers, were thus perceived from the start as anachronistic and primitive. In fact, they represented a significant advance on current methods.

NC was the brainchild of John Parsons, a Michigan manufacturer who was trying to meet the demanding military specifications for helicopter rotor blade templates. He elaborated his ideas when he saw a proposed design for an integrally stiffened wing section of a Lockheed fighter, and subsequently sold them to the Air Force. The Air Force eventually contracted with MIT to build the first NC milling machine and then went on to underwrite the software development, the promotion, and the procurement of the new technology—a bulk order purchase that finally elicited the sustained interest of machine tool and electronic control manufacturers.

It is important to note that, as with the uniformity system, industry generally did not share the Air Force's enthusiasm for the new technology, and for good reason. Although NC was theoretically ideal for complex machining, it was not necessarily ideal for the vast majority of metalworking orders that were not so demanding. NC was also very costly, not only for the hardware but also for the software and computers required to calculate endless amounts of information for the machine controls. NC was also notoriously unreliable, and the programming involved was excessive and time-consuming. With technologies such as record-playback, an analog system that was programmed by manual direction or by following the contour of a pattern, there was no need for computers, programmers, or excessive training of personnel. In addition, the programming was right the first time. In terms of manufacturing needs of the metalworking industry, therefore, it would have been rational to proceed with both technologies, the one for the bulk of metalworking operations, the other (NC) for subsidized military work. But the needs of the Air Force proved hegemonic: when it gave the signal for the development of NC and guaranteed lucrative returns for machine tool and control manufacturers, everyone jumped aboard. All industrial and technical efforts were geared to meet Air Force specifications. Other developments which might have proved more accessible, more practical, and more economical for the metalworking industry as a whole were abandoned.[18] The Air Force wanted highly sophisticated five-axis machines and a complex, expensive software system to go with them (the APT system) and enforced their use through the contract system. Machine manufacturers meanwhile concentrated on the most expensive designs, con-

18. For an extended discussion of alternative production techniques not adopted, see Noble, *Forces of Production*, chap. 7.

fident that for their subsidized customers in the aerospace industry, cost was not a factor.[19]

Performance, command, modern methods: it all made perfect sense for the Air Force. But the economic and social costs were great. Some very promising technical possibilities were foreclosed in the wake of the rush to numerical control. The great expense of the machinery that was manufactured and the overhead requirements of the programming system made the diffusion of NC into the metalworking industry very slow. The contract system fostered concentration in the metalworking industry because it favored the larger shops able to underwrite the expense of NC and APT. In the vast majority of metalworking establishments, there was no gain from the technical advances in automation until the 1970s.

The commercial competitiveness of U.S. machine manufacturers was also undermined by the concentration on NC. In April 1979, at the hearings of the House of Representatives Committee on Science and Technology, Congressman Ritter of Pennsylvania observed that Japan and Germany invested in machinery for commercial rather than military use. Predictably, Presidential Science Advisor Frank Press assured the congressman that defense expenditures "spill over" into commercial use. But Congressman Ritter was onto something quite significant. While American manufacturers were concentrating on highly sophisticated machinery and the APT software system, Japanese and German manufacturers emphasized cheapness, accessibility, and simplicity in their machine designs and software systems. The result is obvious: in 1978 the United States became a net importer of machine tools for the first time since the nineteenth century, and we still can't compete.

Finally, for workers—and this includes the technical personnel as well as production people—modernization orchestrated according to the Air Force objectives has been disastrous, marked by deskilling, downgrading, routinization, and powerlessness. Autonomy and initiative are giving way to precisely prescribed tasks and computer monitoring and supervision. This is happening even though the latest generations of NC machines, equipped with microprocessors at the machine, now make it possible as never before for the operator to program and edit at the machine and to regain control over more so-

---

19. Ibid., chaps. 6–9 and Epilogue.

phisticated technology. The technology is rarely used that way, however, especially in military-oriented plants. There the trend is to integrate these CNC (computer numerical control) machines into a larger DNC (direct numerical control) network under central command. (At a factory in Kongsberg, Norway, for example, workers have successfully struggled to regain control over the editing of machines—except for those who work on the military F-16.) Again, a technical possibility that might mean low costs, higher quality, and better working conditions is foreclosed by military imperatives. Unions are being seriously weakened by the new systems and military strategies. The military requirement of "strategic decentralization" of plants, for example, confronts the unions as runaway shops subsidized by the state. The military demand for interchangeability and communication networks appears to the union as satellite-linked duplicate plants that undermine the power of the strike.

The exaggerated emphasis on capital-intensive methods and automation increases system unreliability (an effect perhaps most obvious in the military itself) while at the same time eliminating irreplaceable human skills—a trend even John Parsons, the inventor of NC, finds insane and shortsighted. (Parsons was trained in manufacturing by an all-around Swedish machinist, precisely the type of person being lost.) In addition, the military imperatives contribute to dislocation and displacement and ultimately to structural unemployment. This ultimate social cost is now being endured by workers everywhere, invisibly and silently. But that low profile will not last long, for it will soon become obvious to all of us that there is simply no place for these people to go—no farms, no factories, no offices. Faith in the inevitable new industry which will absorb them rings hollow. Even the computer and control industries are themselves undergoing automation. Meanwhile the Air Force ICAM Project proceeds apace. Recently, it must be noted, they offered a contract for a study of the social implications of Integrated Computer-Aided Manufacture. But this is no cause for rejoicing. The contract went to Boeing, one of the major users of the latest automated technology.

Conclusion

It should now be obvious that the military has played a central role in industrial development and that this role has left an indelible imprint on that development: the imprint of performance, command, and modern methods. The three examples indicate that while the military influence has not been all bad, neither has it been an unmixed blessing. They suggest rather that we must take a closer look at what is happening under the military aegis. The uniformity system of the Ordnance Department might appear to us as the epitome of progress, but perhaps that is because the people who experienced the disruption of their lives are all dead and gone, their trials forgotten. How rational was the system at the time and what were its social costs? On whose behalf ought the government to foster technological development? Progress for whom? Containerization raises similar questions, only here the survivors are still around to haunt us. What exactly are the presumed economic benefits and what have been the social costs? Again, progress for whom and for what? Finally, the computer revolution in manufacturing: what role has the military played in undermining the competitive viability of U.S. industry? How has it promoted industrial concentration? How has it contributed to the deskilling of workers and the degrading of working conditions, the unreliability of our productive plants, the intensification of management power and control at the expense of workers and unions? And, perhaps most crucial, to what extent has it created a yet-to-be-reckoned-with structural unemployment, our own twentieth century "world turned upside down," to borrow Christopher Hill's phrase.[20] Again, what kind of progress are we talking about here, and progress for whom?

Performance, command, modern methods: these words do not appear in the U.S. Constitution nor are the subtle yet profound and pervasive transformations they imply ever voted upon. The role of the military in shaping our technologies, our productive activities, our social organizations, the power relations between us—in shaping our lives, in short—has gone relatively unnoticed and unrecorded. It is time we gave the matter some serious attention, subjected it to critical scrutiny, brought it under democratic control. It is time we began to answer the fundamental questions: what kind of progress do we want? What kind of progress can we, as a society, afford?

20. Hill, *The World Turned Upside Down* (Penguin Books, 1976).

9

Technology and War: A Bibliographic Essay
Alex Roland

War and technology have influenced every era of man's history; the literature that addresses them is vast. This essay will discuss only those works, mostly books, that have taken as their primary focus (or at least as a major theme) the interrelationship of war and technology. These will be mostly English-language books dealing with the history of the Western world. The purpose is to convey some sense of the range of this field and a sampling of the literature.[1]

The historiography of war and the historiography of technology share many characteristics. Both are seen by many traditional historians as outside the mainstream of historical scholarship, at least in the United States. Neither, for example, appears in the list of traditional "Fields of Specialization" in the American Historical Association *Guide to Departments of History*. Both are viewed with some suspicion in academic circles—military history because things military are distasteful and the history of technology because it appears esoteric or antiquarian or trivial or all of the above. Both are viewed as lacking scholarly rigor and intellectual substance, in part for good reason: some of the early writers in both fields were buffs and enthusiasts, more intent on communicating their own predilections than on surveying their topics critically and analytically.[2] A suspicion attaches to both fields that advocates write their history, a notion akin to assuming that a historian of medicine endorses the plague by studying it. Both fields require specialized knowledge, of the principles of war on the one hand and the nature of technology on the other. Neither requirement is as demanding as it seems, but both contribute to the isolation of the fields. Finally, both fields are growing more popular on American campuses, as war and technology grow more intrusive and important in contemporary life.

In some ways the two fields are different. Military history is older and more established, boasting a richer literature, a larger following, and a greater impact on traditional scholar-

1. Many of the works cited here were first brought to my attention in my graduate colloquium in military history. I am indebted to Kevin Anastas, Jack Atwater, John Bonin, Steve Chiabotti, Winston Choo, David Hogan, Yue Yeong Kwan, Jui Ping Ng, and James Pearson for their contributions. The following colleagues read the entire essay in draft and provided useful criticisms and suggestions: Robert Durden, I. B. Holley, Jr., Thomas P. Hughes, William McNeill, Thomas Misa, Richard Preston, Theodore Ropp, Merritt Roe Smith, and John TePaske.

2. Theodore Ropp chastises me for ignoring the buffs and enthusiasts who, often because of their enthusiasm, made important contributions in both fields. His point is well taken; unfortunately, there have been, I believe, too few of these.

ship. The history of technology, emerging as a distinct field only in the last quarter-century, is still finding its way. Small enough to function more or less coherently as a community of scholars, the field is self-consciously seeking a common agenda and scholarly respectability and influence. One measure of its success is that *Technology and Culture,* the journal of the Society for the History of Technology, has already set a scholarly standard unmatched by journals of military history.

Bibliographic Aids

It is not surprising, therefore, that the best bibliographic introduction to the literature on the history of war and technology may be found in the latter field. Eugene Ferguson's 1968 *Bibliography of the History of Technology* carries very few entries under "Military Technology," but this volume nevertheless opens up the field.[3] In his introduction, Ferguson credits Brooke Hindle for suggesting to him that few books are readily identifiable as military technology because these topics are intertwined throughout the rest of the literature on technology. Military issues appear throughout the history of technology, just as technology appears often in military history. So Ferguson's entire volume is a source for material on the relationship between war and technology.[4] This pattern is also true for the "Current Bibliography in the History of Technology," prepared annually by Jack Goodwin for the April issue of *Technology and Culture,* though there the entries under "Military Technology" are fuller than in Ferguson, partly because they include periodical literature, partly because more work is now being done in the field.

The military side of technology is not entirely devoid of bibliographic aids. Robin Higham's *A Guide to the Sources of British Military History* contains two remarkably informed and detailed essays by W. H. G. Armytage, covering the period up to 1914, and an essay by Ronald W. Clark on the period from 1919 to 1945.[5] Higham's *A Guide to the Sources of United States Military*

---

3. Cambridge, MA: Society for the History of Technology and the MIT Press, 1968.

4. Incidentally, Brooke Hindle's *Technology in Early America: Needs and Opportunities for Study* (Chapel Hill: University of North Carolina Press, 1966) contains some excellent material on military technology.

5. Berkeley, CA: University of California Press, 1971. See W. H. G. Armytage, "The Scientific, Technological and Economic Background to 1815," pp. 167–207; and "Eco-

History contains essays on science and technology by Edward C. Ezell for the nineteenth century and Carroll Pursell for the twentieth century, as does its supplement.[6] Robert G. Albion's *Naval and Maritime History* is especially good on covering military and civilian topics together, a reflection of the nature of the field and the literature it has attracted.[7] Technology is less easy to find in John E. Jessup, Jr. and Robert W. Coakley's *A Guide to the Study and Use of Military History*, but it is there.[8]

Classic Surveys

As is often true in the history of technology, several of the classic works on technology and war have been written by economic historians. Two of these are in a class by themselves. Werner Sombart's *Krieg und Kapitalismus*, volume two of his *Studien zur Entwicklungsgeschichte des modernen Kapitalismus*,[9] argues that war has had a positive influence on the evolution of modern capitalism, industrialization, and technology. War, in Sombart's view, has stimulated invention, investment, production, and innovation, with second-order consequences that spread far beyond military goods and services. John U. Nef attempted to refute Sombart in his *War and Human Progress: An Essay on the Rise of Industrial Civilization*.[10] Real material progress, according to Nef, arises not from physical plants and actual production, but from the emergence of new ideas. These in turn are stimulated by the advance of knowledge, by the free

nomic, Scientific, and Technological Background for Military Studies, 1815–1914," pp. 251–98; and Ronald W. Clark, "Science and Technology, 1919–1945," pp. 542–65. Fill in World War I with Cyril Falls, *War Books: A Critical Bibliography* (London: P. Davies, 1930), dated but still useful for its extensive annotations.

6. Hamden, CT: Archon Books, 1975. See Edward C. Ezell, "Science and Technology in the Nineteenth Century," pp. 185–215; and Carroll W. Pursell, Jr., "Science and Technology in the Twentieth Century," pp. 269–91. See also the comparable sections in Robin Higham and Donald J. Mrozeck eds., *A Guide to the Sources of United States Military History; Supplement I* (Hamden, CT: Archon Books, 1981), pp. 44–55, 69–71.

7. Fourth ed.; rev. and expanded; Mystic, CT: Munson Institute of American Maritime History, 1972. There are, of course, other guides, but none so helpful as these for identifying technological topics in a military context. The *Air University Index*, for example, is especially good on recent military periodical literature, but weak on history, as is William M. Arkin, *Research Guide to Current Military and Strategic Affairs* (Washington, D.C.: Institute for Policy Studies, 1981).

8. Washington, D.C.: United States Army, 1979.

9. Munich: Duncker and Humblot, 1913. This volume has not been translated into English, but its main thesis appears in Waldemar Kaempffert, "War and Technology," *The American Journal of Sociology* 46 (January 1941): 431 ff.

10. Cambridge, MA: Harvard University Press, 1950.

travel and communication of thinkers, by peaceful settings in which ideas can be nurtured and developed. War disrupts all these activities, says Nef, and creates a superficial and false impression of technical advance by using up the accumlated ideas of the past in an orgy of production without renewing the supply. Far from being conducive to technological advance, war is destructive of it, in more important ways than the physical devastation it brings. Neither Sombart nor Nef entirely proves his case, but between them they have posed the most profound question about the relationship between war and technology. Lamentably, the debate has made little progress in the last thirty years.[11]

Another economic historian, at least by training, is Lewis Mumford, whose classic *Technics and Civilization* deals extensively with military topics.[12] This eloquent, idiosyncratic, and provocative book, which still reads well, raises many of the concerns that scholars and intellectuals still have with technology and the military: their authoritarian bent, their dehumanizing influences, and their capacity—indeed power—for tyranny. Mumford also recognized the close, at times symbiotic, relationship between war and technology not only in modern times but throughout Western history. In this regard he anticipated what most others realized only after the experience of World War II.

Quincy Wright, a specialist in international relations, headed a group at the University of Chicago in the 1930s that produced *A Study of War*.[13] Wright developed a theory of the evolution of warfare that saw man move from animal through primitive and historic into what he called modern warfare. The dominant factors in these four periods were, respectively, biological forces or human nature, sociological forces, international law, and finally technology. Wright actually came to believe that in the period after the Renaissance the technology of war was the main, though not the only, factor in determining when, why, and where war would break out and how it would be conducted. His study trained and influenced a whole generation of scholars, like John U. Nef and Bernard Brodie, who

11. An exciting exception to this generalization is the work of Clive Trebilcock. See, for example, his " 'Spin-off' in British Economic History: Armaments and Industry, 1760–1914," *Economic History Review* 22 (December 1969): 474–90.

12. New York: Harcourt, Brace, 1934.

13. Subtitled *With a Commentary on War Since 1942* (2d ed.; Chicago: University of Chicago Press, [1942] 1965).

would bring the same breadth and interdisciplinary approach to their studies of war and technology.

In some respects Wright's hypothesis on the relationship of technology and war had been anticipated by an economist, a Warsaw banker named Ivan S. Bloch. His study, *The Future of War in Its Technical, Economic, and Political Relations,*[14] predicted that the new technology of war, combined with the economic and political resources at the disposal of the modern state, would make war vastly more destructive and pointless than at any other time in human history. He had read the lessons of the nineteenth century, the age of what Wright was to call technological warfare, more clearly than most of his contemporaries, but his warning went largely unheeded.

Lynn White, Jr. belongs in a class by himself. He made the history of technology intellectually respectable with his classic *Medieval Technology and Social Change.*[15] In that volume, and in other studies before and after it, he demonstrated the significance and legitimacy of studying the relationship between war and technology. War was by no means the only context for his investigations of technology, but he demonstrated that technology was worth studying in whatever setting it emerged. Just as war was an important aspect of medieval life, so too was it an important setting for the evolution of technology.

Though not himself trained as an economic historian, William H. McNeill has entered this category with his *The Pursuit of Power: Technology, Armed Force, and Society since A.D. 1000.* Expanding his earlier *The Rise of the West* and *Plagues and People,*[16] McNeill here undertakes to study the political, social, and economic consequences of the ways in which states, especially those in the West, have organized and equipped themselves for war. He attributes the economic and military preeminence of the West to the emergence of free market economies in the late middle ages, replaced in more modern times—especially the twentieth century—by "command technology," the systematic

14. Trans. by R. C. Long (New York: Garland Publishing, [1899] 1972). This is actually a translation of the sixth and last volume of his major study, *Budushchaia Voĭna.*

15. Oxford: Oxford University Press, 1962. See also his works cited in n. 177 below.

16. Chicago: University of Chicago Press, 1982; Chicago: University of Chicago Press, 1963; and Garden City, NY: Doubleday, 1976. Many of the citations appearing in this essay first came to my attention in *The Pursuit of Power.* For the source of one of McNeill's most provocative theses, see Frederic C. Lane, *Profits from Power: Readings in Protection, Rent and Violence-Controlling Enterprises* (Albany: State University of New York, 1979).

manipulation of the sinews of war by the state for the purposes of the state. This provocative and wide-ranging book will likely set the agenda in this field for many years to come. Its rich and erudite argument echoes the tone of early Lewis Mumford, casts serious doubt on the thesis of John U. Nef, and elaborates the argument of Carlo Cippola.

Cippola, another economic historian, is among those who have taken war and technology as the focus of somewhat narrower studies. He argues in *Guns, Sails, and Empires: Technological Innovation and the Early Phases of European Expansion, 1400–1700* that European mastery of cannons and sailing vessels enabled the European explorers to establish hegemony over the entire coastal world.[17] A comparable work, in both theme and quality, is Daniel R. Headrick, *The Tools of Empire: Technology and European Imperialism in the Nineteenth Century.*[18] V. J. Parry and M. E. Yap have collected a set of revealing essays on the evolution of *War, Technology and Society in the Middle East.*[19] Several scholars have addressed these issues from a variety of perspectives in Monte D. Wright and Lawrence J. Paszck, eds., *Science, Technology, and Warfare: Proceedings of the Third Military History Symposium, 8–9 May 1969.*[20] Elting E. Morison has built upon his study of Admiral Sims and naval gunnery (see below) in several insightful essays in *Men, Machines, and Modern Times* and *From Know-How to Nowhere; The Development of American Technology.*[21]

Others have focused on war and included large doses of technology in their analyses. G. N. Clark's *War and Society in the Seventeenth Century* is especially strong on the relationship of war to the emerging scientific movement.[22] In *The Military Revolution, 1560–1660* Michael Roberts examines how a technological revolution in gunpowder weapons led to a thoroughgoing revolution in the methods of conducting war.[23] Joseph P. Smaldone, in *Warfare in the Sokoto Caliphate: Historical and Sociological*

17. New York: Minerva, 1966.

18. New York: Oxford University Press, 1981.

19. London: Oxford University Press, 1975.

20. Washington, D.C.: Office of Air Force History, Headquarters USAF, and United States Air Force Academy, 1971.

21. Cambridge, MA: The MIT Press, 1966; and New York: Basic Books, 1975.

22. Cambridge: Cambridge University Press, 1958.

23. Belfast: Queens University Press, 1956. See also *Essays in Swedish History* (London: Weidenfeld & Nicolson, 1967). See also Geoffrey Parker, "The 'Military Revolution' 1550–1660—a Myth?" *Journal of Modern History* 48 (June 1976): 195–219.

Perspectives, analyzes the effects of changing military technology, including transportation and communications, on the nature of warfare in the Western Sudan from 1790 to 1963.[24] Yigael Yadin engages his own talents as a soldier and archaeologist to explore *The Art of Warfare in Biblical Lands in Light of Archaeological Study,* an analysis rich in artifactual evidence.[25] Klaus Knorr has attempted to evaluate *The War Potential of Nations,* one ingredient of which is technology.[26] J. M. Winter's edited collection *War and Economic Development* delivers more technology than its title suggests.[27]

Some others have produced broad surveys that pay unusual attention to war and technology. Good examples are William H. McNeill, *The Rise of the West: A History of the Human Community*[28] and David S. Landes, *The Unbound Prometheus: Technological Change and Industrial Development in Western Europe from 1750 to the Present.*[29] Both are models of how the history of technology and war may be profitably woven into a survey of a larger topic. Among the surveys of the history of technology, *Technology in Western Civilization* by Melvin Kranzberg and Carroll W. Pursell, Jr. stands out for the strength of its military contributions.[30] This may well be a reflection of initial sponsorship of the project by the United States Armed Forces Institute, which was in search of a text "to explain the critical role of technology in our present society."[31]

The Traditional Weapons Surveys

The most familiar form in which studies of technology and war appear are the survey histories of weapons. Most of these limit themselves to the evolution of the weapons themselves, rather narrowly defined. The best of them analyze the influence of the weapons on war but seldom mention the influence of the weapon on civilian technology. In general these are military

24. London: Cambridge University Press, 1977.

25. New York: McGraw-Hill, 1963.

26. Princeton: Princeton University Press, 1956. See also his *Military Power and Potential* (Lexington, MA: Heath Lexington Books, 1970).

27. Cambridge: Cambridge University Press, 1975.

28. Chicago: University of Chicago Press, 1963.

29. London: Cambridge University Press, 1969.

30. 2 vols.; New York: Oxford University Press, 1967.

31. Ibid., vol. II, p. vi. Sadly and ironically, R. R. Palmer's contribution on military technology is uncharacteristically weak.

histories that pay little attention to the relationship between the military and civilian communities.

Among those that attempt to survey all of Western history, none is entirely satisfactory. The best of the available works is Bernard N. Brodie and Fawn Brodie, *From Crossbow to H-Bomb*.[32] This study is weak on the period before the nineteenth century and concentrates more on science than technology, but it is the most thoughtful and technically informed. Tom Wintringham and J. N. Blashford-Snell's *Weapons and Tactics* is the best of the studies that attempt to link, from the earliest times, the changing nature of warfare to the evolution of weaponry.[33] Though it lacks documentation, it presents an interesting cyclical theory of warfare that has real insights. J. F. C. Fuller's *Armament and History* is in the same category.[34] Trevor N. Dupuy's *The Evolution of Weapons and Warfare* is in a class by itself, a disappointing encyclopedic collection of fascinating information that fails to trace the evolution promised in its title.[35]

Other studies focus more narrowly on shorter time periods or more limited ranges of weapons. Early arms and armor have attracted scores of writers, often buffish and antiquarian. Among the best of these are R. Ewart Oakeshott, *The Archaeology of Weapons: Arms and Armor from Prehistory to the Age of Chivalry*;[36] H. Robinson, *The Armour of Imperial Rome*;[37] Anthony M. Snodgrass, *Arms and Armour of the Greeks*;[38] C. J. Ffoulkes, *Arms and Armament: A Historical Survey of the Weapons of the British Army*;[39] Howard L. Blackmore, *British Military Firearms, 1670–1850*;[40] and O. F. G. Hogg, *Clubs to Cannon: Warfare and Weapons before the Introduction of Gunpowder*.[41]

32. Rev. and enl. ed.; Bloomington, IN: Indiana University Press, 1973. See also P. E. Cleator, *Weapons of War* (New York: Thomas Y. Crowell, 1968).

33. Harmondsworth, Eng.: Penguin Books, 1973. This is an update of Wintringham's *The Story of Weapons and Tactics from Troy to Stalingrad* (Boston: Houghton Mifflin, 1943). See also Jac Weller, *Weapons and Tactics: Hastings to Berlin* (London: Nicolas Vane, 1966); and A. V. B. Norman and Don Pottinger, *A History of War and Weapons, 499 to 1660* (New York: Crowell, 1966).

34. Subtitled *A Study of the Influence of Armament on History From the Dawn of Classical Warfare to the Second World War* (New York: Scribner's Sons, [1945] 1960).

35. Indianapolis: Bobbs-Merrill, 1980. See also Edwin Tumis, *Weapons: A Pictorial History* (Cleveland: World Publishing Co., 1954).

36. New York: Praeger, 1960.

37. New York: Scribner's, 1975.

38. Ithaca, NY: Cornell University Press, 1967.

39. London: G. Harrap & Co., 1945.

40. London: H. Jenkins, 1961.

41. London: Gerald Duckworth & Co., 1968. See also Charles Boutell, *Arms and Ar-*

Some weapons have received individual attention. Artillery has been studied, sometimes in conjunction with fortifications, in Eric W. Marsden's two-volume *Greek and Roman Artillery*;[42] Bryan Hugh St. John O'Neill, *Castles and Cannon: A Study of Early Artillery and Fortification in England*;[43] Warren Ripley, *Artillery and Ammunition of the Civil War*;[44] and O. F. G. Hogg, *Artillery: Its Origin, Heyday, and Decline.*[45] The bow is analyzed in Robert Hardy, *Longbow: A Social and Military History*,[46] which deals with the Hundred Years' War, and more broadly by Victory Hurley in *Arrows Against Steel: The History of the Bow.*[47] The definitive work on the crossbow is still Ralph W. F. Payne-Gallwey, *The Crossbow: Medieval and Modern, Military and Sporting: Its Construction, History, and Management.*[48] Tanks are examined in Richard M. Ogorkiewicz, *Armour*,[49] and in Basil H. Liddell-Hart, *The Tanks: The History of the Royal Tank Regiment and Its Predecessors, Heavy Branch, Machine-Gun Corps, Tank Corps, and Royal Tank Corps, 1914–1945*, the best piece of historical scholarship by one of the great military historians of our time.[50]

Several specialized studies defy categorization. Basil Perronet Hughes, *Firepower: Weapons Effectiveness on the Battlefield, 1630–1850*, examines one of the most important aspects of modern warfare during the period in which it was emerging in its contemporary form.[51] John Ellis, *A Social History of the Machine Gun*, traces the impact of this modern weapon on the people

mour in Antiquity and the Middle Ages (London: Cassell, Petter & Galpin, 1869); Robert Held, *The Age of Firearms: A Pictorial History* (New York: Harper, 1957); James D. Lavin, *A History of Spanish Firearms* (New York: Arco Publishing, 1965); and George C. Stone, *A Glossary of the Construction, Decoration and Use of Arms and Armour in All Countries and All Times* (New York: Jack Brussel, [1934] 1961).

42. Subtitled *Historical Development* (Oxford: Clarendon Press, 1969) and *Technical Treatises* (Oxford: Clarendon Press, 1971). See also Barton C. Hacker, "Greek Catapults and Catapult Technology: Science, Technology, and War in the Ancient World," *Technology and Culture* 9 (January 1968): 34–50; and Ralph W. F. Payne-Gallwey, *A Summary of the History, Construction and Effects in Warfare of the Projectile-Throwing Engines of the Ancients, with a Treatise on the Structure, Power and Management of Turkish and Other Oriental Bows of Medieval and Later Times* (London: Longmans, Green, 1907).

43. Oxford: Clarendon Press, 1960.

44. New York: Van Nostrand, 1970.

45. Hamden, CT: Archon Books, 1970.

46. New York: Arco Publishing, 1976.

47. New York: Mason/Charter, 1975.

48. London: Longmans, 1903.

49. London: Stevens & Sons, 1960.

50. 2 vols; New York: Praeger, 1959.

51. New York: Scribner's, 1975.

who used it and the people it was used against.[52] Robert V. Bruce, *Lincoln and the Tools of War*, describes the unprecedented extent to which this president became involved in the development and employment of weapons and provides an excellent example of how war brings out all manner of geniuses and crackpots with ideas for new and "decisive" weapons.[53] E. T. C. Werner, *Chinese Weapons*, makes use of the linguistic roots of weapons' names to reach conclusions about early Chinese weaponry.[54] Chemist J. R. Partington traces the early development and interrelationship of two important weapons components in his ill-organized but authoritative *A History of Greek Fire and Gunpowder*.[55]

Nonweapons Technology

The technology of war, however, entails more than just weapons and armament. Armies need almost everthing civilian populations do: food, clothing, shelter, medicine, communication, transportation—all of which have peculiar technologies. Furthermore, armies perform some special functions (besides fighting) that may or may not have civilian parallels: engineering, cryptography, chemical and biological warfare, etc.

Logistics comes quickly to mind as a noncombat dimension of military activity that has always been indispensable to success on the battlefield. An overall view of this topic may be found in Hawthorne Daniel, *For Want of a Nail: The Influence of Logistics on War*.[56] A narrower but more scholarly treatment appears in Martin Van Creveld, *Supplying War: Logistics from Wallenstein to Patton*, a well-documented study emphasizing supply and transportation but concentrating more on World War II than its title suggests.[57] Donald W. Engels employs exhaustive scholarship, interdisciplinary research, and a good measure of common sense to unravel the story of *Alexander the Great and the Logistics*

---

52. New York: Pantheon, 1975. See also G. S. Hutchinson, *Machine Guns: Their History and Tactical Employment* (London: Macmillan, 1938).

53. Indianapolis: Bobbs-Merrill, 1956. Merritt Roe Smith recommends Carl L. Davis, *Arming the Union: Small Arms in the Union Army* (Port Washington, NY: Kennikat Press, 1973) and Grady McWhiney and Perry D. Jamieson, *Attack and Die: Civil War Military Tactics and the Southern Heritage* (University, AL: University of Alabama Press, 1982).

54. Shanghai: The Royal Asiatic Society North China Branch, 1932.

55. Cambridge: Heffer, 1960.

56. New York: Whittlesay House, 1948.

57. London: Cambridge University Press, 1977.

of the Macedonian Army, and in the process destroys some myths and lends credence to J. F. C. Fuller's assertion that "supply was the basis of Alexander's strategy and tactics."[58] Equally distinguished scholarship graces Geoffrey Parker's *The Army of Flanders and the Spanish Road, 1567–1659: The Logistics of Spanish Victory and Defeat in the Low Countries Wars,* a model of social history that succeeds in explaining Spanish military experience in the Low Countries without addressing any battles or campaigns.[59] James A. Huston's *The Sinews of War: Army Logistics, 1775–1953* is a volume in the U.S. Army historical series limited to American experience.[60] Richard Goff has studied the logistics of the South in the Civil War in *Confederate Supply,*[61] and R. Arthur Bowler has done the same for the British in the American Revolution in *Logistics and the Failure of the British Army in America, 1775–1783.*[62]

Military transportation has received more attention than most other fields of military technology. Railroads are a special case. Dennis Showalter's *Railroads and Rifles: Soldiers, Technology, and the Unification of Germany* is a model of how effectively war and technology can be integrated in historical studies with findings that reach far beyond the battlefield.[63] Denis Bishop and Keith Dans have studied *Railways and War before 1918,*[64] and D. W. Ronald and J. R. Carter have provided detailed coverage of *The Longmoor Military Railway,* including doctrine, training of operators and maintenance personnel, and technical information on British engines and rolling stock.[65] Nor are these modern studies the only worthwhile books in the field.[66]

---

58. Berkeley: University of California Press, 1978, quote from page 1, citing J. F. C. Fuller, *The Generalship of Alexander the Great* (London: EYRE & Spottiswoode, 1958), pp. 52–53. Some of the statistical assumptions in the study warrant scrutiny, but this hardly compromises the overall value of the work.

59. Cambridge: Cambridge University Press, 1972.

60. Washington, D.C.: Office, Chief of Military History, Department of the Army, 1966.

61. Durham, NC: Duke University Press, 1969.

62. Princeton: Princeton University Press, 1974.

63. Hamden, CT: The Shoe String Press, 1975.

64. New York: Macmillan, 1972.

65. Newton Abbot [Eng.]: David and Charles, 1974.

66. See also H. R. Richardson, *Railroads in Defense and War* (Washington, D.C.: Bureau of Railway Economics Library, 1953); Edwin A. Pratt, *The Rise of Rail Power in War and Conquest, 1833–1914, with a Bibliography* (London: P. S. King and Son, 1916); T. H. Thomas, "Armies and the Railway Revolution," in *War as a Social Institution: The Historian's Perspective,* ed. by J. D. Clarkson and T. C. Cockran (New York: Columbia University Press, 1941).

A more generalized study of military transportation is Forest G. Hill's *Roads, Rails and Waterways: The Army Engineers and Early Transportation,* which is also strong on the relationship between military and civilian activity.[67] More specialized studies are John Maurice Brereton, *The Horse in War,*[68] and Odie B. Faulk, *The U. S. Camel Corps: An Army Experiment.*[69] Other work on military transportation is scattered; for example, R. J. Forbes' treatment of land transport and roads, including Persian and Greek military land communications and the evolution of Roman roads, appears in volume 2 of his *Studies in Ancient Technology.*[70]

Military architecture has received less attention than it deserves, given its influence on civilian architecture. Horst De La Croix, *Military Considerations in City Planning: Fortifications,*[71] and Keith Mallory and Arvid Ottar, *The Architecture of War,*[72] are among the few that have paid attention to the civilian aspects of this issue. For surveys see Sidney Toy, *A History of Fortifications from 300 B. C to A. D. 1700;*[73] Ian V. Hogg, *Fortress: A History of Military Defense;*[74] and James Quentin Hughes, *Military Architecture.*[75] More specialized treatments include Christopher Duffy's two complementary volumes, *Siege Warfare* and *Fire and Stone,*[76] and two revisionist studies: Byron Tsangadas's *The Fortifications and Defense of Constantinople*[77] and Vivian Rowe's *The Great Wall of France: The Triumph of the Maginot Line.*[78] Some of the original classics in the field can also be

67. Norman, OK: University of Oklahoma Press, 1957. See also Hill's "Formative Relations of American Enterprise, Government and Science," *Political Science Quarterly* 75 (September 1960): 400–419.

68. Newton Abbot [Eng.]: David and Charles, 1976.

69. New York: Oxford University Press, 1976.

70. 9 vols.; Leiden: E. J. Brill, 1955–1964.

71. New York: G. Braziller, 1972.

72. New York: Pantheon Books, 1973. This volume, which has interesting material on concrete, the geodetic dome, prefabrication, and high-speed road networks, is unfortunately limited to the first half of the twentieth century.

73. London: Heinemann, 1955.

74. London: MacDonald and Jones, 1975.

75. London: Evelyn, 1974. See also the chapter on fortifications in Albert Neuberger, *The Technical Arts and Sciences of the Ancients,* trans. by Harry L. Brose (London: Methuen, 1930).

76. Subtitled, respectively, *The Fortress in the Early Modern World, 1494–1660* (London: Routledge & Kegan Paul, 1979) and *The Science of Fortress Warfare, 1660–1860* (Newton Abbot [Eng.]: David and Charles, 1975).

77. East European Monographs, No. 71 (New York: Columbia University Press, 1980).

78. London: Putnam, 1959.

rewarding; see, for example, Sebastien Le Prestre Vauban, *A Manual of Siegecraft and Fortification,*[79] and Eugene Emmanuel Viollet-Le-Duc, *Military Architecture.*[80]

Cryptography has attracted much attention in recent years, largely because of the revelation that the Allies had broken the German code during World War II. In this case not only did a technology influence the conduct of the war, but knowledge of the technology is also altering the historiography of the event. David Kahn's *The Codebreakers: The Story of Secret Writing* provides a general historical survey.[81] On the breaking of the German code in World War II, see F. W. Winterbotham, *The Ultra Secret*;[82] Reginald Victor Jones, *The Wizard War: British Scientific Intelligence, 1939–1945*;[83] and Ronald Lewin, *Ultra Goes to War.*[84] Chemical and biological warfare is another topic currently in the news; the best work is Frederick J. Brown, *Chemical Warfare: A Study in Restraints.*[85]

Other topics that have received noteworthy historical treatment are as varied as the nature of war and the preparation for war. In the field of weaponry are Malvern Lumsden's *Incendiary Weapons,*[86] and Constance McLaughlin Green, Harry C. Thompson, and Peter C. Roots, *The Ordinance Department: Planning Munitions for War,* one of the more thoughtful and analytical volumes in the Army series on World War II.[87] Alfred Price's *Instruments of Darkness: The History of Electronic Warfare* addresses an important but secrecy-enshrouded topic.[88] David MacIsaac has analyzed the effects of strategic bombing on military and civilian targets in *Strategic Bombing in World War Two:*

79. Trans. by A. Rothrock (Ann Arbor: University of Michigan Press, 1968).

80. Trans. by M. Macdermott (London: James Parker, 1879).

81. New York: Macmillan, 1968.

82. New York: Harper and Row, 1974.

83. New York: McCann and Geoghegan, 1978.

84. New York: McGraw-Hill, 1978.

85. Princeton: Princeton University Press, 1968. See also Steven Rosse, *CBW: Chemical and Biological Warfare* (Boston: Beacon Press, 1969); Samuel P. Jones, "From Military to Civilian Technology: The Introduction of Tear Gas for Civil War Control," *Technology and Culture* 19 (April 1978): 151–68; and John H. Perkins, "Reshaping Technology in Wartime: The Effect of Military Goals on Entomological Research and Insect-control Practices," ibid., pp. 169–86.

86. Cambridge, MA: The MIT Press, 1975.

87. Washington, D.C.: Department of the Army, 1955. Other volumes in the series on Technical Services are rich in materials on technology, though not all are of this caliber.

88. London: MacDonald and Jones, 1977.

The Story of the Strategic Bombing Survey.[89] Brian Pearce suggests the potential of a sadly neglected topic in "Elizabethan Food Policy and the Armed Forces."[90] Much more work is needed in this area and in others like medicine, engineering, sanitation, clothing, communication, and electronics.

Naval and Air Forces

Navies and air forces have a special relationship with technology, for the vehicles in which they conduct their missions are complex machines, usually embodying the most sophisticated technology of their day. Men can fight on land with the most primitive equipment, in fact with no equipment at all, but fighting on or under the sea or in the air requires technical support. Thus it is that navies, and later air forces, have always been more alive to technology than their land-based counterparts, which does not mean they have necessarily been more progressive.[91] Furthermore, ships and planes may carry both guns and butter: many advances in maritime and aeronautical science and engineering affect both civilian and military applications. Ideas flow more freely between the two realms, and many craft often find use in both peace and war. Thus the institutions that foster technological progress at sea and in the air often mix civilian and military purposes.

Naval warfare has been conducted in three great eras, defined by the ships that dominated them: galley, sail, and steam. Some histories cover all, or most, of these periods. Bjorn Landström's *The Ship*[92] is the best of these, but it may be profitably supplemented with Philip Cowburn, *The Warship in History*.[93] On galley warfare see R. C. Anderson, *Oared Fighting*

---

89. New York: Garland Publishing, 1976. *The United States Strategic Bombing Survey* (339 vols.; Washington, D.C., 1945-1947), on which Dr. MacIssac based his study, is itself a remarkable resource for historians investigating the resistance of modern industrial society to the effects of conventional weapons.

90. *Economic History Review* 12 (1942): 39–45.

91. See, for example, Lance C. Buhl, "Marines and Machines: Resistance to Technological Change in the American Navy, 1865–1869," *Journal of American History* 61 (December 1974): 703–27, and the works by Elting Morison cited below.

92. Garden City, NY: Doubleday, 1961.

93. New York: Macmillan, 1965. Viking craft were a special case; see A. W. Brøgger and Haakon Shetelig, *The Viking Ships: Their Ancestry and Evolution*, trans. by Katherine John (Oslo: Dreyers Forlag, 1953).

Ships, from Classical Times to the Coming of Steam[94] and John F. Guilmartin's brilliant and stimulating *Gunpowder and Galleys: Changing Technology and Mediterranean Warfare at Sea in the Sixteenth Century*,[95] a work to place beside Frederic C. Lane's classic *Venetian Ships and Shipbuilding of the Renaissance*.[96] Romula Anderson and R. C. Anderson have treated sail in all ages in *The Sailing Ship: Six Thousand Years of History*,[97] while its greatest exploitation in war has been addressed by E. H. H. Archibald in *The Wooden Fighting Ship in the Royal Navy, A.D. 879–1860* and by C. N. Longridge, *The Anatomy of Nelson's Ships*.[98] The transition to the age of steam is analyzed in James P. Baxter's classic *The Introduction of the Ironclad Warship*.[99] The best survey is Bernard Brodie, *Sea Power in the Machine Age*,[100] to be complemented by Edgar C. Smith, *A Short History of Naval and Marine Engineering*, which is especially strong on the relation of civil to military developments.[101] Richard G. Hewett and Francis Duncan have extended the story into the *Nuclear Navy, 1946–1962*.[102]

Specialized studies in naval technology abound. In the vast literature of submarines, mines, and torpedoes, see especially J. S. Cowie, *Mines, Minelayers, and Minelaying*;[103] Alex Roland, *Underwater Warfare in the Age of Sail*;[104] and two biographies of

94. London: P. Marshall, 1962. See also John W. Morrison and R. T. Williams, *Greek Oared Ships, 900–322 B.C.* (London: Cambridge University Press, 1968); and Lionel Casson, *The Ancient Mariners: Seafarers and Sea Fighters of the Mediterranean in Ancient Times* (New York: Macmillan, 1959).

95. London: Cambridge University Press, 1974.

96. Baltimore: Johns Hopkins Press, 1934. See also Lane's "The Crossbow in the Nautical Revolution of the Middle Ages," *Explorations in Economic History* 7 (Fall 1969–1970): 161–71; this too sheds light on the transition from galley to sail.

97. London: George G. Harrap, 1926.

98. London: Blanford Press, 1968; London: Percival Marchall, 1955. See also G. J. Marcus, *Heart of Oak: A Survey of British Seapower in the Georgian Era* (London: Oxford University Press, 1975); and Howard I. Chapelle, *The History of the American Sailing Navy: The Ships and Their Development* (New York: Norton, 1949).

99. Cambridge, MA: Harvard University Press, 1933. See also Frank M. Bennett, *The Monitor and the Navy Under Steam* (Boston: Houghton, Mifflin, 1900).

100. Princeton: Princeton University Press, 1941.

101. Cambridge: Printed for Babcock and Wilcox Ltd. at the University Press, 1937. See also Brian Ranft, ed., *Technical Change and British Naval Policy, 1800–1939* (London: Hodder and Stoughton, 1977); Oscar Parkes, *British Battleships "Warrior" 1860 to "Vanguard" 1950: A History of Design, Construction and Armament* (rev. ed.; London: Seeley Service, 1966); and Stanley Sandler, *The Emergence of the Modern Capital Ship* (Newark: University of Delaware Press, 1979).

102. Chicago: University of Chicago Press, 1974.

103. London: Oxford University Press, 1949.

104. Bloomington, IN: Indiana University Press, 1978.

key inventors: Edwin Gray, *The Devil's Device: The Story of Robert Whitehead, Inventor of the Torpedo*,[105] and Richard K. Morris, *John P. Holland, 1841–1914: Inventor of the Modern Submarine*.[106] Navigation is treated from Ulysses to Captain Cook in E. G. R. Taylor, *The Haven-Finding Art*,[107] and more narrowly in David W. Waters, *The Art of Navigation in England in Elizabethan and Early Stuart Times*.[108] The insatiable appetite of sailing navies for good wood and the effects of this on civilian economies are treated in Robert G. Albion's classic *Forests and Sea Power: The Timber Problem of the Royal Navy, 1652–1862*[109] and in Paul W. Bamford, *Forests and French Sea Power, 1660–1789*.[110] Naval arms and armament are covered broadly in P. Padfield, *Guns at Sea*,[111] and more narrowly in Michael Lewis's revisionist study, *Armada Guns: A Comparative Study of English and Spanish Armaments*.[112] J. J. Keevil has addressed an otherwise neglected topic in *Medicine and the Navy: 1200–1900*,[113] as has Sir Arthur Hezlet in *The Electron and Sea Power*.[114] The collection of pieces by Ken J. Hagan and others on *Naval Technology and Social Modernization in the Nineteenth Century* fits no particular category but is representative of the best scholarship that is currently being done.[115]

The literature on the technology of military flight is more vast than profound. More than any other military field save heraldry, this one is still dominated by buffs and tail-number counters. Still, there are enough significant exceptions to this

105. London: Seeley, 1975.

106. Annapolis, MD: United States Naval Institute Press, 1966. See also John D. Alden, *The Fleet Submarine in the U.S. Navy: A Design and Construction History* (Annapolis, MD: United States Naval Institute Press, 1979).

107. New York: Abeland-Schulman, 1957.

108. New Haven, CT: Yale University Press, 1958. See also Rupert T. Gould, *The Marine Chronometer: Its History and Development* (London: The Holland Press, 1960); and Humphrey Quill, *John Harrison: The Man Who Found Longitude* (London: Baker, 1966).

109. Cambridge, MA: Harvard University Press, 1926.

110. Toronto: University of Toronto Press, 1956.

111. New York: St. Martins, 1974. See also Frederick L. Robertson, *The Evolution of Naval Armament* (London: Constable, 1921).

112. London: George Allen and Unwin, 1961. Lewis concludes it was seamanship and logistics that gave victory to the English, not guns, which did comparatively little damage. See also Herman T. Wallinga, *The Boarding Bridge of the Romans: Its Construction and Its Function in the Naval Tactics of the First Punic War* (Groningen, Neth.: J. B. Wolfers, 1956), for an instance of a revolutionary, once secret, and often decisive weapon.

113. 4 vols.; London: E&S Livingstone, 1957.

114. New York: Stein and Day, 1975.

115. Manhattan, KS: *Military Affairs* and the American Military Institute, 1976.

sad rule to provide a useful introduction to the field. The best of the general surveys is Ronald Miller and David Sawers, *The Technical Development of Modern Aviation*, in spite of its emphasis on civilian aviation.[116] Complement this with John D. Anderson, Jr., *Introduction to Flight: Its Engineering and History* (a technical text with brief historical sketches);[117] *Research and Development Contributions to Aviation Progress (RADCAP): Joint DoD-NASA-DoT Study*;[118] and J. L. Nayler and E. Ower, *Aviation: Its Technical Development*, for the British view.[119] Robert Schlaifer and R. D. Heron, *Development of Aircraft Engines and Aviation Fuels*,[120] remains the best work on aviation propulsion, to be supplemented with L. J. K. Setright, *The Power to Fly: The Development of the Piston Engine in Aviation*,[121] and Edward Constant's thoughtful and analytical *The Origins of the Turbojet Revolution*.[122] Institutions that have fostered the technical development of aviation are treated in George W. Gray, *Frontiers of Flight: The Story of NACA Research*[123] and Percy B. Walter, *Early Aviation at Farnborough: The History of the Royal Aircraft Establishment*.[124] Monte Wright's *Most Probable Position: A History of Aerial Navigation to 1941* does for flying what Taylor and Waters have done for sailing.[125]

Procurement

Nowhere does the military have a greater impact on technology—including civilian technology—than in procurement. It is

116. New York: Praeger, 1970.

117. New York: McGraw-Hill, 1978.

118. Washington, D.C.: Department of Defense, National Aeronautics and Space Administration, Department of Transportation, 1972. Though this volume reaches just the conclusions one would expect from its sponsors, it is probably accurate nonetheless and has the virtue of postulating a list of the major advances in aeronautical technology.

119. Philadelphia: Dufours Editions, 1965.

120. Subtitled *Two Studies of Relations between Government and Business* (Boston: Division of Research, Graduate School of Business Administration, Harvard University, 1950; Elmsford, NY: Maxwell Reprint Co., 1970).

121. London: George Allen & Unwin, 1971.

122. Baltimore: Johns Hopkins University Press, 1980.

123. New York: Alfred A. Knopf, 1948. See also Alex Roland's *Model Research: The National Advisory Committee for Aeronautics, 1915–1958* (2 vols.; Washington, D.C.: NASA, 1985).

124. 2 vols.; London: MacDonald, 1971–1974. For a glimpse into the world of the aircraft designer, a crucial but little-known figure in aviation development, see E. H. Heinemann and Rosario Rausa, *Ed Heinemann: Combat Aircraft Designer* (Annapolis, MD: United States Naval Institute Press, 1980).

125. Lawrence, KS: University of Kansas Press, 1972.

here that military needs and specifications determine what and how the civilian economy will produce. Sometimes the military will simply buy what is available on the civilian market; more often it will insist upon custom-made materials, produced either by civilian firms under contract or by its own arsenals. The research to develop new and better products for military use is supported in the same two ways. In any case, the military often acts as the major purchasing and subsidizing arm of the national government, developing and buying technology on a scale that dwarfs most private enterprise.

Studies of arms manufacture comprise the classic form of military history in this field. Charles J. Ffoulkes's richly informed studies of gunfounding in Europe have served as something of a model for this kind of study and have held up well.[126] A more recent study, employing the latest rubrics of scholarship and addressing questions of contemporary concern to historians of technology, is Melvin H. Jackson and Carel de Beer, *Eighteenth Century Gunfounding: The Verbruggens at the Royal Brass Foundry: A Chapter in the History of Technology.*[127] I. B. Holley's monumental *Buying Aircraft: Matériel Procurement for the Army Air Forces,* another volume of official history in the Army series on World War II, is a model of meticulous research and exhaustive analysis of one of the most arcane yet crucial facets of modern military experience.[128] J. A. Stockfisch provides a more popular, broad-ranging survey of the hazards of modern military procurement in *Plowshares into Swords: Managing the American Defense Establishment.*[129]

Perhaps the most interesting issue in military procurement is the choice between contracting out to civilians and producing materials directly in government arsenals. On this topic generally, see M. M. Postan, D. Hay, and J. D. Scott, *The Design and Development of Weapons: Studies in Government and Industrial*

---

126. *The Armourer and His Craft from the 11th to 15th Century* (New York: B. Blom, [1917] 1967); and *The Gun-Founders of England, with a List of English and Continental Gun-Founders from the XIVth to the XIXth Centuries* (Cambridge: Cambridge University Press, 1937).

127. Washington, D.C.: Smithsonian Institution Press, 1974. See also Claude Gaier, *Four Centuries of Liège Gunmaking* (London: Eugène Wahle and Sotheby Parke Bernet, 1977) on an early-modern center of European arms manufacture; and Fritz Redlich, *The German Military Enterpriser* (2 vols.; Weisbaden: F. Steiner, 1964) on mercenaries, with some attention to arms manufacture and sale.

128. Washington, D.C.: Office of the Chief of Military History, Department of the Army, 1964.

129. New York: Mason and Lipscomb, 1973.

Organization;[130] and Tibor Scitovsky, Edward Shaw, and Lorie Tarshis, *Mobilizing Resources for War: The Economic Alternatives*.[131] Arsenals (and much else besides) are addressed in one of the most distinguished studies in the history of technology in recent years, Merritt Roe Smith's *Harpers Ferry Arsenal and the New Technology: The Challenge of Change*,[132] which may be profitably complemented by Edward Ames and Nathan Rosenberg, "Enfield Arsenal in Theory and History,"[133] and by several European studies: O. F. G. Hogg, *The Royal Arsenal: Its Background, Origin, and Subsequent History*;[134] H. A. Young, *The East India Company's Arsenals and Manufactories*;[135] and P. M. J. Conturie, *Histoire de la fonderie nationale de Ruelle, 1750–1940, et des anciennes fonderies de canons de fer de la marine*.[136] On contracting see Philip Noel-Baker, *The Private Manufacture of Armaments*.[137] Traditional studies of specific experiences, like Felicia Johnson Deyrup, *Arms Making in the Connecticut Valley*;[138] John Anderson Miller, *Men and Volts at War: The Story of General Electric in World War II*;[139] and Frank E. Vandiver, *Ploughshares into Swords: Josiah Gorgas and Confederate Ordnance*,[140] should be supplemented with investigations of war profiteering [141] and histories of industries with close ties to the military.[142] Closely related to contracting are the entrepeneurs, whose careers make for fascinating biography. In addition to the numerous

130. London: HMSO, 1964.

131. New York: McGraw-Hill, 1951. See also Arthur Forbes, *A History of the Army Ordnance Services* (3 vols.; London: Medici Society, 1929).

132. Ithaca, NY: Cornell University Press, 1977. See also Russell I. Fries, "British Response to the American System: The Case of the Small Arms Industry after 1850," *Technology and Culture* 16 (July 1975): 377–403.

133. *Economic Journal* 78 (December 1968): 827–42.

134. 2 vols.; London: Oxford University Press, 1963.

135. Oxford: Clarendon Press, 1939.

136. Paris: Impr. nationale, 1951.

137. 2 vols.; New York: Oxford University Press, 1937.

138. Subtitled *A Regional Study of the Economic Development of the Small Arms Industry, 1798–1870* (York, PA: Shumway, 1970).

139. New York: Whittlesay House, McGraw-Hill, 1947.

140. Austin: University of Texas Press, 1952.

141. For example, Richard F. Kaufman, *The War Profiteers* (Garden City, NY: Doubleday, 1972); and Berkeley Rice, *The C-5A Scandal: An Inside Story of the Military Industrial Complex* (Boston: Houghton Mifflin, 1971).

142. For example, Arthur Pine VanGelder and Hugo Schlater, *History of the Explosives Industry in America* (New York: Arno Press, [1927] 1972); and Alan P. Cartwright, *The Dynamite Company: The Story of African Explosives and Chemical Industries Limited* (Cape Town: Purnell, 1964).

studies of the Krupp and Vickers dynasties,[143] see Charles B. Dew, *Ironmaker to the Confederacy: Joseph R. Anderson and the Tredegar Iron Works*;[144] Leonard A. Swann, Jr., *John Roach, Maritime Entrepeneur: The Years as Naval Contractor, 1862– 1866*;[145] and Thomas P. Hughes's model study of *Elmer Sperry: Inventor and Engineer,* whose remarkable career moved in and out of military contracting.[146]

World War II and the Cold War

The influence of technology on war (not to be confused with the influence of war on technology) has undergone three great revolutions in the course of Western history. The gunpowder revolution is treated in the weapons surveys cited above. The second came with the industrial revolution and played itself out between the Napoleonic wars and World War II. During this period machine weapons increased long-range firepower dramatically, expanded the size of the battlefield, gave an advantage to the defensive, and turned large-scale conflicts into wars of industrial mobilization. World War II, itself a conflict of industrial attrition, precipitated a third revolution in the technology of war by instituting an era in which the quality of military technology, more than the quantity of industrial production, was widely viewed as the most important predictor of success in the next war. The technological era that Quincy Wright saw beginning with the industrial revolution did not really come into full flower until his study was published. After World War II technology really did become a prime determinant of how and why wars would start, and traditional military conservatism toward new weapons was transferred almost overnight into an enthusiasm for new and better weapons and a technological arms race of such rapid pace that obsolescence became its hallmark.

---

143. Gert von Klass, *Krupp: The Story of an Industrial Empire,* trans. by James Cleugh (London: Sidgwick and Jackson, 1954); William Manchester, *The Arms of Krupp, 1587– 1968* (Boston: Little, Brown, 1968); Bernhard Menne, *Blood and Steel: The Rise of the House of Krupp,* trans. by G. H. Smith (New York: L. Furman, 1938); Wilhelm Bardrow, *The Krupps: 150 Years of Krupp History* (Berlin: P. Schmidt, 1937); J. D. Scott, *Vickers: A History* (London: Weidenfeld and Nicolson, 1962); and Clive Trebilcock, *The Vickers Brothers: Armaments and Enterprise, 1854–1914* (London: Europa, 1977).

144. New Haven, CT: Yale University Press, 1966.

145. Annapolis, MD: United States Naval Institute Press, 1965.

146. Baltimore: Johns Hopkins University Press, 1971. See also Alden Hatch, *Remington Arms: An American History* (New York: Rinehart, 1956); and H. W. Dickinson's short but suggestive *John Wilkinson: Ironmaster* (Ulverstone: Hume Kitchin, 1914).

The sources of the revolution may be found in the history of World War II itself. Alan Milward treats the war "as an economic event" in *War, Economy and Society, 1939–1945*,[147] and argues that in modern, capital-intensive wars like WWII, economics—and by extension technology—is decisive. The story of traditional industrial mobilization is presented in such studies as Michael M. Postan, *British War Production*,[148] and Robert Howe Connery, *The Navy and Industrial Mobilization in World War II.*[149] But the real difference in this war occurred in the systematic harnessing of science and technology to the needs of the state, as described in James Phinney Baxter, III, *Scientists Against Time*;[150] James G. Crowther and R. Widdington, *Science at War*;[151] and Guy Hartcup, *The Challenge of War: Britain's Scientific and Engineering Contribution to World War II.*[152] Studies of individual developments may be found in such books as Louis F. Fieser, *The Scientific Method* (napalm);[153] Ralph B. Baldwin, *The Deadly Fuze: Secret Weapon of World War II* (proximity fuse);[154] and Robert Morris Page, *The Origins of Radar.*[155] Of course, the great weapons revolution of World War II, the one that really set the tone for the postwar world, was the development of the atomic bomb, described most ably in two official histories: Richard G. Hewlett and Oscar E. Anderson, Jr., *The New World, 1939–1946*,[156] and Margaret Gowing, *Britain and Atomic Energy, 1939–1945.*[157]

147. Berkeley: University of California Press, 1979.

148. History of the Second World War, United Kingdom civil series (London: HMSO, 1952).

149. Princeton: Princeton University Press, 1951; New York: Capo Press, 1972.

150. Boston: Little, Brown, 1946.

151. New York: Philosophical Library, 1948.

152. New York: Taplinger, 1970. See also Leslie E. Simon, *Secret Weapons of the Third Reich: German Research in World War II* (2d ed.; Old Greenwich, CT: We Inc., 1971); and Ronald W. Clark, *The Rise of the Boffins* (London: Phoenix House, 1962). Compare these with Carol S. Gruber, *Mars and Minerva: World War I and the Uses of the Higher Learning in America* (Baton Rouge; Louisiana State University Press, 1975).

153. New York: Distributed by Reinhold Publishing, 1964.

154. San Rafael, CA: Presidio Press, 1980.

155. Garden City, NY: Doubleday, 1962. See also David Kite Allison, *New Eye for the Navy: The Origin of Radar at the Naval Research Laboratory*, NRL Report 8466 (Washington, D.C.: Naval Research Laboratory, 1981); and Albert Percival Rowe, *One Story of Radar* (Cambridge: Cambridge University Press, 1948); both of which focus on the institutional setting in which these developments took place and the relationship of those institutions with the scientific community in academia. And see Reader's Digest Association, *The Tools of War 1939/1945, and a Chronology of Important Events* (Montreal: RDA, 1969).

156. Vol. 1 of A History of the Atomic Energy Commission (University Park: Pennsylvania State University Press, 1962).

157. London: Macmillan, 1964. See also Henry deWolf Smyth, *Atomic Energy for Mili-*

With the topic of atomic energy, the story of the relation between war and technology slides quickly into the period of the Cold War.[158] The literature increases exponentially as experts in science, technology, government, political science, international relations, national security, management, economics, and public policy join historians in analyzing the relationship between war and technology. This essay cannot hope to sample even the best of this literature, let alone provide a comprehensive survey. It can, however, mention two areas that have attracted some of the best scholarship and suggest (in the following section) some of the most interesting themes that are emerging in the literature and prompting reexamination of previous eras in the light of contemporary experience.

The handmaiden of nuclear weapons has been the missile, which transformed these unprecedented devices of destruction into virtually unstoppable ones. The roots of this story and the technology at work can be traced in Eugene M. Emme, ed., *The History of Rocket Technology: Essays on Research, Development, and Utility.*[159] The critical role played by Wernher von Braun is best recorded by Frederick I. Ordway III and Mitchell R. Sharpe in *The Rocket Team.*[160] The development of this technology into a military weapon is thoughtfully analyzed in Edmund Beard, *Developing the ICBM: A Study in Bureaucratic Politics,*[161] and Harvey M. Sapolsky, *The Polaris System Development: Bureaucratic and Programmatic Success in Government.*[162] Ernest J. Yanarella has investigated the integration of technology into national policy in *The Missile Defense Controversy: Strategy, Technology, and Politics, 1955–1972.*[163]

How to get the maximum military advantage from the new

---

tary Purposes: The Official Report on the Development of the Atomic Bomb under the Auspices of the United States Government, 1940–1945 (Princeton: Princeton University Press, 1945); Leslie R. Groves, *Now It Can Be Told: The Story of the Manhattan Project* (New York: Harper, 1962); David Irving, *The German Atomic Bomb: The History of Nuclear Research in Nazi Germany* (New York: Simon and Schuster, 1968); and Arnold Kramish, *Atomic Energy in the Soviet Union* (Stanford: Stanford University Press, 1959).

158. See, for example, the sequels to the volumes of official history cited above: Richard G. Hewlett and Francis Duncan, *The Atomic Shield, 1947–1952* (Chicago: University of Chicago Press, 1974); and Margaret M. Gowing, *Independence and Deterrence: Britain and Atomic Energy, 1945–1952* (2 vols.; New York: St. Martin's Press, 1974).

159. Detroit, MI: Wayne State University Press, 1964.

160. New York: Crowell, 1979.

161. New York: Columbia University Press, 1972. See also Ernest G. Schwiebert, *A History of the U.S. Air Force Ballistic Missiles* (New York: Praeger, 1965).

162. Cambridge, MA: Harvard University Press, 1972.

163. Lexington:University Press of Kentucky, 1977.

technology of nuclear weapons and missiles has attracted the widest variety of scholarship. Among the better overviews are Bernard Brodie, *Strategy in the Missile Age*;[164] John Erickson, *The Military-Technical Revolution: Its Impact on Strategy and Foreign Policy*;[165] and Stefen Possony and J. E. Purnelle, *The Strategy of Technology: Winning the Decisive War*.[166] A unique perspective is provided in Herbert York's lively and insightful *Race to Oblivion: A Participant's View of the Arms Race*;[167] his forthcoming history of weapons development in the Cold War may prove to be the definitive work in the field. An observation by York prompted the title of Mary Kaldor's *The Baroque Arsenal*, a biased but penetrating study of the built-in obsolescence and dysfunction of today's most sophisticated weapons.[168]

Military Themes

Military historians have achieved no explicit consensus on the important themes in the study of technology and war. Most who address this topic treat it tangentially. Those who take it as their primary focus have emphasized the effect of weapons on combat. Only a handful have dealt with topics like logistics, industrial mobilization, procurement, and innovation, at least for the period before World War II. For the postwar period, a flood of studies have swamped the field without yet carving out a clearly defined new landscape. This general lack of consensus makes the few themes that have emerged all the more striking.

The military-industrial complex is the most familiar theme, and in some ways the most representative, for it provides a clear example of postwar interests being projected back into earlier periods to reveal new insights. Numerous studies have examined the topic in its postwar setting: for example, Stephen Rosen, ed., *Testing the Theory of the Military-Industrial Complex*,[169]

164. Princeton: Princeton University Press, 1959. Brodie was one of the first to see clearly the consequences of the atomic bomb; see his "War in the Atomic Age" in Bernard Brodie, ed., *The Absolute Weapon* (New York: Harcourt Brace Jovanovich, 1946).

165. New York: Praeger, 1966.

166. New York: Dunellan, 1970. This is the most extreme work, arguing a kind of technological determinism.

167. New York: Simon and Schuster, 1970. See also his "Military Technology and National Security," *Scientific American* 221 (August 1969): 17–29.

168. New York: Hill and Wang, 1981.

169. Lexington, MA: Lexington Books, D. C. Heath, 1973.

and Carroll W. Pursell, Jr., ed., *The Military-Industrial Complex.*[170] Others like Benjamin Franklin Cooling and Paul A. C. Koistinen have convincingly demonstrated that the phenomenon had a long history before President Eisenhower, in his 1960 farewell address, gave it the currency it now enjoys.[171] Many of the studies are highly critical of the military-industrial complex in its modern form and take on the air of exposés: Ralph E. Lapp, *Arms Beyond Doubt: The Tyranny of Weapons Technology,*[172] and H. L. Nieburg, *In the Name of Science*[173] are in this category, as is Berkeley Rice, *The C5-A Scandal: An Inside Story of the Military Industrial Complex.*[174] Others like Kenneth S. Davis, W. Henry Lambright, and Jacques S. Gansler, view the topic more dispassionately.[175] At least one author has essayed a defense of the military-industrial complex.[176] The issue seems to be turning not on whether the military-industrial complex exists, but on whether it works very well, whether it produces security commensurate with its cost, and whether it is in any event too powerful and subversive a force to be in the long-term best interests of the republic. The same questions may be profitably asked of other societies and other times in which close cooperation has grown up between the state and the manufacturers of arms.[177]

170. New York: Harper & Row, 1972.

171. Benjamin Franklin Cooling, *War, Business, and American Society: Historical Perspectives on the Military Industrial Complex* (Port Washington, NY: Kennikat Press, 1977); and *Gray Steel and Blue Water Navy: The Formative Years of America's Military-Industrial Complex, 1881–1917* (Hamden, CT: Shoe String Press, 1979); Paul A. C. Koistinen, *The Military-Industrial Complex: A Historical Perspective* (New York: Praeger, 1980).

172. New York: Crowell, 1970. See also his *The Weapons Culture* (New York: W. W. Norton, 1968).

173. Chicago: Quadrangle Books, 1966.

174. Boston: Houghton Mifflin, 1971. See also Seymour Melman's *Pentagon Capitalism: The Political Economy of War* (New York: McGraw Hill, 1970) and *The Permanent War Economy: American Capitalism in Decline* (New York: Simon and Schuster, 1974).

175. Davis, *Arms, Industry and America* (New York: H. H. Wilson Company, 1971); Lambright, *Shooting Down the Nuclear Airplane* (Indianapolis: Bobbs Merrill, 1967); and Gansler, *The Defense Industry* (Cambridge, MA: The MIT Press, 1980). See also J. A. Stockfisch, *Plowshares into Swords.* The F-111 evoked similar studies: Robert F. Art, *The TFX Decision: McNamara and the Military* (Boston: Little, Brown, 1968); and Robert Coulam, *Illusions of Choice: The F-111 and the Problem of Weapons Acquisition Reform* (Princeton: Princeton University Press, 1977).

176. John Stanley Baumgartner, *The Lonely Warriors: Case for the Military-Industrial Complex* (Los Angeles: Nash Publishing, 1970).

177. Of course, this issue is closely tied to procurement, discussed above. See the works cited there, and the modern classic, Merton J. Peck and Frederic M. Scherer, *The Weapons Acquisition Process: An Economic Analysis* (Boston: Division of Research, Graduate School of Business Administration, Harvard University, 1962).

No topic in the historiography of technology and war has been dominated so thoroughly by one scholar as has doctrine by I. B. Holley, Jr. In his classic *Ideas and Weapons,* he virtually invented the field, demonstrating in a study of American aircraft in World War I that new technologies will not be exploited fully until a doctrine is developed prescribing their use in war.[178] His work is now widely cited, especially in the aviation literature, and his ideas have been incorporated in many recent studies. Of course other works have treated this topic independently of Dr. Holley's example,[179] but much remains to be done.

Military engineering antedates civilian engineering by centuries, and engines of war gave the profession its name; yet studies of military engineers are sadly lacking. The ambitious U.S. Army Corps of Engineers Historical Program is beginning to fill this gap for the United States, including in its agenda a forthcoming *Biographical Dictionary of the U.S. Army Corps of Engineers, 1775–1980.* Yet most of the studies done and projected are institutional and programmatic histories, and the other services have no such undertaking in train. The richness of this neglected field has been suggested by Forest G. Hill's *Roads, Rails and Waterways: The Army Engineers and Early Transportation* and Russell F. Weigley's *Quartermaster General of the Union Army: A Biography of M. C. Meigs.*[180] More has been done on naval engineering, such as Edward W. Sloan, III, *Benjamin Franklin Isherwood, Naval Engineer: The Years as Engineer in Chief, 1861–1869;*[181] Harold G. Bowen, *Ships, Machinery, and Mossbacks: The Autobiography of a Naval Engineer;*[182] and Elting E. Morison, *Admiral Sims and the Modern American Navy.*[183] Among the important issues deserving further study are the creation of

---

178. Subtitled *Exploitation of the Aerial Weapon by the United States during World War I: A Study in the Relationship of Technological Advance, Military Doctrine, and the Development of Weapons* (New Haven, CT: Yale University Press, 1953; Hamden, CT: Archon Books, 1971). His thesis may be profitably compared with Alfred D. Chandler's structure-strategy concept, presented in *The Visible Hand: The Managerial Revolution in American Business* (Cambridge, MA: Belknap Press of Harvard University Press, 1977).

179. Doctrine is insightfully addressed in Liddell-Hart, *The Tanks*; and Ronald and Carter, *The Longmoor Military Railway,* to say nothing of such classics as Vauban, *A Manual of Siegecraft and Fortification.*

180. New York: Columbia University Press, 1959.

181. Annapolis, MD: United States Naval Institute Press, 1965.

182. Princeton: Princeton University Press, 1954.

183. Boston: Houghton Mifflin, 1942.

technical schools to train career officers in engineering[184] and the relations between the engineers and the line combat officers.

Military officers were traditionally viewed as technologically conservative, often preparing to fight the last war with yesterday's weapons.[185] Since World War II they have been seen as technological enthusiasts, trading yesterday's weapons for tomorrow's without exploiting the former or understanding the latter.[186] Neither behavior is especially surprising, for the military profession has always been a life-and-death business that values the tools it knows over those that have yet to prove themselves in the test of battle, and the Cold War has upset that proclivity with a conviction that the next year will be decided by the most advanced technology. But these stereotypes need to be tested more thoroughly than they have been, and the notion of "decisive weapons" needs further scrutiny.

The influence of international law and the unwritten rules of war on the introduction and use of new military technology has received just enough scholarly attention to suggest how fruitful a field it is for further study. Maurice Keen's model study of *The Laws of War in the Late Middle Ages*[187] traces this theme in the period when modern international law was in the making. Frederick Brown focuses on it in his equally fine *Chemical Warfare: A Study in Restraints.*[188] Alex Roland has examined its application to an exotic field of weaponry in *Underwater Warfare in the Age of Sail.*[189] The names that military men gave to new weapons often suggested the moral issues surrounding their use, as is demonstrated in Edwin Gray's *The Devil's Device*[190] and Milton F. Perry's *Infernal Machines.*[191] The literature to date suggests that military communities have shied away from

184. For now see John P. Lovell, *Neither Athens nor Sparta: The American Service Academies in Transition* (Bloomington, IN: Indiana University Press, 1979).

185. See, for example, Cowie, *Mines, Minelayers, and Minelaying*; Morris, *John P. Holland*; and Morison, *Admiral Sims and the Modern American Navy*.

186. See, for example, Beard, *Developing the ICBM*; Frederic A. Bergerson, *The Army Gets an Air Force: Tactics of Insurgent Bureaucratic Politics* (Baltimore: Johns Hopkins University Press, 1980); and the works cited above for the military-industrial complex.

187. London: Routledge & Kegan Paul, 1965.

188. Princeton: Princeton University Press, 1968.

189. Bloomington, IN: Indiana University Press, 1978.

190. London: Seeley, 1975.

191. Subtitled *The Story of Confederate Submarine and Mine Warfare* (Baton Rouge: Louisiana State University Press, 1965). See also Lumsden, *Incendiary Weapons*.

weapons they view as unmanly or unfair, but that once one side employs them, usually in desperation, the other side feels compelled to accept them as well, ratcheting ever upward the technology of violence. Established powers tend to outlaw the radical technological innovations of the aspiring nations.

The military need for secrecy imposes special constraints on the development of the technology of war. Greek fire is a classic example of a decisive weapon owing at least part of its success to secrecy. So well kept was the secret of its composition that scholars today are still unsure of its makeup, in spite of the vast amount of scholarship it has attracted.[192] Another early case study embroiled in scholarly debate is Herman T. Wallinga, *The Boarding Bridge of the Romans: Its Construction and Its Function in the Naval Tactics of the First Punic War,* which addresses among other topics the first use of a new weapon on an unsuspecting and unprepared enemy.[193] Modern examples include Simon, *Secret Weapons of the Third Reich*[194] and Baldwin, *The Deadly Fuze.*[195] Edward Constant has dealt with the extent to which excessive secrecy between different units of the same national project can retard development in *The Origins of the Turbojet Revolution;*[196] the issue appears as well in the histories of the Manhattan Project.

Several other themes that have received less attention in the literature suggest areas military historians may investigate with profit. Systems engineering and operational analysis introduced in World War II have had far-reaching effects in both military and civilian sectors, yet have received little historical analysis.[197] The moral and political position of the scientist and the engineer in the service of the state has received some attention, but far more needs to be done.[198] A special aspect of this

192. Begin with Partington, *A History of Greek Fire and Gunpowder;* and see also D. Ayalon, *Gunpowder and Firearms of the Mamluk Kingdom* (London: Frank Cass and Co., 1956).

193. Groningen, Neth.: J. B. Wolfers, 1956.

194. 2d ed.; Old Greenwich, CT: We Inc., 1971.

195. San Rafael, CA: Presidio Press, 1980.

196. Baltimore: Johns Hopkins University Press, 1980.

197. Brodie and Brodie treat this topic in *From Crossbow to H-Bomb.* See also I. B. Holley, Jr., "The Evolution of Operations Research and Its Impact on the Military Establishment; the Air Force Experience," in Wright and Paszek, eds., *Science, Technology, and Warfare,* pp. 89–109; and the commentary by Robert L. Perry, ibid., pp. 110–21.

198. See, for example, Baxter, *Scientists Against Time;* Bruce, *Lincoln and the Tools of War;* Clark, *The Rise of the Boffins;* and R. W. Reid, *Tongues of Conscience: Weapons Research and the Scientists' Dilemma* (New York: Walker, 1969).

problem, the argument by weapons developers that they will make war horrible in order to eliminate it, has become especially important in the nuclear era, though it has deep and largely unexplored historical roots.[199] In fact, virtually all the modern issues surrounding the development and employment of the technology of war have historical antecedents worthy of study.

Technology Themes '

The themes and issues historians of technology have come to focus upon lend themselves to treatment of military technology. Thomas Hughes has done more than any other scholar in the field to identify and define these themes.[200] A sampling of these issues and the military literature that pertains to them will suggest how fruitful additional research may prove.

The role of systems and institutions in technological development is nowhere more evident than in the technology of war. In fact we owe our modern appreciation of the importance of systems to the emergence of weapons systems development during and since World War II, and of course the military has always been an institutional mold forming that technology. Weapons systems are as old as warfare; Lynn White has shown how a single technological innovation like the stirrup can upset the entire military structure and usher in a revolution in warfare like the shift from foot to cavalry, with other components of the system—saddle, armor, lance, sword, and comunication—undergoing changes in turn.[201] The English longbow, gunpowder weapons, and the Swiss halberd all contributed to the termination of the resulting cavalry cycle and introduced a new tactical paradigm in which these weapons were integrated in different but still coherent fighting systems. Modern weapons systems are simply more self-conscious and more sophisticated in achieving the same ends.

Governments have always subsidized technological development by being the primary institution to stimulate the introduction of new weapons. Though most studies of this phenomenon

199. See Roland, *Underwater Warfare in the Age of Sail.*

200. See especially Thomas P. Hughes, "Emerging Themes in the History of Technology," *Technology and Culture* 20 (October 1979): 697–711; and "Convergent Themes in the History of Science, Medicine, and Technology," ibid., 22 (July 1981): 550–58.

201. White, *Medieval Technology and Social Change.*

focus on the twentieth century, there is evidence that research on earlier periods can be just as rewarding.[202] *R & D Contributions to Aviation Progress (RADCAP): Joint DoD-NASA-DoT Study*[203] was consciously designed to show how its sponsoring agencies contributed to the advance of American aviation, both military and civilian. Numerous other studies have focused on specific government agencies in an attempt to trace their impact on weapons development. Among these might be singled out Merritt Roe Smith's *Harper's Ferry Armory and the New Technology*,[204] remarkable for its contrast of two different arsenals, and David Allison's *New Eye for the Navy*, which takes the institutional influence on technology as one of its major themes.[205]

Differences in national and regional style can be seen in almost any survey of international weapons development, from early swords and body armor to modern aviation and space development. Carlo Cippola has contrasted Western developments in ships and ordnance with those in the rest of the world in *Guns, Sails, and Empires*.[206] John F. Guilmartin has neatly isolated the peculiarities of Mediterranean naval warfare and their effect on the technology of the region in *Gunpowder and Galleys*.[207] Ken Hagan and his coauthors have compared and contrasted the naval technologies of Russia, China, and the United States in *Naval Technology and Social Modernization in the Nineteenth Century*.[208] Melvin Jackson and Carel de Beer have demonstrated how so minor a factor as the sulphur content in the soil can alter a region's military technology in *Eighteenth*

202. See, for example, Baxter, *Scientists Against Time*; Kendall E. Bailes, *Technology and Society under Lenin and Stalin* (Princeton: Princeton University Press, 1978); Hartcup, *The Challenge of War*; and Postan, Hay, and Scott, *The Design and Development of Weapons*. A. Rupert Hall has demonstrated the influence of guilds and governments on artillery development in *Ballistics in the Seventeenth Century: A Study in the Relations of Science and War with Reference Principally to England* (Cambridge: Cambridge University Press, 1952).

203. Washington, D.C.: Department of Defense, National Aeronautics and Space Administration, Department of Transportation, 1972.

204. Ithaca, NY: Cornell University Press, 1977.

205. NRL Report 8466 (Washington, D.C.: Naval Research Laboratory, 1981). See also Gray, *Frontiers of Flight*; Green, Thomson, and Roots, *The Ordnance Department*; Hewlett and Anderson, *The New World*; Hogg, *The Royal Arsenal*; Walter, *Early Aviation at Farnborough*; and Young, *The East India Company's Arsenal and Manufactories*.

206. New York: Minerva, 1966.

207. London: Cambridge University Press, 1974.

208. Manhattan, KS: *Military Affairs* and the American Military Institute, 1976.

Century Gunfounding.[209] Michael Lewis has traced the effects of differing approaches to guns and tactics in *Armada Guns.*[210] Richard Ogorkiewicz's *Armor*[211] examines differences in weapons and tactics among nine countries. These studies, and many others like them, reveal how fruitful a topic national and regional style can prove to be because military technology is by definition one in which contrasting styles will find their way into direct confrontation with each other. What really needs more attention is the extent to which the transfer of ideas (another theme in the history of technology) between arms makers has produced homogenization of international weaponry and military techniques. Some work has been done in this field, but more is needed.[212]

The relation of science to technology in war and preparation for war is another fruitful theme that has received less attention than it deserves. A. Rupert Hall's *Ballistics in the Seventeenth Century* finds the science and technology of artillery, at least in that period, difficult to separate.[213] Brooke Hindle has examined the influence of war on science and science on war in the context of the American Revolution.[214] In more modern times, the Manhattan Project has proved to be a remarkably revealing case study of scientists taking the lead in weapon development—from first conception through ultimate use. As science and technology grow more dependent on each other in the contemporary world, the dividing line between their contributions will continue to blur.[215]

The Carthaginians learned the importance of appropriate

209. Washington, D.C.: Smithsonian Institution Press, 1974.

210. London: George Allen and Unwin, 1961.

211. London: Stevens & Sons, 1960.

212. See, for example, Lynn White, Jr., "Jacopo Aconcio as an Engineer," *Medieval Religion and Technology: Collected Essays* (Berkeley: University of California Press, 1978), pp. 149–73; and "The Crusades and Western Technology," ibid., pp. 277–96; Partington, *A History of Greek Fire and Gunpowder*; and Faulk, *The U.S. Camel Corps.* For an interesting departure from this theme, see Arnold Krammer, "Technology Transfer as War Booty: The U.S. Technical Oil Mission to Europe, 1945," *Technology and Culture* 22 (January 1981): 68–103.

213. See also Hall's essay "Science, Technology, and Warfare, 1400–1700," in Wright and Paszek, eds., *Science, Technology, and Warfare*, pp. 3–29.

214. Hindle, *The Pursuit of Science in Revolutionary America* (Chapel Hill: University of North Carolina Press, 1956).

215. See Fieser, *The Scientific Method*; and Charles Susskind, "Relative Roles of Science and Technology in Early Radar," *Actes*, 12th International History of Science Congress, Paris, 1965.

technology when they first faced the Roman corvus at Mylae.[216] Chance favors not the "superior" technology nor the most sophisticated technology, but rather the technology best suited to the resources and circumstances at hand. The French were subjected to the same lesson under fire from the longbow in the Hundred Years' War[217] and behind their Maginot Line in 1940.[218] Americans encountered this problem in Vietnam, and in the eyes of some analysts are up against it now in strategic weapons.[219] But the currency of the issue should not be allowed to obscure the fact that there are countless historical examples of this phenomenon that have yet to receive adequate investigation by scholars.

Technological momentum is a common theme in modern weaponry,[220] from the overall strategic arms race itself to the endless refinement of specific weapons like the fighter plane and the tank. Similar examples are available throughout recorded history, from the accretion of layers to the fortifications around Constantinople to the proliferation of artillery types in early modern Europe. Some recent examples have received scholarly treatment, as in Edward Constant's *The Origins of the Turbojet Revolution*[221] and Thomas Hughes's insightful "Technological Momentum in History: Hydrogenation in Germany, 1893–1933,"[222] but more research is needed on earlier examples of this phenomenon and the motivations of the decison-makers of the time.

The influence of policy on technological development is seldom more clear than in military activities. Warlike nations committed to conquest and expansion naturally adopted weapons and techniques suited to their purposes. Their neighbors often responded with fortification or imitation. Such decisions dictated the technologies that would be employed and shaped not only the nature of warfare, but also, in many cases, the nature

216. Wallinga, *The Boarding Bridge of the Romans.*

217. Hardy, *Longbow.*

218. Rowe, *The Great Wall of France.*

219. Kaldor, *The Baroque Arsenal.* See also York, "Military Technology and National Security."

220. Thomas P. Hughes defines technological momentum as "loosely connected, mutually reinforcing components that constitute a system of vested interests involving people, institutions, ideas, and artifacts. The system has a momentum that tends to resist change and softly determine the course of events." (Personal communication.)

221. Baltimore: Johns Hopkins University Press, 1980.

222. *Past and Present* (August 1969): 106–32.

of the society itself. Many studies investigate policy and some examine technology, but not enough analyze the relationship between the two. Bruce's *Lincoln and the Tools of War*[223] suggests the rewards of this type of investigation, as do David MacIsaac's *Strategic Bombing in World War II*[224] and Michael Armacost's *The Politics of Weapons Innovation.*[225] Comparable studies of earlier periods are needed.

Generally, in studies of major themes in the history of technology as they apply to warfare, more good work has been done on the contemporary period than on earlier times. No doubt this reflects in part the youth of the field and the fact that modern technological growth helped bring it into existence. Surely the history of contemporary military technology will continue to attract productive scholars who will contribute significantly to our understanding of this complex phenomenon. But in the plethora of modern topics, the historical community should not lose sight of the rewarding and revealing issues from earlier periods that await attention. These too have important lessons to teach, not only about the evolution of war and technology, but also about the contemporary problems we face in these intertwined fields.

223. Indianapolis: Bobbs-Merrill, 1956.
224. New York: Garland Publishing, 1976.
225. New York: Columbia University Press, 1969.

Contributors

David K. Allison is Historian of Navy Laboratories in the U.S. Naval Material Command. He is the author of *New Eye for the Navy: The Origin of Radar at the Naval Research Laboratory* (1981).

Peter Buck is Associate Professor of the History of Science in MIT's Program in Science, Technology and Society. He is the author of *American Science and Modern China* (1980) as well as several articles on the history of the social sciences.

Susan Douglas teaches at Hampshire College and is finishing a book on the early history of radio. Her research focuses primarily on the organizational contexts of technological innovation.

David A. Hounshell is Associate Professor of History at the University of Delaware and curator of technology at the Hagley Museum and Library in Wilmington. He has written *From the American System to Mass Production, 1800–1932: The Development of Manufacturing Technology in the United States* (1984).

Thomas J. Misa is a doctoral candidate in the Department of History and Sociology of Science at the University of Pennsylvania, where he is completing a dissertation under Thomas P. Hughes. His research interests include the history of modern electronics and the history of science and technology in America.

David F. Noble is Associate Professor of History at Drexel University. His most recent book is *Forces of Production: A Social History of Industrial Automation* (1984).

Charles F. O'Connell, Jr., is Assistant Chief of the Tactical Air Command History Office, Langley Air Force Base, Virginia. His essay in this volume presents an overview of a 1982 doctoral dissertation completed at The Ohio State University.

Alex Roland is Associate Professor of History at Duke University and current Secretary for the Society for the History of Technology. His writings include *Underwater Warfare in the Age of Sail* (1978) and *Model Research: A History of the National Advisory Committee for Aeronautics, 1915–1958* (1985).

Merritt Roe Smith is Professor of the History of Technology in MIT's Program in Science, Technology and Society. His book *Harpers Ferry Armory and the New Technology* received the 1977 Frederick Jackson Turner Award of the Organization of American Historians and the 1978 Pfizer Award of the History of Science Society.

Index

Accounting procedures, 17, 93–94,
98, 107
and uniformity system, 11, 56–59
Air Force
bibliography, 363–64
and numerical control, 9, 331,
340–45
Aitken, Hugh, 14, 120, 130, 171
Allison, David K., 15, 29, 289–328
Allport, Gordon, 206, 237
American Marconi Company, 135,
140, 168
American Soldier, The, 211, 220, 226,
227, 228, 232–34, 236, 238, 239,
241–42, 243, 245, 246, 249, 250–
51, 252
"American system" of manufacturing, 28, 41–86. *See also* Interchangeable manufacture;
Uniformity system
Ames Manufacturing Company, 77,
78
Anthropology. *See* Social research
Archer, Gleason, 118–19
Architecture, bibliography, 359–60
Armories, 337
administration of, 12, 83–86
in diffusion of new techniques, 76–
78
Harpers Ferry, 44, 51, 53–54, 56,
64, 68, 81–82, 337–38
inspection procedures at, 59–61,
67–69
monitoring operations of, 57–59
Springfield, 44, 51, 54, 55, 60, 68,
80, 81, 82, 338

and uniformity system, 51–84,
337–38
Army Corps of Engineers, 11
bibliography, 372–73
and managerial innovation, 11, 88–
116
Army Ordnance Department, 11–
12, 70–72, 74, 335–38
and "American system" of manufactures, 41–86
and Christian stewardship/pater-
nalism, 79–80
creation and authority of, 43–44
and interchangeable manufacture,
59, 61–64
and metallurgy, 18–19
open door policy of, 76–77
reorganization of, 66–67
and uniformity system, 49–84
Army Signal Corps
and miniaturization of equipment,
263–64, 268
and transistor development, 253–
55, 262–87
and walkie-talkie devices, 262
Art, Robert, 10, 25
Artillery
bibliography, 377
and Ordnance Department reorganization, 67
uniform system of, 45, 49–50, 70–
71
Ashton, Thomas S., 31
Austin, Louis, 156
Authoritarian technics, 334–35

Badger, Rear Admiral Charles J., 159, 163
Baltimore & Ohio Railroad Company, 97–98, 101–6
Barber, Commander Francis M., 133–37, 141–42, 144, 146
Bell Telephone Laboratories, 254, 256, 258–62, 264–70, 272–75, 277–79, 281–83, 285–86, 322
Bethlehem Steel Company, 8, 19
Bibliography
 architecture, 359–60
 artillery, 377
 aviation, 363–64
 British military history, 349
 classic surveys on war and technology, 350–54
 Cold War, 369–70
 cryptography, 360
 doctrine, 372
 engineering, 372–73
 food supply, 361
 historiography of war and technology, 348–50
 military–industrial complex, 370–71
 moral/political position of scientists, 374–75
 naval technology, 361–63
 nonweapons technology, 357
 nuclear weapons, 370
 operational analysis, 374
 policymaking, 378–79
 procurement, 364–67
 rules of war, 373–74
 secrecy, 374
 subsidies, 375
 systems engineering, 374
 technology themes, 375–79
 traditional weapons survey, 354–57
 transportation, 358–59
 warfare, 377
 World War II, 367–68
Blanc, Honoré, 47
Bliss, George, 108, 109
Blood, Wallace B., 255
Bomford, Colonel George, 28, 52–53, 56, 60–61, 64, 66–67, 82–83, 335, 337
Booth, James C., 73
Boston & Albany Railroad, 110
Boston & Worcester Railroad, 107, 110
Bowen, Admiral Harold, 28, 195

Bown, Ralph, 259, 264
Bradford, Admiral R. B., 128, 129, 136, 139
Brattain, Walter, 256, 258
Braun-Siemens-Halske, 133, 135, 141
Bridges, Harry, 340
British gun carriage construction, 46–47, 48
Brown, Harold, 303
Buck, Peter, 14, 203–52
Buckley, Oliver E., 264
Bush, Vannevar, 257, 291

Calhoun, John C., 91, 92, 93, 96
Cantril, Hadley, 215, 240
Carlucci, Frank, 307, 308
Carnegie Steel Company, 19
Chandler, Alfred D., Jr., 87, 167
Clark, George, 148, 155
Classic surveys on war and technology, bibliography, 350–54
Cold War, bibliography, 369–70
Collins Radio, 323
Colt Patent Fire Arms Manufacturing Co., 11, 78
Command and control, 6, 13–14, 17, 36, 69, 94, 170, 320–21, 333
 of containerization, 339
 numerical control, 331, 340–45
Computer-aided manufacturing, 341, 345
Conant, James B., 237, 247, 257
Constant, Edward, 18
Containerization of cargo, 338–40
Contractors
 bibliography, 365–66
 and cost effectiveness, 9
 and military contracting practices, 8–9, 63–64, 201, 331
 for Naval research and development, 309–10, 314–27
 and Office of Scientific Research and Development, 290
 of small arms, 26
 and uniformity system, 61–66
Cooley, Michael, 332
Costs, 332, 334
 and design, 7
 of new technologies, 8–9
 and performance, 8
 of radio operations, 129–31, 142, 146

in railroad management, 102–3, 105, 107
and uniformity system, 58, 86
Cottrell, Leonard, 240
Craft knowledge, 56, 60, 64, 79, 180, 202, 335
Craven, Commander, 157, 158
Crocker-Wheeler Company, 168
Cryptography, bibliography, 360

Daniels, Josephus, 24, 179, 181, 183–84, 195, 198, 201
De Forest, Lee, 119, 132, 142–44, 146, 256
De Forest Radiophone, 136–38, 149, 169
Defense Department, 16, 299, 307, 308, 334
Defense Reorganization Act of 1958, 300
Defense System Acquisition Review Council, 305–7, 312
Designs, execution of, 5–10
Discipline. *See* Labor, discipline of
Dissemination of new technologies, 5–8, 76–78, 120–21, 255, 267–68, 284
Doctrine, bibliography, 372
Douglas, Susan J., 25–27, 117–73
Ducretet, 133–35
Dumas, Lloyd J., 33–34
Duncan, Francis, 6
du Pont, Lammot, 12
DuPont Company, 12, 19

Eagle boat project, 24, 175, 179–202
Economic growth, and technological innovation, 33, 36, 40
Eisenhower, Dwight D., 290, 291
Emergency Fleet Corporation, 168–69, 181–82
Entrepreneurship, 1–2, 75–76, 118, 255
bibliography, 366–67
Ezell, Edward, 10

Fairley, Lincoln, 340
Fessenden, Reginald, 132, 139–44
Firearms
Blanc's muskets, 47
continuous-aim firing, 23, 26
Hall's rifle, 61–62
inspection of, 60
interchangeable parts in, 42

Model 1816 musket, 53
Model 1821 musket, 61
Model 1841 percussion rifle, 63, 65
Model 1842 percussion musket, 63, 65
small arms contractors, 26
uniformity system of manufacture of, 51–84
Fiske, Commander Bradley A., 119
Fitch, Charles H., 12
Food supply, bibliography, 361
Ford, Henry, 24, 177, 179–80, 182
Ford Instrument Company, 323
Ford Motor Company, 175, 178–202
Fort Pitt Iron Works, 19, 72
Fosdick, Raymond, 216
French military tradition, and uniformity system, 44–49, 73
Fullingwider, Admiral E. G., 338
Furth, Admiral F. R., 321

General Dynamics Corporation, 309
General Electric Company, 169, 170, 270, 275, 318
General Regulations for the Army, 93–94, 100–1
George S. Lincoln & Company, 78
Globe-Union Corporation, 263
Gribeauval, General Jean-Baptiste de, 46–49
Groves, General Leslie, 28
Guthrie, Edwin, 216

Hall, John H., 61–63, 335
Harvard University, 206, 210, 236
Haupt, Herman, 111–14
Hauser, Philip, 215
Hewlett, Richard, 6
Historiography of war and technology, bibliography, 348 50
Hitch, Charles J., 16, 302–3
Holley, I. B., 16
Hooper, Stanford C., 27, 154–68, 172
Horwitch, Mel, 9
Hough, R. R., 282
Hounshell, David A., 24, 175–202
Hudgins, Lieutenant J. M., 149
Huger, Captain Benjamin, 74
Hughes Aircraft, 323, 331
Hurley, Edward N., 181

Inspection, of armories and arsenals, 46, 59–60, 64–69
instrumentation in, 61–62
Instrumentation, 17
in inspection of firearms, 61–62
Interchangeable manufacture, 42, 47, 59, 60, 62–64, 78, 335–38
See also Mass production; Uniformity system

Japanese-American relocation camps, 208–10
Japanese semiconductor industry, 286
Jeffers, Captain William N., 8, 28

Kaiser, Henry J., 201
Keegan, John, 226, 227
Kelly, Mervin J., 256, 274
King, Ernest J., 290–91
Kluckhohn, Clyde, 206–8, 209, 210, 230, 236–37
Knight, Jonathan, 96, 98, 102, 104–6
Knudsen, William S., 183
Korean War, 236–37, 248, 266

Labor
discipline of, 80–83, 85
regulation of, 17, 83–86
relations with management, 79–80, 83–84, 332–33, 338, 339–40
Laird, Melvin, 305
Large-scale technological programs, 9–10
Lark-Horovitz, Karl, 256–57
Latrobe, Benjamin H., 106
Lazarsfeld, Paul, 215, 227
Lee, Roswell, 51, 54–57, 60–61, 64, 66, 80
Leighton, Alexander, 207–10, 230–32, 234, 249–52
Liberty ships, 201–2
Liebold, E. G., 181, 184, 191
Little, J. B., 269
Lockheed, 331, 343
Logistics, bibliography, 357
Long, Lieutenant Colonel Stephen H., 98, 102–3

MacFarland, Jacob Corey, 77
Machine tool manufacture, 5, 9, 77–78, 342, 344
MacNair, Walter A., 283

Macomb, Colonel Alexander, 95–96, 98, 100
MacWilliams, W. H., Jr., 265
Maddox, Ensign C. H., 156
Magruder, T. P., 148–49
Mahan, A. T., 123
Management, 10–17, 332–33
accountability, 69, 92, 94, 100–1, 104–5, 107
accounting procedures. *See* Accounting procedures
Army administration, 90–93, 96–97, 114
Army influence on innovations in, 87–116
bureaucratic procedures, 25, 27, 89, 92, 94, 96–97, 101, 122
centralization, 11, 17, 21, 25, 92, 299–302
civilian adoption of military procedures, 98, 100–2, 105–6, 109, 116
command and control. *See* Command and control
division of management functions, 112
hierarchy in, 101, 104, 109, 121, 243
influence of military procurement on, 331
and innovation, 171–73
labor's relations with, 79–80, 83–84, 332–33, 338, 339–40
manual of military management, 93, 94
Navy administration, 121–26, 152–53, 290–95
and new technologies, 160–67, 170–73
operations research, 15–16, 374
reporting procedures. *See* Reporting procedures
of research and development, 292–308, 327–29
scientific, 12–13
and social research, 14–15
systems analysis, 15–16
time studies, 13–14
Marconi, Guglielmo, 126–27, 129–31, 134, 138
Market forces, 9, 332
invention and marketing, 254n2
and managerial innovation, 11, 132
Marshall, General S. L. A., 227
Mass production, 175, 178–202

automobile v. shipbuilding, 183, 185, 191, 197, 201–2
and craft knowledge, 180, 202
design v. production priorities, 200–1
and Ford Motor Company, 178–79
model construction in, 186, 188
techniques used in, 193–95, 197, 199
Massachusetts Institute of Technology, 256, 321, 343
Mayo, William B., 182–84, 188
McLean, William B., 315–16
McNamara, Robert S., 15, 302, 304–5, 312
McNeill, Lieutenant Colonel Stephen H., 106–8
McNeill, Captain William G., 99, 100–3, 106–8, 110, 114–15
McNeill, William H., 32, 176, 177
McNemar, Quinn, 215
McRae, James, 264, 266
Mechanization, 64, 79, 82, 86
Melman, Seymour, 276, 331
Metallurgy, 5, 18
in transistor development, 268–72
uniform foundry practices, 71–74
Metcalfe, Henry, 12
Midvale Steel Company, 8, 13, 20
Military enterprise, 1–29, 330–31, 345–46
and command, 333
effects of, 32–37, 330–31
externality of, 330
and industrial might, 4
and management, 10–17
and modern methods, 333, 334
and performance, 7–8, 332, 340
Military-industrial complex, 291
bibliography, 370–71
Misa, Thomas J., 34, 253–87
Missile systems, 267, 278–80, 282–83, 300, 314–20
Modern methods, 333–34
Moral/political position of scientists, bibliography, 374–75
Mordecai, Major Alfred, 72–75, 335
Morfit, Campbell, 73
Morison, Elting E., 23, 27, 126, 171
Morton, Jack A., 259, 274, 277, 279–80, 282
Mumford, Lewis, 11, 28–31, 176, 330, 334

Murray, Henry, 206, 207, 209, 223, 236

Nashua Iron and Steel Works, 19
National Defense Research Committee, 290
National Electric Signaling Company, 139
National Military Establishment, 291, 299, 315
National Research Council Committee on Amphibious Operations, 338
National Resources Planning Board, 214
National Science Foundation, 257, 297, 308
National Security Act of 1947, 291
Naval Electronics Laboratory, 323, 324
Naval Experimental Battery, 19
Naval Material Support Establishment, 301
Naval Ordnance Test Station, 315–18, 320
Naval Radio Service, 164–66
Naval Research Laboratory, 277, 310
Naval Tactical Data System, 320–27
Naval Torpedo Station, 19
Navy, 121, 170
administration of, 122–26, 152–53, 290–95, 299–302
bibliography of, 361–63
Bureau of Construction and Repair, 24, 181, 183–84, 196–98
Bureau of Equipment, 125, 128, 130, 136, 148–49, 152
Bureau of Navigation, 148–49, 157, 165
Bureau of Ships, 323
Bureau of Steam Engineering, 152, 155, 157, 165
Bureau of Weapons, 300
bureau system of, 294–96, 298, 300
Chief of Naval Development, 304, 309
Chief of Naval Operations, 166–67, 294, 297, 301–2, 322–23, 325–26
and containerization, 338–40
Deputy Chief of Naval Operations, 300
Director of Naval Laboratories, 312

Navy, *cont.*
Eagle boat project, 24, 175, 179–202
Office of Communications, 167
Office of Naval Material, 300, 301, 312
Office of Naval Research, 257, 295, 297, 302, 304, 309–10, 321–23, 338
Ordnance Bureau. *See* Navy Ordnance Bureau
and patents, 143–44, 146
radio network of, 25–26, 117, 126–73
research program post World War II, 289–328
System Acquisition Review Council, 307
systems commands, 301
Navy Ordnance Bureau, 8, 19, 23, 26, 300
and Sidewinder missile, 315–18
and transistor applications, 265
Navy Yards, 19, 135, 150–53
Naylor and Company, 19
Nef, John U., 29, 31
Newton, Lieutenant Commander J. T., 128
Noble, David F., 6, 9, 35, 329–46
North, Simeon, 63, 335
Nuclear energy and weapons
bibliography, 370
civilian use of, 6–7
Numerical control, 9, 331, 340–45

O'Connell, Charles F., Jr., 11, 87–116
Office of Management and Budget, 305, 306, 326
Office of Scientific Research and Development, 257, 290, 295
Office of Strategic Services, 207–10, 223
Office of War Information, 230
Osborn, Frederick, 214–15, 235–36, 248, 251–52
Osterhaus, Rear Admiral Hugo, 159, 163

Pacific Semiconductor, 275
Parsons, Admiral William S., 317
Parsons, John, 342, 345
Parsons, Talcott, 206, 237
Patents, 153

automatic soldering system, 263
on firearms, 61
radio, 143–44, 146
on semiconductors, 261
Pennsylvania Railroad, 111, 112
Performance, and costs, 7–8, 332, 340
Personality
and culture, 206, 209, 211, 239
of inventors, 131–35
of military personnel, 131–32
and railroad management, 102
and resistance to innovation, 23–5, 171
and technological change, 27–28
PERT (computerized program evaluation technique), 16–17, 28
Philco, 318
Planning, Programming and Budgeting System, 302, 304, 307, 312
Polaris submarine project, 16
Policymaking
by Army Ordnance Department, 76–77
bibliography, 378–79
and social research, 212–22, 235–36, 247, 249
Process-oriented technologies, 6, 8
Procurement, bibliography, 364–67
Project Lamp Light, 321
Propaganda, 209, 229, 249
Providence Tool Company, 78
Psychology. *See* Social research
Purdue University, 256, 257, 258

Quantity production and consumption, 176, 178, 182–83. *See also* Mass production
Quarles, Donald, 264

Raborn, Admiral William F., 28
Radio
contracts awarded by Navy, 137–39, 141–42, 145–46, 148, 168–69
De Forest Radiophone, 136–38, 146
Fessenden's equipment, 139–43
French and German equipment, 133–35, 137, 141, 148
installation and use of, 143, 148–52, 156, 158, 160–63, 168
Marconi's wireless, 126–30, 138
Navy's radio network, 25–26, 117, 126–73

and organizational realignment, 160–67, 170–73
patents, 143–44, 146
regulation of, 153–54
resistance to, 119–20, 130–32
and Stanford Hooper, 155–73
tactical signaling, 157–58, 164
testing by Navy, 126–29, 135–37, 139–41
transistors in, 274, 281–82
Radio Corporation of America, 118, 170, 270, 275
Railroads, 87–116, 358–59. *See also specific companies*
Rationalization of production, 11, 36, 79, 82, 334
Raytheon, 274, 275, 332
Reagan, Ronald, 306–7
Remington Rand, 323
Remington & Sons Company, 11, 78
Reporting procedures, 94–96, 98, 100, 105, 107
development concept paper, 304–5
in research and development, 295–96
Research and development
budgeting oversight by Congress, 297
centralization of authority over, 299–302, 309
current Naval projects, 314–15
in-house military laboratories, 310–12, 320
Naval authority over, 292–302, 327–28
Naval management of, 302–8
Naval Tactical Data System, 320–27
Navy program, 289–328
Office of Scientific Research and Development, 290, 295
planning documents in, 295, 296
private industry in, 309–10, 314–27
Sidewinder, 315–20
Research Branch of Special Services Division, 14–15
Research and Development Board, 266, 291
Rickover, Admiral Hyman, 6, 15, 28
Robbins & Lawrence, 78
Rochefort, 133, 134, 135
Rodman, Lieutenant Thomas J., 74
Roland, Alex, 347–79

Roosevelt, Theodore, 122, 125
Roosevelt, Theodore, Jr., 200
Rules of war, bibliography, 373–74

Sapolsky, Harvey, 16
Scaff, J. H., 268, 269
Schwimmer, Roskia, 177
Scott, Brigadier General Winfield, 93, 100
Secrecy, bibliography, 374
Sellers, William, 12, 13
Sharps Rifle Company, 78
Sherwin, Chalmers, 310, 312
Shipbuilding
  Eagle boat project, 24, 175, 179–202
  Liberty ships, 201–2
Shockley, William, 256, 258, 259, 269
Shop floor, control of, 6, 13, 17, 20, 36, 332–35
Shriver, James, 95, 96
Sicard, Captain Montgomery, 8
Sidewinder missile, 315–20
Sims, William S., 23–24, 119
Slaby-Arco Company, 133, 134, 135, 137, 138, 141, 148
Smith, Admiral Levering, 28
Smith, Merritt Roe, 1–86, 332, 337
Social engineering, 205, 212, 220, 228, 250
Social and human needs, 334–35
Social organization, and technological change, 24–26
Social process, technological change as, 3, 23, 35, 80
Social product, technological change as, 2, 21, 35, 330–35
Social research
  into adjustment and adaptation, 242–43
  adjustment to military life study, 212, 214
  attitude surveys, 14
  autonomy of social scientists, 205, 219–21
  into belief systems, 208–10, 229–31
  into combat behavior and leadership, 226–28
  Committee on Social Adjustment of the SSRC, 212–15, 228, 242
  into cultural stereotypes, 231–35

Social research, *cont.*
into cultural stereotyping, 245–46, 248
into demobilization, 217–18, 240–41, 246
development of new theoretical resources, 206–7, 213, 220–21, 239
during 1941–1950, 203–52
into education and social class, 242
into Hiroshima survivors, 231
and human development, 213–14, 251–52
intelligence testing, 14
for managerial purposes, 14–15
organizational sociology, 243–44
OSS assessments of recruits, 207–10, 223
into people under stress, 208–9
and policymaking, 212–22, 235–36, 247, 249
and propaganda, 209, 229, 249
and psychological warfare, 209
scaling, 14
small-group sociology, 212, 227
and social theory, 211, 212, 248
at relocation camps, 208, 210
into technological change, 244–45
into war's meaning for soldiers, 222–23, 226–27, 228–30, 232, 235–38, 241
Sombart, Werner, 4, 29–32, 176
South Boston Iron Company, 19
Sperry Rand, 323
Spinoff of military technologies to civilian use, 6–7, 34–36, 177, 286–87, 330
Standardization, 5–10, 17, 47, 150, 176, 182, 285. *See also* Uniformity system
Steel industry, 8, 13, 18–20
Stocker, Captain Robert, 183, 200–1
Stouffer, Samuel, 15, 206–7, 210–12, 214–19, 221–22, 237–38, 240–42, 250–52
Stubblefield, James, 51, 53–54, 56, 61, 64
Submarine patrol boats, 24, 175, 179–202
Submarines, 300, 309, 314
bibliography, 362
Subsidies by military, 8–9
bibliography, 375–76
of transistor production, 272–76

Swift, Captain William H., 107, 108, 109, 110, 114–15
Sylvania, 275
Systems engineering, 374

Talcott, Colonel George, 67–68, 82, 84–85, 335
Taylor, Admiral David W., 183, 200–1
Taylor, Frederick Winslow, 12, 13
Teal, Gordon, 269
Technological change, 1–2, 330
adaptation to, 22–27, 205
bibliography, 375–79
catalysts to, 27–29
design and dissemination of new technologies, 5
institutional contexts of, 3
as knowledge, 2
and managerial innovation, 10–17, 87–116
and order, 20–21, 69
resistance to, 22–24, 119–20, 171
and social choice, 3
as social force, 2
and social organization, 24–26
as social process, 3, 23, 35, 80
as social product, 2, 21, 35
and unemployment, 334
and war, 176
Technological convergence, 78
Telefunken, 141, 143, 144, 169
Telephone systems, 274, 277
Testing, 17–18
Emery Testing Machine, 19
in steel industry, 18, 19, 20
Theuerer, H. C., 268, 269
Thomas, Philip E., 98, 100, 102
Thomson, J. Edgar, 111–13
Tinus, W. C., 282
Todd, Lieutenant Commander D. W., 157, 158
Tousard, Major Louis de, 45, 46
Towne, Henry R., 12
Traditional weapons survey, bibliography, 354–57
Transistors, 253–87
applications projects, 265–68, 274
and Army Signal Corps, 253, 254, 255, 262–87
cultural context of industry, 284–87
diffusion technology, 281–82, 285

dissemination of information about, 267, 268, 284
invention and development of, 254, 255–62, 264
military applications of, 277–83
production of, 272–76
standardization of specifications, 285
technological innovations in development of, 268–72
"Translators," 27–28, 120–21, 133, 146, 171–72
Transportation, bibliography, 358–59
Triumph Electric Company, 168
Tyler, Daniel, 66

Uniformity system, 7, 20–22, 42, 332, 335, 337. *See also* Interchangeable manufacture
and accounting procedures, 57–58
of artillery, 45, 49–50, 70–71
codification of regulations, 69–70
costs of, 86
disseminating new techniques, 76–78
and foundry practices, 71–74
inspection procedures under, 46, 59–61, 64n34, 64–66, 68–69
and interarmory cooperation, 54, 57
and interfactory cooperation, 63
and inventory control, 57
and mechanization, 64
muskets developed under, 53–54, 61–63, 65
origins of, 44–47
of small arms manufacture, 51–84
work rules under, 80–86
United States Shipping Board, 181, 182
University of Chicago, 215, 216

Values, 6, 21, 85
Vietnam War, 227

Wade, William, 72
Wadsworth, Colonel Decius, 49–54, 335
Walbach, Lieutenant Louis A. B., 73
War Department, 43, 91–94, 96
Foreign Morale Analysis Division, 208–10, 230–31, 249–51

Research Branch of the Information and Education Division, 210–12, 214, 216–17, 219–22, 229–30, 232, 238–41, 242–44, 246, 247–52
War of 1812, 42, 44, 90–91
Warfare
bibliography, 361–62, 377
displacements caused by, 206, 208
and human development, 29–32
and mass production, 176
psychological, 209
simulated warfare computer, 265
Weaponry, bibliography, 360
West Point Foundry, 19
West, Charles C., 193, 195, 197
Western Electric Company, 169, 275, 278, 283
Western Railroad Corporation, 107–10
Wever, Caspar, 96, 98, 102–4
Whistler, George W., 108, 110
Whitney, Eli, 47, 49, 51
Wilson, Woodrow, 181, 182
Winner, Langdon, 6
World War I, 177, 178
World War II, 320
bibliography, 367–68
social research by military during, 203–35, 249–52

Young, Donald, 215
Young, Kimball, 215–16

Zahl, Harold A., 264